State-of-the-Art Reviews on Energetic Ion-Atom and Ion-Molecule Collisions

Interdisciplinary Research on Particle Collisions and Quantitative Spectroscopy

ISSN: 2315-4233

Series Editor: Dževad Belkić

Vol. 1: Fast Ion-Atom and Ion-Molecule Collisions
edited by Dževad Belkić

Vol. 2: State-of-the-Art Reviews on Energetic Ion-Atom and Ion-Molecule Collisions
edited by Dževad Belkić, Igor Bray and Alisher Kadyrov

Interdisciplinary Research on Particle Collisions
and Quantitative Spectroscopy
Volume 2

State-of-the-Art Reviews on Energetic Ion-Atom and Ion-Molecule Collisions

Editors

Dževad Belkić
Karolinska Institute, Stockholm, Sweden

Igor Bray
Curtin University, Australia

Alisher Kadyrov
Curtin University, Australia

World Scientific

NEW JERSEY • LONDON • SINGAPORE • BEIJING • SHANGHAI • HONG KONG • TAIPEI • CHENNAI

Published by

World Scientific Publishing Co. Pte. Ltd.
5 Toh Tuck Link, Singapore 596224
USA office: 27 Warren Street, Suite 401-402, Hackensack, NJ 07601
UK office: 57 Shelton Street, Covent Garden, London WC2H 9HE

British Library Cataloguing-in-Publication Data
A catalogue record for this book is available from the British Library.

Interdisciplinary Research on Particle Collisions and Quantitative Spectroscopy — Vol. 2
STATE-OF-THE-ART REVIEWS ON ENERGETIC ION-ATOM AND ION-MOLECULE COLLISIONS

ISBN 978-981-121-160-7

For any available supplementary material, please visit
https://www.worldscientific.com/worldscibooks/10.1142/11588#t=suppl

Aims and Scope: Mission Statement of This Book Series

Vision on interdisciplinarity

It is instructive to inquire about the real meaning of interdisciplinary research within a given context, such as book series. In general, besides obvious obstacles such as evident differences in the attitudes to problem solving and terminology, there is also the more subtle issue of mentality and barriers when experts from different research fields envisage to join forces. It would not be a rare event if a theoretical physicist would wish, from the start, to venture to tell e.g. a potentially collaborating biologist what is actually important to study in biology. A biologist might attempt to view physicists or mathematicians alike as those who merely provide the needed service to biology. Both approaches are wrong, as they evidently stem from the lack of genuine appreciation of the special aspects and potential of the other branches of research. Some research physicists might even be tempted to think of going to another hardly necessary extreme of studying e.g. all of biomedicine or *vice versa*, as if this were a pre-requisite for a successful interdisciplinary cooperation.

A more appropriate approach to a good cooperation of experts from different fields would seem to be striking a balance in acquiring the necessary minimum information about each others' expertize within the concrete problems under study, then to ask the right questions and finally to know what is at stake. It is within this general and realistic realm of viewing interdisciplinarity that this series is planned to encourage cooperative approaches to complex problems to be reviewed through different Volumes of varying levels of complexity.

Specifically, this series is aimed at giving expert coverage to cutting-edge advances in studies of collisions and spectroscopy with a focus on prospects for interdisciplinary applications in the sense to which we alluded above. These two main strategies for investigating the structure of matter on various fundamental levels are deeply and intrinsically intertwined through a panoply of similar or common concepts as well as via mathematical and computational methods of both a deterministic and stochastic nature.

The notion of spectroscopy naturally emerges from resonant scattering phenomena when unstable, decaying states are involved. Regarding collisions, the series will cover both light and heavy particle laser-free and laser-assisted collisions from low to high energies. This is rooted in the key interdisciplinary significance of such collisions in wide applications, including plasma physics, astrophysics, thermonuclear fusion research, radiotherapy, etc. The series will encourage contributions from basic as well applied sciences, the joint home of both collisions and spectroscopy.

Over a long period of time, collisions and spectroscopy from basic research in physics and chemistry have made gigantic strides across interdisciplinary fields, including the life sciences. For example, nuclear magnetic resonance spectroscopy from fundamental research in physics and chemistry is viewed by experts to possess the potential of revolutionizing particularly early cancer diagnostics, molecular image-guided surgery and radiotherapy as well as follow-up.

Likewise, energetic collisions involving multiply-charged heavy particles are of paramount importance, as they are nowadays going through a veritable renaissance in several leading strategies, such as hadron therapy, fusion research, etc. High-energy light ion beams from protons to oxygen nuclei, as a powerful part of hadron therapy of deep-seated tumors, are increasingly in demand worldwide, and this highly motivates construction of hospital-based accelerators. Positron-emitting secondary ions generated from collisions of primary projectiles with the treated tissue provide the possibility for special tomographic verifications of dose depositions at the targeted sites. Such information is of utmost importance, since any inaccuracy in dose delivery to the tumor relative to the prescribed dose planning system can be used to update the input data to the algorithmic codes for energy losses in order to provide the corrected doses in the subsequent fractionated treatments of the patient. This adaptive dose delivery and the proper inclusion of the so-called relative biological effectiveness are among the major constituents of biologically-optimized and, indeed, personalized radiotherapy which is adapted to the specific needs of each patient.

Collision physics involving ions is topical again in the quest for new energy sources, as greatly boosted by the International Thermonuclear Reactor, ITER, which is currently being built in France and shall be put into operation in 2020. This type of energy stems from high-temperature fusion of light nuclei and the main significance of this achievement is not just in a commercially profitable energy gain, but, most importantly, in having an incomparably safer energy production than the one from nuclear fission, due to minimum radioactive waste and no runaway nuclear reactions. This was made possible by tremendous intra- and inter-disciplinary cooperative efforts involving ion collision physics, plasma physics, engineering and technology. For example, the most critical to steady energy production is plasma stability. Plasma, being comprised of hot charged ions, can readily be neutralized and, as such, lost for generation of the current. Neutralization can occur through collisions of ions with the reactor walls via electron capture from the material contained in the tokamak. Knowledge from atomic collision physics about the largest cross sections for electron capture for certain elements enabled elimination of such materials from the walls of the tokamak and thereby contributed substantially to plasma stabilization.

The series explores both direct and inverse problems in scattering and spectroscopy. The primary emphasis is on methodologies of the proven validity simultaneously based on fundamental theoretical concepts and principles as well as on indispensable and thorough comparisons with experiments. There is a growing concern among e.g. medical physicists that critical input from theories of basic sciences is needed for going beyond phenomenological and empirical, data-driven modeling with no predictive power. Such models are still overwhelmingly in use in e.g. radiobiology of cell survival as well as in cross section data bases for Monte Carlo simulations of energy losses of ions during their passage through tissue. Critical and expert state-of-the-art reviews of the existing progress and stringent assessments of the available methods in comparative analyses of theories versus experiments will be strongly encouraged.

Although physics was the main branch for studying particle collisions as well as for the major three spectroscopies (sub-nuclear, nuclear and atomic), mathematics was and still is pivotal to furthering these scientific strategies. This coupling resulted in the creation of a special field in mathematics *per se*, called "Scattering Theory". The themes of this series will nurture the ever needed inspiration from mathematics, both computational and fundamental, ranging from integral transforms through differential equations to inverse problems. The most interesting/important areas in theoretical and experimental studies across inter-disciplinary fields are within inverse problems of finding the causes from the observed effects. Only the simplest experiments in e.g. physics deal with direct measurements of the sought observables. By contrast, in more fundamental physics experiments such observables are directly inaccessible, so that resorting to the concept of inverse problems is unavoidable by making inferences about the causes for the recorded effects. Likewise, and practically by definition, virtually the whole of medicine is about solving inverse problems. This is of paramount significance in practice and the series will embrace inverse scattering and spectroscopy problems that are plagued by notorious mathematical ill-conditioning and ill-posedness due to the lack of the models' continued dependence upon the analyzed data. Such inherent difficulties are further exacerbated by the presence of unavoidable noise.

Ubiquitous noise in experimental data and computationally generated spuriousness are invariably considered as a nuisance and detrimental to analysis and synthesis. This series will pay special attention to the modern advances in inverse collisional and spectroscopic problems where, surprisingly, noise can advantageously be exploited to reliably identify the sought physical and biochemical information and, thus, separate noise from true signal (stochastic resonances, Froissart doublets, etc).

The series' interest in basic structure and overall interactions is aimed to offer the possibility for perceiving kindred relationships among apparently diverse systems in vastly different fields. This naturally includes system theory and optimization. System theory perceives the system's interactive dynamics as a whole, rather than merely its isolated constituent parts. Optimization comes into play to synergize the favorable and minimize the adverse influences of the surrounding environment of the system. This approach has already brought fruits to biology and medicine. For instance, optimization is a key to conformal radiotherapy which selectively targets the diseased tissue, while simultaneously sparing the healthy parts of the treated organ. In the like spirit of cross-fertilization, the self-contained field of signal processing is currently itself benefiting from the recently imported advances in quantum-mechanical spectral analysis. Here, regarding e.g. medical diagnostics, mathematical optimization is also the most important, since experimentally encoded data are frequently uninterpretable in the direct domain in which measurements are performed.

There is not a single book or book series or journal which unifies spectroscopy and collisions into a common framework, as envisaged here. This series is designed to bridge such a gap as a publication of the outlined unique profile. Such a scope is indispensable given the inevitable scatter of information across interdisciplinary fields. The need to make bridges across various sources of spectroscopy and collisions is urgent, given the demonstrated potential of these universal methodologies for studying the structure of matter on vastly different levels/complexities. One of the rationale for establishing this type of an interdisciplinary book series is to avoid duplication of methods and results. The lack of cross-disciplinary fertilization often leads to "discovering" some methodologies or techniques in one branch only to subsequently realize that it was merely a reinvention of the already known results from another branch of research and development.

The logo of this series (shown on the front cover page) is a symbol consisting of three intertwined rings that are known as Borromean rings in mathematics and Effimov states in physics. The latter symbolize a bound system of e.g. three atoms where none of the three two-body sub-systems is bound. As such, the compound tripartite system breaks apart as soon as any single ring is removed. In the logo, the three rings represent symbolically mathematics, physics and medicine, as the main backbone of interdisciplinarity of the series. What is meant by this symbolism is that both physics and medicine are substantially enriched by their linkage to mathematics, without which progress of either science would be most severely hampered, to say the least. Modern hospitals could not function without X-rays, computerized tomography, ultrasound, positron emission tomography, magnetic resonance imaging and magnetic resonance spectroscopy, all of which were brought to medicine by physics and mathematics. The logo of the series places emphasis on the synergism achievable when these three disciplines are tightly bound together. This triple link is beneficial to mathematics, as well. For example, the most difficult problems and questions raised particularly within physics are already known as being capable of spurring fundamental progress in mathematics. Critical advances on deeper levels in biomedical sciences also rely heavily upon mathematics due to the complexity of the problems and systems in the life sciences. It is likely that in the future, biomedical problems might also initiate research in yet unexplored pure mathematical avenues. Here, it pertinent to cite an article by E. Wigner, "The unreasonable effectiveness of mathematics in the natural sciences", *Commun. Pure & Appl. Math.* **13**, 1 – 14 (1960).

Overall, this series will focus on the multifaceted interdisciplinary character of collisions and spectroscopy. This is motivated by the existing wide consensus that the most profound progress in modern sciences of strategic importance for the society at large is critically guided by inter- and multi-disciplinary approaches to complex research problems.

By its versatile nature, the intended audience for this series is vast, including researchers and graduate students in physics, chemistry, biology, physicians of various levels of training, and the associated mathematical/computational sub-branches alongside the separate field of signal processing. The series will be published through independent books by one or more invited authors or edited volumes with chapters by different expert contributors.

Editor-in-Chief: Dževad Belkić
Professor of Mathematical Radiation Physics
Nobel Medical Institute, Karolinska Institute, Stockholm, Sweden

Preface

This book, as Volume 2 of the Book Series "Interdisciplinary Particle Collisions and Quantitative Spectroscopy", is based upon a part of the invited and contributing talks at the 25th International Symposium on Ion-Atom Collisions (ISIAC), held in Palm Cove, Queensland, Australia. ISIAC is a biennial satellite of the International Conference on Photonic, Electronic and Atomic Collisions (ICPEAC). In 2017 both events were held in Australia: ISIAC took place on July 23-25, 2017 right before ICPEAC. Usually, ISIAC attracts about 60 participants from all over the world, giving them a convenient opportunity to present and exchange detailed information and ideas at a smaller meeting of specialists in the field. The topics covered by ISIAC encompass basically all types of ion-atom collisions (excitation, ionization, charge transfer and their combinations) from low through intermediate to high (non-relativistic as well as relativistic) impact energies. Also included are interactions of ions with surfaces, molecule and clusters.

More specifically, the main themes covered by ISIAC are:

– Fundamental processes in ion-atom and ion-molecule collisions,

– Ion-surface and ion-solid interactions,

– Fragmentation of molecules and clusters,

– Ion-biomolecule interactions,

– Laser-assisted collisions,

– Atomic and molecular collision processes,

– New development techniques and instrumentation,

–Versatile applications, including cross-disciplinary branching to plasma physics, astrophysics to biomedicine with hadron therapy.

Some of these themes are also present in this ISIAC-2017 book by way of the following 11 chapters:

Chapter 1 (Experiment), I. Madesis, A. Laoutaris, T.J.M. Zouros, S. Nanos, and E.P. Benis, "Projectile electron spectroscopy and new answers to old questions: Latest results at the new atomic physics beamline in Demokritos, Athens"

Chapter 2 (Experiment), R.T. Zhang and C.C. Havener, "Soft X-ray emission following charge exchange at solar wind velocities"

Chapter 3 (Experiment), L.C. Tribedi, "Ionization of large biomolecules by fast heavy ions"

Chapter 4 (Theory), R.D. Rivarola, C.A. Tachino, O.A. Fojón and M.F. Ciappina, "Young-type interference effects in heavy-ion collisions with diatomic molecules"

Chapter 5 (Experiment and Theory), L. Nagy, F. Járai-Szabó, S. Borbély, T. Arthanayaka, B.R. Lamichhane, A. Hasan and M. Schulz, "Wave packet scattering in intermediate-energy p-He collisions"

Chapter 6 (Theory), M.J. Ambrosio and L.U. Ancarani, "Double ionization of helium by fast electrons"

Chapter 7 (Theory), M.S. Gravielle, J.E. Miraglia and G.A. Bocan, "Interaction potentials for atoms in front of a LiF(001) surface"

Chapter 8 (Theory), R. Rodríguez-Segundo, D.J. Arismendi-Arrieta and R. Prosmiti, "A bottom-up approach for ion-water interactions: from clusters to bulk"

Chapter 9 (Theory), J.J. Bailey, I.B. Abdurakhmanov, A.S. Kadyrov and I. Bray, "The stopping power of hydrogen for protons and antiprotons"

Chapter 10 (Theory), S. Guatelli, D. Bolst, Z. Francis, S. Incerti, V. Ivanchenko and A.B. Rosenfeld, "Physics models for Monte Carlo simulations in carbon ion therapy"

Chapter 11 (Theory), Dž. Belkić, "Theory of heavy ion collision physics for hadron therapy".

The plan of the book runs as follows. The individual contents of each chapter are shown together, one after another, so as to give the first insight into the specific topics covered by the book. This is followed by the list of the joint abbreviations. Subsequently, every chapter begins with an abstract, continues with its own contents and proceeds with the main presentation, accompanied with illustrations and discussion. Such a chapter-by-chapter exposition is finalized by conclusions and bibliography. The book itself ends with the subject Index.

To aid the general reader, all the authors tried to present their chapters in the context of the development of the addressed particular themes and the underlying major ideas and intricacies. Some chapters contain new results that have not been previously published elsewhere. Whenever possible, the authors made their attempts to connect the basic research in atomic/molecular collision physics with some important applications in other branches of physics as well as across the physics borders. It is hoped that the material presented in this book will be interesting and useful to the beginners and specialists alike. The contents and expositions are deemed to be helpful to the beginners in assessing the potential overlap of some of the presented material with their own research themes and this might provide motivations for possible further upgrades. Likewise, specialists could take advantage of these reviews to see where the addressed themes were and where they are going, in order to acknowledge the fruits of the efforts made thus far and actively contribute to tailoring the directions of future research.

Overall, this book is truly interdisciplinary. It judiciously combines experiments and theories within particle collision physics on atomic and molecular levels. It presents state-of-the-art fundamental research in this field. It addresses the possibilities for significant and versatile applications outside standard atomic and molecular collision physics ranging from astrophysics, surface as well as cluster physics/chemistry, hadron therapy in medicine and to the chemical industry. It is then, as Volume 2, fully in the spirit of the "Aims and Scope" of this Book Series by reference to its "Mission Statement".

Editors of Volume 2: *Dževad Belkić, Igor Bray and Alisher Kadyrov*

Acknowledgments

Thanks are due to the Swedish Funding Agencies: The Jubilee Gustav the 5th Research Fund (Radiumhemmet, Karolinska University Hospital, Stockholm) and the Research Fund of Stockholm County Council (FoUU — Forskning för Utbildning och Utveckling: Research for Education and Development). Essential help for language editing was received from Adjunct Professor Dr. Karen (Edinger) Belkić, MD, clinical researcher. LaTex processing of the final, camera-ready manuscript was assisted by Associate Professor Dr. Nenad Milojević, theoretical atomic physicist. Indispensible advice and expertise of Mr. D. Rajesh Babu (from World Scientific Publishing) in the finalization process of this book are highly appreciated. Also invaluable were the continued support and patience of Ms. Kim Tan and Ms. Yolande Koh from World Scientific Publishing.

Contents

Abbreviation	Description
3D	Three-dimensional
AAA	Advanced adiabatic approach
A-CDW	Atomic-continuum distorted wave
ADC	Autoionizing double capture
AI	Adiabatic ionization
AO	Atomic orbital
AOCC	Atomic orbital close coupling
APAPES	Atomic physics with accelerators & PES
B	Becke
B-CDW	Bragg-continuum distorted wave
BIC	Binary light ion cascade
BP	Bragg peak
BSSE	Basis set superposition error
CB1	First Born with correct boundary conditions
CBC	Correct boundary conditions
CC	Charge-charge
CCC	Convergent close coupling
CD	Charge-dipole
CCSD(T)-F12	Correlated coupled-cluster single, double & perturbative triples
CDW-EIS	Continuum distorted wave - eikonal initial state
CDW-HF	Continuum distorted wave – Hartree-Fock
CEM	Channel electron multiplier
CNDO	Complete neglect of differential overlap
COB	Classical over-barrier
COLTRIMS	Cold target recoil ion momentum spectroscopy
CPU	Central processing unit
CR	Clementi-Roetti
CR-RHF	Clementi-Roetti-Roothaan-Hartree-Fock
CTMC	Classical trajectory Monte Carlo
CTMC-COB	Classical trajectory Monte Carlo – classical over barrier
CX	Charge exchange
DD	Dipole-dipole
DDCS	Double differential cross section
DEA	Dissociative electron attachment
DF-MP2	Density-fitting – Møller-Plesset of second order
DFT	Density functional theory
DNA	Deoxyribonucleic acid
DW	Distorted wave
ECC	Electron capture to continuum

Abbreviation	Description
ECP	Effective core potential
ECR	Electron cyclotron resonance
ECRIA	Electron cyclotron resonance ion accelerator
EL	Electron loss
EM	Electromagnetic
END	Electron nuclear dynamics
ET	Electron transfer
FAD	Fast atom diffraction
FBA	First Born approximation
FBDR	Forward backward DDCS ratio
FC	Faraday cup
FDCS	Fully differential cross section
FEDVR	Finite element discrete variable representation
FLUKA	Fluktuierende kaskade (Fluctuating cascades)
FWHM	Full width at half maximum
GEANT4	Geometry and tracking, version # 4
GEM	Generalized evaporation model
GGA	Generalized gradient approximation
GIFAD	Grazing-incidence fast atom diffraction
GPS	Gas post-stripper
GSI	Gesellschaft für Schwerionenforschung (Helmholtz Center for Heavy Ion Research)
GSF	Generalized Sturmian function
GTO	Gaussian-type orbital
GTS	Gas terminal stripper
HB	Hydrogen bond
HDA	Hemispherical deflector analyzer
HEP	High energy physics
HF	Hartree-Fock
HIMAC	Heavy ion medical accelerator in Chiba
HP	High precision
IAM	Independent atom model
ICRU	International Commission on Radiation Units & Measurements
IEM	Independent electron model
IHB	Ion hydrogen bond
IMRT	Intensity modulated radiotherapy
INCL	Liège intranuclear cascade
IR	Infrared

Abbreviation	Description
KE	Kinetic energy
LCAO	Linear combination of atomic orbitals
LDA	Local density approximation
LEED	Low-energy electron diffraction
LLPB3LYP	Lee-Lee-Parr+Becke+3-coefficient-Lee-Yang-Parr
LYP	Lee-Yang-Parr
MAE	Mean absolute error
MC	Monte Carlo
MCNPX	Monte Carlo N-particle extended
MCS	Multiple Coulomb scattering
MEP	Minimum energy path
MIRF	Multi-charged ion research facility/ORNL
MO	Molecular orbital
MOCC	Molecular orbital close coupling
MP2	Møller-Plesset of second order
MP4	Møller-Plesset of fourth order
MRS	Magnetic resonance spectroscopy
MSP	Multi-sphere plate
NCSRD	National Center for Scientific Research "Demokritos", Athens
NIRS	National Institute of Radiological Science
NTE	Non-resonant transfer excitation
OCE	One center expansion
OCE-CDW	One center expansion – continuum distorted wave
OER	Oxygen enhancement ratio
OFHC	Oxygen free high conductivity
OO	Object orientated
ORNL	Oak Ridge National Laboratory
PA	Pairwise additive
PAW	Projector-augmented wave
PAW-PBE	Projector-augmented wave + Perdew-Burke-Ernzerhof
PAW-PBE-vdW	Projector-augmented wave + Perdew-Burke-Ernzerhof + vdW
PBE	Perdew-Burke-Ernzerhof
PCS	Photoionization cross section
PES	Potential energy surface
PHITS	Particle and heavy ion transport
PSD	Position sensitive detector
PTCOG	Particle therapy co-operative group
PWB	Plane wave Born

Abbreviation	Description
QMD	Quantum molecular dynamics
QMOCC	Quantum molecular orbital close coupling
RBE	Relative biological effectiveness
RBED	Relativistic binary encounter Bethe
RGA	Residual gas analyzer
RHF	Roothaan-Hartree-Fock
RIMS	Recoil ion momentum spectrometer
RNA	Ribonucleic acid
RTE	Resonant transfer excitation
SCA	Semi-classical approximation
SCAN	Strongly constrained and appropriately normed
SCF	Self consistent field
SDCS	Single differential cross section
SHIELD-HIT	Shield-heavy ion transport
SI	Single ionization
SIMION®	Ion and electron optics simulator
SIVR	Surface initial-value representation
SOBP	Spread out Bragg peak
STO	Slater-type orbital
TANDEM	The 5.5 MV tandem Van de Graaff accelerator/NCSRD
TCBGM	Two center basis generator method
TCS	Total cross section
TDCC	Time-dependent close-coupling
TDCS	Triply differential cross section
TDSE	Time-dependent Schrödinger equation
TE	Transfer excitation
TEC	Two-effective center
TI	Transfer ionization
ToF	Time-of-flight
TPS	Treatment planning system
TTM	Thole-type model
vdW	van der Waals
WET	Water equivalent thickness
WF	Wave function
XDM	Exchange-hole dipole moment
ZAPS	Zero-degree Auger projectile spectroscopy

Chapter 1

Projectile electron spectroscopy and new answers to old questions: Latest results at the new atomic physics beamline in Demokritos, Athens

I. Madesis[1,3], A. Laoutaris[1,3], T.J.M. Zouros[1,3*], S. Nanos[2] and E.P. Benis[2]

[1]*Department of Physics, University of Crete, Voutes University Campus, GR-70013 Heraklion, Crete, Greece*

[2]*Department of Physics, University of Ioannina, GR-45110 Ioannina, Greece*

[3]*Tandem Accelerator Laboratory, Institute of Nuclear and Particle Physics, National Center for Scientific Research "Demokritos", GR-15310 Aghia Paraskevi, Greece*

A new experimental station has been setup at the Athens 5.5 MV tandem accelerator to perform zero-degree Auger projectile electron spectroscopy (ZAPS). In addition, two systems of gas and foil terminal- and post-strippers have also been installed to vary the metastable $1s2s\,^3S$ beam component in the He-like ion beams. After a short introduction to the ZAPS technique and our new experimental station, we present new results on the production of the $C^{3+}(1s2s2p\,^{2,4}P, 1s2p^2\,^2D)$ and $C^{4+}(2s2p\,^{3,1}P)$ states in collisions of mixed $C^{4+}(1s2s\,^3S,\ 1s^2)$ ions with He. The $2s2p\,^{3,1}P$ states can be produced by direct $1s \rightarrow 2p$ excitation, electron-electron interaction plus spin exchange and/or transfer-loss processes. The $1s2s2p$ states can be produced either by direct electron capture to the $1s2s\,^3S$ component or by transfer-excitation from the $1s^2$ ground state. A new technique allows for the separate determination of the contributions from the ground and metastable beam components. The ratio $R_m = \sigma_m(^4P)/\sigma_m(^2P)$ of transfer to the metastable component leading to the production of the $^4P,^2P$ states is determined and compared to other older measurements on B^{3+} and F^{7+} similarly analyzed, giving $R_m \simeq 2$ in agreement with spin statistics. No convincing evidence of cascade feeding of the $1s2s2p\,^4P$ state is found, in clear dis-

*Corresponding author. Email: tzouros@physics.uoc.gr

agreement with previous experimental and theoretical results. Clearly, more work is needed to resolve this puzzle.

Contents

1. Introduction

Over the last few decades, considerable progress has been made to obtain information on both the atomic structure and collision dynamics of multiply excited atomic states using high resolution Auger electron spectroscopy [1, 2]. This interest has been generated to a large degree in the fields of plasma physics, thermonuclear fusion research, astrophysics and radiotherapy with light ions (hadron therapy) [3], where the collisional properties of highly stripped ions play an important role. The determination of highly accurate excitation energies, transition rates and lifetimes combined with production cross section information obtained from line intensity measurements lead to a better overall understanding of the dominant processes at play.

The technique of zero-degree Auger projectile electron spectroscopy (ZAPS), described next, has been highly successful in enabling measurements of LS-resolved cross sections as a function of collision energy. The

ZAPS technique offers an experimental window, particularly for ions with low projectile atomic number (Z_p) where the Auger yields are more favorable, for an improved understanding of ion-atom collision mechanisms at the state-selective level and provides stringent testing of theory.

A state-of-the-art ZAPS experimental station has recently been setup in the Athens 5.5 MV tandem Van de Graaff accelerator (TANDEM) at the National Center for Scientific Research "Demokritos" under the Atomic physics with accelerators: projectile electron spectroscopy (APAPES) [4] initiative. Recent progress and new results are presented here.

2. The technique of zero-degree Auger projectile spectroscopy

High resolution Auger projectile spectroscopy refers to the experimental energy- and angular-resolved electron spectroscopic technique in which Auger electrons ejected from a moving emitter (usually a projectile ion beam) are energy analyzed and detected at or close to the observation angle $\theta = 0°$ with respect to the beam direction. This technique is, therefore, known as ZAPS [1].

The ZAPS technique refers mostly to high energy resolution ($< 0.5\%$) measurements of Auger KLL electrons with characteristic energies in the range of 50-600 eV (in the projectile rest frame). It also includes electrons emitted in autoionization, autodetachment as well as photo-ionization processes of ions. In the future, ZAPS could possibly include even high energy [5] conversion [6–8] and beta-electron [9, 10] emission processes measured by magnetic zero-degree spectrometers. All these emitted electrons have characteristic energies and, therefore, carry important atomic structure information such as state energy, binding energy, line width, etc. The particular laboratory detection angle of $\theta = 0°$ offers the optimal kinematic conditions for attaining the highest possible resolution. Equally important, within the category of state-selective measurements, the ZAPS technique also allows for the determination of LS-resolved electron production cross sections. This provides important information about the projectile dynamics in atomic collisions, such as single electron capture, a process of primary interest to the APAPES [4] investigations.

2.1. *Kinematics of fast emitters*

Since the measured electrons are emitted from fast projectile ions, it is important to understand some of the basics of projectile electron kinematics.

A detailed analysis can be quite complicated [11]. However, in the case of energetic collisions of a few MeV/u or more, projectile ions are scattered through very small angles (~mrads) resulting in negligible energy loss and influence on the projectile trajectory. Thus, a simple velocity vector addition model is sufficient for the determination of most kinematic effects [1]. The velocity $\boldsymbol{v}$ of the Auger electron in the laboratory frame is obtained by adding the projectile velocity $\boldsymbol{V}_p$ to the velocity $\boldsymbol{v}'$ of the electron in the projectile rest frame, as shown in Fig. 1. Using the notation with primed quantities in the projectile rest frame, the electron kinetic energy ε in the laboratory frame can be related to the corresponding rest frame ε' as [1]:

$$\varepsilon = \frac{1}{2} m\boldsymbol{v} \cdot \boldsymbol{v} = \varepsilon' + t_p + 2\sqrt{\varepsilon' t_p} \cos\theta', \tag{1a}$$

or its more accurate relativistic counterpart

$$\varepsilon = \gamma_p \varepsilon' + t_p + \sqrt{(1+\gamma')(1+\gamma_p)\varepsilon' t_p} \cos\theta'. \tag{1b}$$

Here, $t_p = (m/M_p)E_p$ is the reduced projectile energy (the kinetic energy of an isotachic electron, also known as the Cusp energy), E_p and M_p are the kinetic energy and mass of the projectile, respectively, while m is the electron mass and $\gamma_p(= 1 + t_p/(mc^2))$ and $\gamma'(= 1 + \varepsilon'/(mc^2))$ are the usual relativistic γ-parameters. Both the second and the third term in Eqs. (1a) and (1b) imply a substantial energy shift from the rest frame energy (the Auger electron energy) ε' to either higher or lower laboratory energies ε depending on the value of θ'. This latter angle determines whether the resulting laboratory electron speed v is larger or smaller than the speed of the electron v′ in the projectile rest frame. The third term in Eqs. (1a) and (1b) introduces a stretching effect of the spectra that can be used experimentally with the advantage of achieving high energy resolution measurements. In addition, for fast emitters, there is a limiting laboratory observation angle θ_{max}, as per Fig. 1 [left], beyond which no projectile electrons can be observed. Finally, the relativistic effects at a few MeV/u beam energies are just a small, but observable correction (a few hundred meV) to the measured laboratory electron energy ε. This helps in the more accurate energy calibration and identification of the Auger lines.

For an electron spectrometer with a finite acceptance angle $\Delta\theta'$ (with $\Delta\theta$ in the lab frame), it is simple to show [1], by direct differentiation of Eq. (1) with respect to θ', that the following proportionality holds true:

$$\frac{d\varepsilon}{d\theta'} \propto \sin\theta'. \tag{2}$$

Such a dependence on the finite acceptance angle of the spectrometer induces a kinematic broadening, which is a substantial limiting factor at non-zero observation angles. However, at the observation angle $\theta = 0°$, corresponding to $\theta' = 0°$ or $180°$, it follows that $d\varepsilon/d\theta'$ is zero. The implication is that only the second order contributions in $\Delta\theta'$ or smaller are expected. A simple calculation shows that for a practical electron spectrometer acceptance angle of $\Delta\theta \simeq 0.57°$ or $\simeq 10$ mrad, a substantial, almost two-orders of magnitude improvement becomes attainable. This enables high resolution ($\Delta\varepsilon \simeq 100$ meV) electron spectra measurements at realistic count rates. At

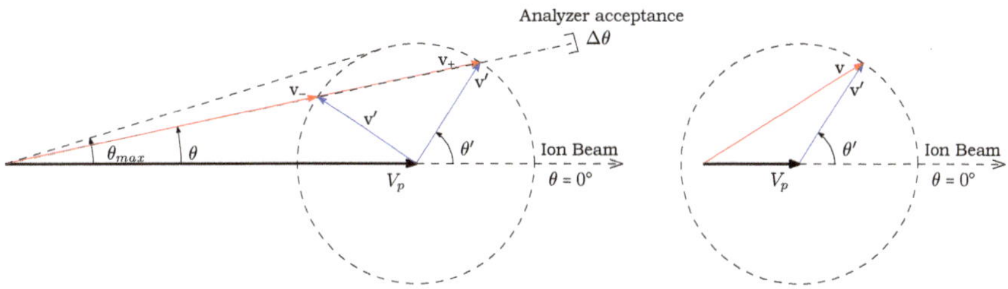

Fig. 1. Velocity addition diagrams. The electron velocity $\boldsymbol{v}'$ in the projectile rest frame is transformed to that in the laboratory frame according to the vector addition rule $\boldsymbol{v} = \boldsymbol{V}_p + \boldsymbol{v}'$, where $\boldsymbol{V}_p$ is the velocity of the projectile emitter. Left: $V_p > \mathrm{v}'$ leading to a maximum possible laboratory emission angle θ_{max}. Right: $V_p < \mathrm{v}'$, with all observation angles θ being possible. The ZAPS technique sets the electron analyzer at $\theta = 0°$, where kinematic broadening effects due to the finite analyzer acceptance angle $\Delta\theta$ are minimized (see the main text).

$\theta = 0°$, the angle θ' can be either $0°$ or $180°$, according to Fig. 1. This results in a corresponding addition or subtraction of velocities. In such a case, the ensuing laboratory electron energy can be readily obtained from Eq. (1) by setting $\theta' = 0°$ or $180°$ to be:

$$\varepsilon_{\pm}(\theta = 0°) = (\sqrt{\varepsilon'} \pm \sqrt{t_p})^2, \tag{3a}$$

or its more accurate relativistic counterpart

$$\varepsilon_{\pm}(\theta = 0°) = \frac{1}{2}\left[\sqrt{(1+\gamma_p)\varepsilon'} \pm \sqrt{(1+\gamma')t_p}\right]^2. \tag{3b}$$

In Fig. 2, we show typical high resolution zero-degree electron spectra, where the + sign in Eq. (3) is applicable. The projectile lines corresponding to KLL-Auger transitions are well resolved. The particular states can be readily identified by comparison to the available atomic structure calculations [12].

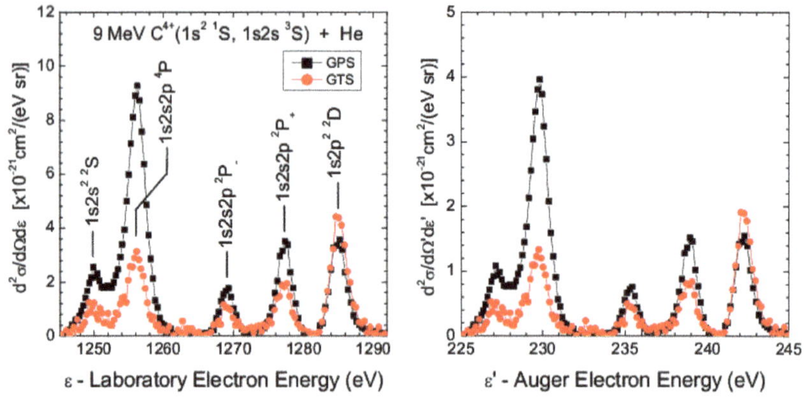

Fig. 2. Typical ZAPS electron spectra obtained with our new apparatus at Demokritos for 9 MeV C^{4+} collisions with helium gas. Left: Laboratory frame spectra. Right: The corresponding projectile rest frame spectra. The spectra were obtained by: Gas Post-Stripping (GPS) the delivered C^{3+} beam to the final C^{4+} beam (squares), Gas Terminal Stripping (GTS) of the C^- beam in the terminal of the accelerator (circles). As can be seen, these different stripping methods produce different ratios of the ${}^4P/{}^2D$ line intensities. This implies different amounts of pre-excited metastable fraction $1s2s\,{}^3S$. The energy axes are related by Eq. (1), while the axes of double differential cross sections have been transformed according to Eq. (4).

In addition to the transformation of the energy axis, there is also a corresponding transformation of the double differential cross section (DDCS) on the y-axis such that:

$$\frac{d^2\sigma}{d\Omega' d\varepsilon'} = \sqrt{\frac{\varepsilon'}{\varepsilon}}\frac{d^2\sigma}{d\Omega d\varepsilon}. \tag{4}$$

In the case of Fig. 2, since $\varepsilon' < \varepsilon$, the transformation of Eq. (4) to the rest frame shows that the spectra in the laboratory frame are relatively enhanced by the factor $\sqrt{\varepsilon/\varepsilon'}$, which is an additional help.

As illustrated in Fig. 1, for fast emitters ($V_p > \mathrm{v}'$ or equivalently $t_p > \varepsilon'$), there can be two possible solutions $\varepsilon_\pm$ (with corresponding velocities $\mathrm{v}_\pm$) for the same laboratory emission angle θ. For slow emitters ($V_p \leq \mathrm{v}'$ or equivalently $t_p \leq \varepsilon'$), there is only one solution. On the other hand, for fast emitters, this results in a restriction on the emission angle θ to within a maximum value:

$$\theta_{max} = \arcsin\left(\frac{\mathrm{v}'}{V_p}\right), \tag{5}$$

or its more accurate relativistic counterpart resulting from Eq. (1b)

$$\theta_{max} = \arcsin\left\{1 + \gamma_p^2\left[(V_p/\mathrm{v}')^2 - 1\right]\right\}^{-1/2}. \tag{6}$$

2.2. *The ZAPS experimental setup of the APAPES project*

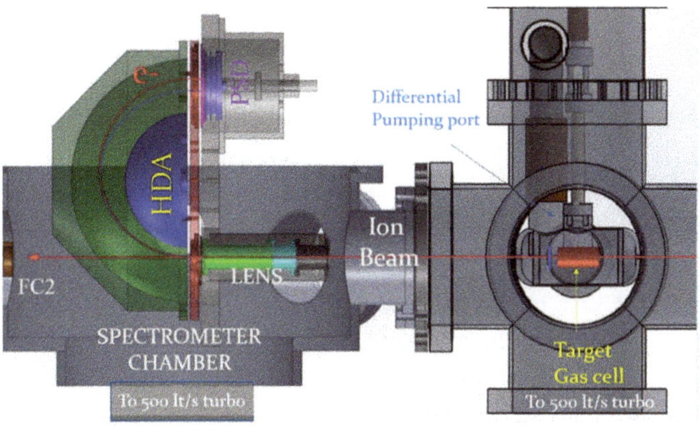

Fig. 3. Our experimental setup consists (from right to left) of a doubly differentially pumped target gas cell, a 4-element entry lens, a large HDA with a 40 mm diameter two-dimensional PSD. The length of the target gas cell is $L_c = 50$ mm and the distance of its center to the lens entry is $s_0 = 289$ mm (see also Fig. 8).

The ZAPS experimental setup of the APAPES project is hosted on beam line L45 at the Demokritos TANDEM measurement hall dedicated to atomic collision physics and is shown in Fig. 3. The mixed state C^{4+} ion beam supplied by the TANDEM is directed into a doubly differentially pumped gas cell where it collides with the target gas. With this system, a stable analyzer chamber vacuum of 10^{-6} Torr can be maintained for typical gas cell pressures in the mTorr range. The emitted Auger electrons are analyzed in energy with a maximum resolution $\Delta\varepsilon/\varepsilon \simeq 0.1\%$ using an electrostatic paracentric hemispherical deflector analyzer (HDA) [13–17] equipped with a 4-element focusing/deceleration entry lens and a two-dimensional (2D) position sensitive detector (PSD) [18, 19]. Since the energy resolution is fixed by the spatial characteristics of the analyzer, higher absolute resolution $\Delta\varepsilon$ is achieved by pre-retarding the Auger electrons. A pre-retardation factor of $F = 4$ is usually sufficient to resolve the Li-like KLL-Auger lines presented here. Finally, the analyzed electron spectra are normalized to the number of ions collected in the final Faraday Cup (FC2) located at the exit after the HDA (see Fig. 3).

The most critical part of the spectrometer is the four-element electrostatic lens which focuses the emitted electrons from the gas cell to the entry of the HDA. This provides a much smaller virtual slit of less than a millimeter compared to the much larger (6 mm) diameter of the physical HDA entry aperture. Such an arrangement has been found very successful [18] since it allows for the unobstructed passage of the ion beam through the entire lens and the HDA without the production of spurious background electrons. At the same time, it provides the appropriate conditions for high transmission and high energy resolution electron optics in the combined lens plus HDA system. Thus, a single-stage arrangement can be used, as opposed to the conventional two-stage arrangement. This facilitates the use of a 2D PSD instead of the conventional slit exit to increase the overall counting efficiency by at least two orders [20]. The entire system has been simulated [21] in detail (see Fig. 4) by the ion and electron optics simulator – SIMION [22], which provides an alternative, well-tested [15] way to determine the appropriate voltages on both lens and the HDA for optimal operation. The resulting full width at half maximum (FWHM) energy resolution as a function of the pre-retardation parameter F is shown in Fig. 5.

Following the successful installation and testing both the terminal gas stripper and foil/gas post-strippers, we have been able to obtain Auger spectra with variable amounts of metastable beam fraction as already shown in Fig. 2. We then apply the new technique recently developed for the accurate determination of the electron capture ratio R_m [23] (see section 4.4 under the technical restrictions of our method).

3. Ion beam stripping considerations — production of pre-excited ions

The number of electrons carried into the collision by the projectile is important since it also determines the resulting atomic state after the collision, depending on whether a net ionization, excitation or capture process takes place. Highly-charged ions typically carry just a few electrons, resulting in easier interpretable collision spectra which, in turn, can provide more stringent tests of theory [1].

The production of highly-charged ions is, therefore, of great interest in ion-atom collisions. Such ions are typically produced by passing a low charge state beam through a thin foil or gas, where numerous electrons can

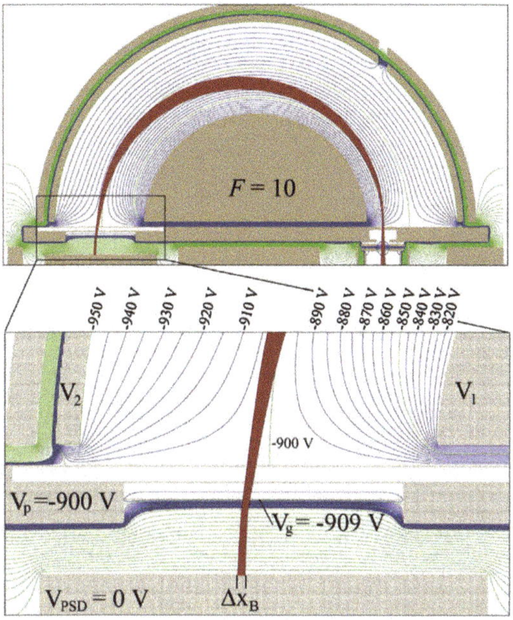

Fig. 4. Details of SIMION schematic of the HDA electrodes, equipotentials and trajectories in the symmetry plane $y = 0$. The insert shows the PSD region at the exit of the HDA for pre-retardation factor $F = 10$ (adapted from Ref. [21]).

be stripped from the ion, thus increasing its charge state. These strippers are found in the terminal of all tandem Van de Graaff accelerators since they are needed to convert the initially negatively charged ion beam to a subsequently positively charged beam, which is further energy boosted in the second stage of acceleration. The desired charge state and energy are typically selected by the analyzing magnet and sent on to the experiment. Depending on the energy of the ion beam during the stripping process, a particular Gaussian-like charge state distribution emerges as being centered around the mean charge state. The higher the energy of the beam, the higher the mean charge state [24–26]. Thus, to produce more intense few-electron or even bare ion beams, additional stripping points are provided after the beam exits the accelerator. These are known as post-strippers.

Few-electron ionic beams, depending on the number of electrons, can be delivered in mixed states including ground state and other long-lived metastable components. The lifetimes of the He-like beam metastable components $1s2s\,^3S$ and $1s2s\,^1S$ have been mostly calculated and are seen in

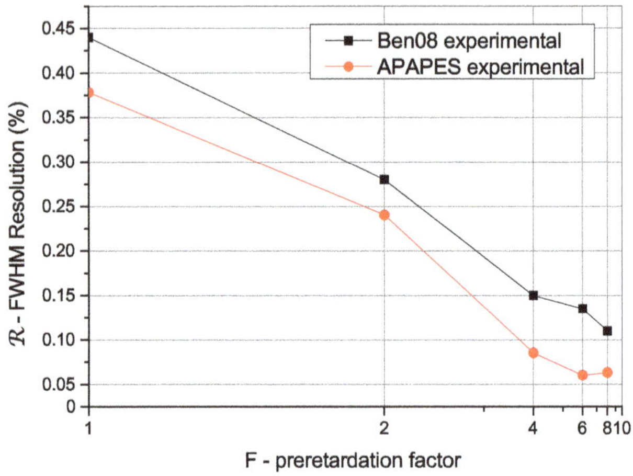

Fig. 5. The FWHM energy resolution is plotted as a function of $1/F$ for pre-retardation factor F. The FWHM from the older work [15] (squares). The improved FWHM obtained by lens voltages optimized by SIMION [21] (Circles).

Fig. 6(a) to drop rapidly with projectile atomic number Z_p. Assuming the states are statistically produced in a $3:1$ ratio with a 30% content of $1s2s\,^3S$ at the point of production, we show in Fig. 6(b) the surviving fraction from either the terminal stripper in the accelerator tank (at a distance $s_1 = 25.4$ m) or the post-stripper (at a distance $s_2 = 10$ m), away from the target interaction region of our experiment, as a function of Z_p. Clearly, because of its much longer lifetime, the $1s2s\,^3S$ fraction remains practically intact over the whole range of $Z_p = 3-9$, while the $1s2s\,^1S$ fraction is considerably reduced (in the present case of the C^{4+} beams) to below 5%. Therefore, the $1s2s\,^1S$ fraction has been assumed to have a negligible effect on the analysis.

An additional requirement of the APAPES project is to also have a variable (and controllable) amount of metastable beam component ($1s2s\,^3S$). To date, there are no codes that can generally predict the fraction of metastable ions. However, previous studies have shown that the amount of metastable beam components varies mainly depending on whether a foil or a gas stripper is used. The metastable amount of $1s2s\,^3S$ typically reaches a maximum of ~30% for He-like ions, when using a foil stripper [28–33]. We have recently installed a gas terminal stripper inside the TANDEM which produces the He-like beam predominantly in the ground state. By performing two different measurements with beams of appreciably different

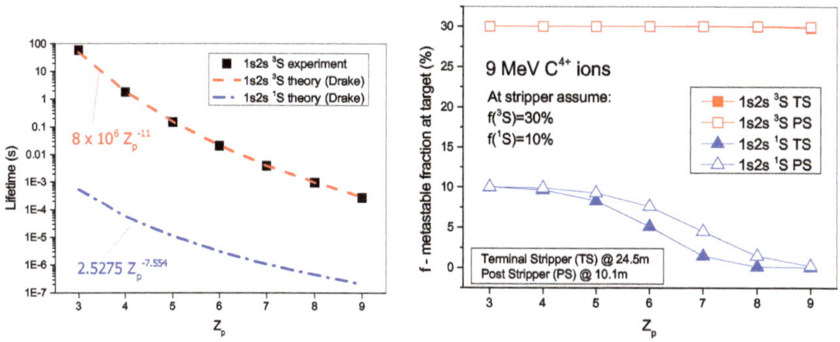

(a) Lifetimes of states. (b) Survival of metastable fractions.

Fig. 6. Parameters of the $1s2s\,^3S$ and $1s2s\,^1S$ levels as a function of projectile atomic number Z_p. (a) Lifetimes calculated by Drake [27]. The explicit Z_p dependencies are the results of least square fits (this work). (b) Survival of metastable fractions from their point of production either at the TANDEM tank terminal stripper or at the post-stripper (see the main text). The two square symbols overlap.

metastable fractions [33, 34], we can extract the contributions from either the ground state or the metastable state. This technique has been presented in detail in a recent publication [23].

4. Single-electron capture

In ion-atom collisions one of the more interesting processes that can be investigated is electron capture to excited atomic states. In single electron capture, an electron is transferred from the target atom to the projectile ion. Details of the electron transfer process depend on the collision system investigated. The information about the transfer to excited states of the projectile can be inferred spectroscopically by observing various transitions (either radiative [35, 36] or Auger [37]) from the excited atomic states to lower lying states.

4.1. *Capture to the metastable $1s2s\,^3S$ beam component*

Recently, interest has focused on electron capture to pre-excited states of ions [37–41], i.e. to collisions in which the incoming ion is already in some excited state. This allows for the investigation of excited states not readily produced in any other way. Thus, for example, the spectra shown in Fig. 2 were obtained using the two-electron C^{4+} ion beams in a mixture of pre-excited $1s2s\,^3S$ and $1s^2$ ground states. The $1s2s\,^3S$ state is metastable (long-lived) since it can only decay by a forbidden spin changing transition. The $1s2s\,^3S$ state is produced in the ion stripper inside the tandem accelerator terminal together with the ground state ions in variable amounts 5-30%, depending on the stripping medium (gas or foil) and stripping energy [29, 31]. Capture to the ground state $1s^22s$ of the formed Li-like ion cannot further de-excite so this transition is not accessible spectroscopically. However, electron capture to the $1s2s\,^3S$ metastable state leads to a spectroscopically rich spectrum:

$$C^{4+}(1s2s\,^3S) + e^- \rightarrow C^{3+}[(1s2s\,^3S)n\ell] \qquad (n \geq 2,\, 0 \leq \ell \leq n-1). \qquad (7)$$

The excited states formed in the case of $n = 2$ (according to electron spin-orbit coupling rules) are identified in the spectra of Fig. 2 from their characteristic Auger energies ε'_A:

$$\begin{array}{ll} C^{3+}[1s2s2\ell\,^{2S+1}L_J] & (L = \ell = 0, 1,\ S = 3/2, 1/2,\ \boldsymbol{J} = \boldsymbol{L} + \boldsymbol{S}) \\ \quad\downarrow & \\ \quad C^{4+}(1s^2) + e^-_A(\varepsilon'_A). & \end{array} \qquad (8)$$

Thus, the production of the $1s2s^2\,^2S$ and $1s2s2p\,^{2,4}P$ lines observed in the Auger spectra can be readily produced by direct single $2s$ or $2p$ electron capture into the $1s2s\,^3S$ component of the ion beam [40, 42]. However, the line labeled as $1s2p^2\,^2D$ cannot be produced by single electron capture to the $1s2s\,^3S$ state since it requires an additional $2s \rightarrow 2p$ excitation. In the energy range of these collisions, another related two-electron process involving the transfer of a target electron to the $2p$ orbital simultaneously with $1s \rightarrow 2p$ excitation from the ground state is quite probable. This process, known as transfer-excitation (TE) [43, 44], can populate the other two observed ^2P lines (through a simultaneous $2s$ transfer/$1s \rightarrow 2p$ excitation), but not the ^4P line. This again has to do with active spin selection rules which would require a low probability spin flipping TE transition to occur from the ground state. Therefore, according to our present understanding [23], the ^4P line is produced only from the metastable state by direct single electron capture [40, 42], while the ^2D line stems dominantly from the ground state through the aforementioned TE process.

Finally, the inherent long lifetime (ns-μs depending on projectile atomic number Z_p and total angular momentum J) of the $1s2s2p\,^4P_J$ metastable states (with $J = 1/2, 3/2, 5/2$ lines unresolved) leads to an additional difficulty in the quantitative analysis of the measured $1s2s2p\,^4P$ production cross section [34, 42]. In zero-degree measurements, the electron spectrometer lies in the direct path of the ion beam. Therefore, the metastable projectile states decay all along the ionic projectile path towards and through the spectrometer. Thus, one has to correctly take into account both the decay of these states along the path of observation and the increase of the spectrometer acceptance solid angle, as the emission point approaches the entry of the spectrometer. This can result in a considerable correction to the measured metastable electron yield. This correction has been treated in the literature, either in a purely geometrical approach [34, 42] or very recently, for our measurements using an HDA with entry lens, in a Monte Carlo (MC) electron trajectory simulation approach within the SIMION [22] charged particle optics software [45]. In this latter approach, kinematic effects, particular to Auger emission from fast moving projectile ions (e.g. kinematic line broadening and solid angle limitations due to θ_{max}) are included for the first time. This allows for a more accurate and realistic line shape modeling [45] of both metastable and prompt (much shorter lifetimes less than a few ps) Auger lines. The APAPES project has embarked on an isoelectronic study of electron transfer to the $1s2s\,^3S$ state of He-like ions to shed more light on these processes.

4.2. *Separation of contributions from the $1s^2$ ground state and the $1s2s\,^3S$ metastable state component*

The processes that predominantly contribute to the production of the five KLL states of interest are shown in Fig. 7. In general, we may write for the single differential yield of each level x (for $x : {}^2S, {}^2P_{\pm}, {}^4P, {}^2D$):

$$\frac{dY[x]}{d\Omega'} = (1 - f_{^3S})\left(\frac{d\sigma_g[x]}{d\Omega'}\right) + f_{^3S}\left(\frac{d\sigma_m[x]}{d\Omega'}\right), \tag{9}$$

where $d\sigma_g[x]/d\Omega'$ and $d\sigma_m[x]/d\Omega'$ represent the differential production cross sections for all possible processes that can populate the state x from either the ground state (g) or metastable states (m). The single differential yield is the yield normalized to the total number of ions in the beam N_I (no matter with which beam components these ions are associated):

$$\frac{dY}{d\Omega'}[x] = \frac{N^e[x]}{\kappa\, G_\tau[x]\,\xi[x]}. \tag{10}$$

Here, $\xi[x]$ is the Auger yield and $N^e[x]$ is the total number of electrons in the Auger peak x divided by the overall absolute "normalization" constant

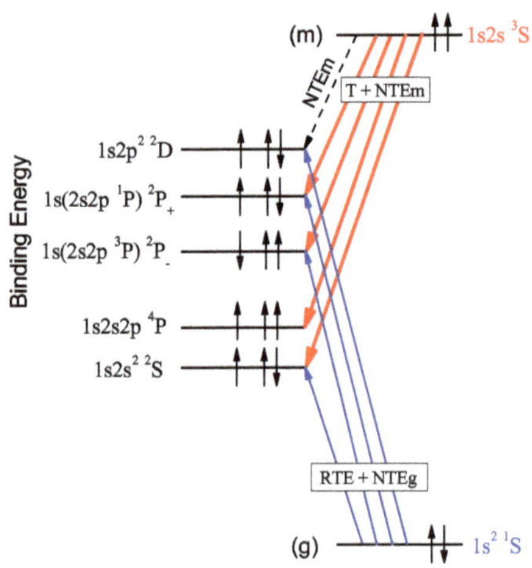

Fig. 7. Energy level diagram showing the dominant mechanisms for the production of the $1s2l2l'\,^{(2S+1)}L$ doubly excited states formed in collisions of energetic He-like $(1s^2\,^1S, 1s2s\,^3S)$ beams with low Z_t targets (e.g. H_2). Single $2p$ or $2s$ electron transfer (ET) or non-resonant transfer excitation (NTEm) to the metastable state (m), simultaneous transfer and excitation [both resonant (RTE) and non-resonant (NTEg)] from the ground state (g) (for more details, see the main text).

κ for each Auger line given by:

$$N^e[x] \equiv \int_{[x]} N^e(\varepsilon')\frac{d\varepsilon'}{\Delta\varepsilon} = \int_{[x]} \left(\frac{\varepsilon'}{\varepsilon}\right) N^e(\varepsilon)\frac{d\varepsilon}{\Delta\varepsilon}, \tag{11}$$

$$\kappa \equiv N_I\, n\, L_c\, T\, \overline{\eta}\, \overline{\Delta\Omega_0}, \tag{12}$$

$$\overline{\Delta\Omega}[x] = G_\tau[x]\,\overline{\Delta\Omega_0}, \tag{13}$$

where $\overline{\eta}$ is the detection efficiency averaged over the PSD spectrum energy range. Quantity $\overline{\Delta\Omega}[x]$ is the effective solid angle of the measurement, while G_τ is a spectrometer-dependent solid angle correction factor accounting for a possible dependence on the lifetime $\tau[x]$ of the particular state x (more correctly on the product $V_p\tau[x]$, see Refs. [46, 47]). Recall that the doublet states are all prompt, so that $G_\tau \simeq 1$, implying practically the same solid angle $\overline{\Delta\Omega_0}$. On the other hand, the ^4P state is metastable for which $G_\tau \equiv G_\tau[^4P]$ is significantly different from 1. We have recently computed G_τ for the HDA used in this work [45]. Computations of this correction factor for the two-stage parallel plate analyzer (tandem analyzer) have also been reported in Refs. [1, 34, 43, 48]. The exact value of G_τ is an important correction when measuring the Auger electron yield from the ^4P state for ions with low Z_p. This is further discussed below. Finally, κ is seen to include the values of all the experimental parameters of the measurement and the accuracy of each of these parameters determines the overall systematic uncertainty of the cross section. All the experimental parameters except for G_τ have been collected together into the overall absolute normalization parameter κ, which for our HDA has been computed and shown to have an overall systematic uncertainty $(\Delta\kappa)/\kappa \lesssim 0.17$ [49].

Let us now assume, by reference to Fig. 7, that the ^2S and $^2P_\pm$ levels are populated from both the ground state (g) and the metastable (m) state. Further, let the ^4P level be populated only from the metastable state (by capture). Moreover, let the ^2D level stem from the ground state alone (by TE). Then, it can be shown [23] that the production cross sections for the five levels can be obtained from the single differential yields dY_i (short notation for $dY_i/d\Omega'$) in two measurements having different metastable fractions. These cross sections are given by the following equations:

$$\frac{d\sigma_m[x]}{d\Omega'} = \frac{dY_2[x]\,dY_1[^2D] - dY_1[x]\,dY_2[^2D]}{dY_1[^2D] - dY_2[^2D]}, \tag{14}$$

$$\frac{d\sigma_g[x]}{d\Omega'} = \frac{dY_2[x]\,dY_1[^4P] - dY_1[x]\,dY_2[^4P]}{dY_1[^4P] - dY_2[^4P]}. \tag{15}$$

Note that Eqs. (14-15) satisfy the main assumption $d\sigma_g[^4P]/d\Omega' = d\sigma_m[^2D]/d\Omega' = 0$.

4.3. *Correction for the long lifetime of the* 1*s*2*s*2*p* ^4P *state - the solid angle correction factor* G_τ

The effective solid angle for electrons emitted from some long-lived Auger projectile states, $\overline{\Delta\Omega}$, can be determined adopting the geometry of Fig. 8. A projectile ion beam of velocity V_p is moving along the $+z$ axis and collides with a gas cell target of length L_c, the center of which is at a distance s_0 from the entry opening of the spectrometer. The projectile is collisionally excited to a long-lived J-state, having a lifetime τ_J, and subsequently Auger decays along the path to and through the spectrometer.

Then, the effective solid angle averaged over the length of the gas cell, $\overline{\Delta\Omega_J}$, can be determined from pure geometrical considerations using the double integral:

$$\overline{\Delta\Omega_J}(L, V_p\tau_J, L_c) \equiv \frac{1}{L_c}\int_{z'=0}^{L_c} dz' \int_{z=0}^{L-z'} dz\, \frac{e^{-z/V_p\tau_J}}{V_p\tau_J}\, \Delta\Omega_0(L - z' - z), \tag{16}$$

where L is the maximal distance along the ion trajectory over which significant contributions to the Auger electron line profile can be made [45]. The solid angle $\Delta\Omega_0(s)$ corresponds to point source emission from a distance s away from the spectrometer entry. For the HDA under study, the solid angle $\Delta\Omega_0(s)$ for point source emission (subtended by the lens entry aperture of radius r at distance s) is given by $\Delta\Omega_0(s) = 2\pi\left(1 - s/\sqrt{r^2+s^2}\right)$.

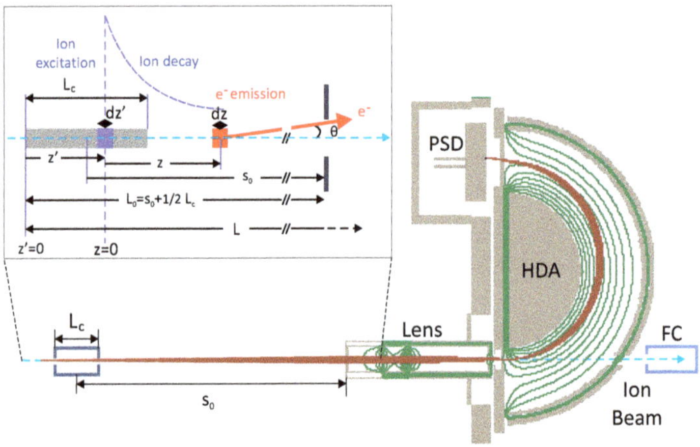

Fig. 8. Schematic geometries for the HDA spectrometer [45]. The ion beam propagates from the gas cell to and through the spectrometer to be finally collected by the FC2. Electron trajectories emitted at the 0° angle with respect to the ion beam are energy analyzed and detected at the two-dimensional PSD. Equipotential lines are also indicated. The insert shows the parameters relevant to the integrations of Eq. (16). Their values are: $L_c = 50$ mm, $r = 2$ mm and $s_0 = 289$ mm [21].

Regarding Eq. (16), it becomes clear that $\overline{\Delta\Omega_J}$ accounts for two competing effects: (i) the gain in the solid angle as the ions approach the spectrometer and (ii) the loss due to the temporal decay along the ion path towards and beyond the spectrometer. A more detailed examination of these effects is given in Ref. [45].

Applying Eq. (16) to a prompt state in the limit of $V_p\tau_J \to 0$, the effective solid angle becomes:

$$\overline{\Delta\Omega_0}(s_0, L_c) \equiv \frac{1}{L_c}\int_{z'=0}^{L_c} dz' \Delta\Omega_0(L_c/2 + s_0 - z'). \qquad (17)$$

Thus, we can define the effective solid angle correction factor for the J-state as the ratio of the effective solid angle of the long-lived state to that of the same, but for prompt emission, i.e.:

$$G_{\tau_J} = G_{\tau_J}(L, V_p\tau_J, s_0, L_c) \equiv \frac{\overline{\Delta\Omega_J}(L, V_p\tau_J, L_c)}{\overline{\Delta\Omega_0}(s_0, L_c)}. \qquad (18)$$

To account for the J sub-levels, since they are not energy resolved experimentally, a statistical averaging over all the J-states leads to the J-averaged effective solid angle correction factor G_τ given by [45]:

$$G_\tau = \frac{\sum_J (2J+1)\xi_J G_{\tau_J}(L, V_p\tau_J, s_0, L_c)}{\sum_J (2J+1)\xi_J}. \tag{19}$$

4.3.1. *Determination of G_τ: SIMION simulations*

In Ref. [45], we evaluated G_τ for our HDA using MC-type simulations realized within the SIMION 8.1 ion optics package in the following way. For a number N_0 of initially (t=0) populated projectiles for each J-state, the number of the observed Auger electrons will be:

$$N^e_{obs_J} \sim \xi_J N_0 \overline{\Delta\Omega_J}(L, V_p\tau_J, L_c). \tag{20}$$

However, the true number of detected electrons should be:

$$N^e_{true_J} \sim \xi_J N_0 \overline{\Delta\Omega_0}(s_0, L_c), \tag{21}$$

so that

$$\frac{N^e_{obs_J}}{N^e_{true_J}} = \frac{\overline{\Delta\Omega_J}(L, V_p\tau_J, L_c)}{\overline{\Delta\Omega_0}(s_0, L_c)} \equiv G_{\tau_J}. \tag{22}$$

Therefore, the ratio of the number of observed electrons of a long-lived J-state to the number of true electrons, which corresponds to a prompt state of the same initial population N_0, essentially describes the effective solid angle correction factor G_{τ_J}. The J-averaged effective solid angle correction factor G_τ is then estimated through Eq. (19).

4.3.2. *Experimental determination of G_τ using Be-like carbon beams*

The experimental determination of G_τ was based on the fact that the Be-like ions produced in tandem Van de Graaff accelerators are delivered in mixed ionic states, including the ground $1s^2 2s^2\ ^1S$ and the metastable $1s^2 2s2p\,^3P_J$ states with lifetimes in the μs-to-s range (depending again on atomic number Z_p and angular momentum J) [50–52]. In collisions with the H_2 targets, the $1s$ electron of the $1s^2 2s2p\,^3P$ state is needle ionized [53] (similar to the situation encountered in photo-ionization of Be-like ions [52, 54 57]). This results in the production of the $1s2s2p$ configuration, which in the LS coupling scheme gives rise to the $1s2s2p\,^4P$ and $1s(2s2p\,^3P)^2P$ states (or $^2P_-$ for short). These two states are used in our experimental determination of the effective solid angle correction factor G_τ. As already mentioned, the ^4P state is a long-lived state which Auger decays to the $1s^2$ ground state. On the other hand, the $^2P_-$ state Auger decays promptly to the $1s^2$ ground state having lifetimes ranging from tens to hundreds of fs.

In addition, the production population statistics of the ^4P and $^2P_-$ states should result in the ratios $\sigma(^4P) : \sigma(^2P_-) = 2 : 1$, as follows from the multiplicity of the states. A direct measurement of this ratio requires the necessary correction due to the metastability of the ^4P state and, therefore, the determination of the effective solid angle correction factor G_τ. Thus, we have:

$$\frac{\sigma(^4P)}{\sigma(^2P_-)} = \frac{Y(^4P)/(\xi_{^4P}G_\tau)}{Y(^2P_-)/\xi_{^2P_-}} = 2, \tag{23}$$

and, therefore,

$$G_\tau = \frac{1}{2}\frac{Y(^4P)}{Y(^2P_-)}\frac{\xi_{^2P_-}}{\xi_{^4P}}, \tag{24}$$

where Y denotes the measured normalized electron yields obtained from the fitted areas, while the values of the Auger yields are adopted from Ref. [58].

The production of the ^4P and $^2P_-$ states, by $1s$ ionization of the $1s^22s2p\,^3P$ beam component has an advantage crucial to the correct determination of the G_τ correction factor: it is free from additional deleterious effects on their ratio which can modify its spin statistics value of 2. As mentioned earlier, the ^4P and $^2P_-$ states can also be populated in collisions of He-like ions delivered in mixed $(1s^2, 1s2s\ ^3S)$ states via single capture to the $2p$ state. However, capture to higher $n\ell$-states is known to be strong [41, 59] and, as such, can lead to cascade feeding of these states [60], which could alter their expected spin statistics population ratio [61].

In Fig. 9, we present high resolution electron spectra measured with our HDA in 17.5 MeV collisions of O^{4+} with H_2. The electrons were recorded at pre-retardation factor F=8 for higher resolution. As can be seen, the small asymmetry in the peak near 425 eV is due to the additional low intensity $1s2s^22p\,^3P$ Auger line, which was also reported for similar spectra in Ref. [62]. This state can be formed from the $1s^22s^2\ ^1S$ ground state via $1s \to 2p$ excitation, and it will promptly decay to the $1s^22p$ final state. It is important to account for this extra peak area for the accurate determination of G_τ. Therefore, the line spectra were fitted with the Voigt profiles, except for the ^4P peak whose asymmetric profile, due to its metastability, was considered in the fitting process. The $1s2s^22p\,^3P$ peak area was found to be about 17% of the total area of both peaks in the vicinity of 425 eV.

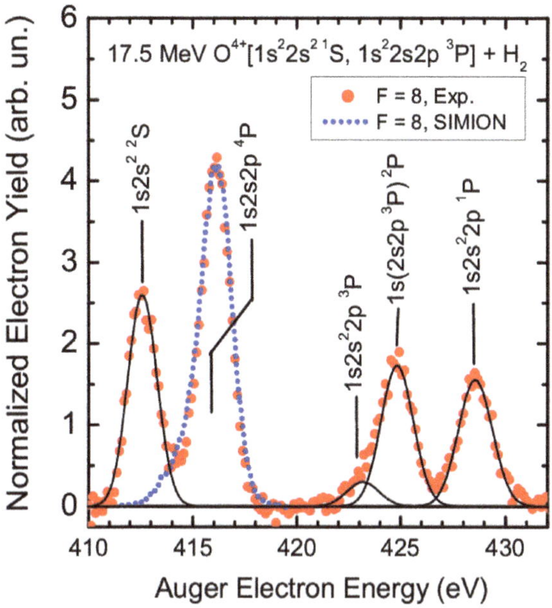

0.2

Fig. 9. Oxygen Auger KLL spectra obtained in collisions of 17.5 MeV O^{4+} with H_2 using our HDA spectrometer. The SIMION simulations of the ^4P spectral distributions are shown by the dotted lines in excellent agreement with the measurements. The solid lines correspond to the Voigt profile least square fits of the peaks. The small peak at the low energy shoulder of the $1s(2s2p\,^3P)\ ^2P$ peak, was identified as arising from the $1s2s^22p\,^3P$ intermediate state [62]. Its contribution was carefully accounted for and separated from that of the $^2P_-$ peak of primary interest (see the main text).

Our results for the collision system $O^{4+} + H_2$ at 17.5 MeV with two different deceleration factors ($F = 4$ and $F = 8$), as well as for the collision system $C^{2+} + H_2$ at 6.6 MeV with the deceleration factor of $F = 4$ are presented in Table 1.

For the case of oxygen, the G_τ values obtained with SIMION are comparable to the estimated experimental uncertainty of the measured values, for the lifetimes reported in Ref. [63], while they deviate by an additional ~ 20% for the lifetimes reported in Ref. [58]. However, in the case of carbon, the SIMION results are well within the experimental uncertainties. Further, it is seen from Table 1, that G_τ depends sensitively on the lifetimes and the involved Auger yields, as evident from the ≃ 15% differences in the estimated values of G_τ using the lifetimes reported in Refs. [46] and [58],

Table 1. Results for the effective solid angle correction factor G_τ determined experimentally (denoted as "Exp.", referred to Eq. (24)) and by the MC-type SIMION simulations.

State	E_p MeV	F	G_τ Exp.	G_τ SIMION
C^{2+}	6.6	4	2.0(4)	1.92[a]
				2.41[b]
O^{4+}	17.5	4	1.9(4)	2.47[a]
				2.80[b]
		8	1.5(4)	2.08[a]
				2.33[b]

[a] See Ref. [46], [b] see Ref. [58].

both for oxygen and carbon. Clearly, this sensitivity on the lifetimes can be a source of considerable systematic theoretical uncertainty in the estimation of G_τ. Thus, overall, we may claim that an acceptable agreement was found between our SIMION computations and the measured values of G_τ.

4.4. *Capture to metastable ratios R_m and r_m*

Basic quantum mechanics requires the spin coupling of the $2p$ electron to the $1s2s\,^3S$ state yielding $1s2s2p\,^4P$ quartet and $1s2s2p\,^2P$ doublet states to be in the ratio of $2:1$, i.e.:

$$R_m \equiv \frac{\frac{d\sigma_m[^4P]}{d\Omega'}}{\frac{d\sigma_m[^2P_+]}{d\Omega'} + \frac{d\sigma_m[^2P_-]}{d\Omega'}}, \tag{25}$$

while the ratio r_m

$$r_m \equiv \frac{\frac{d\sigma_m[^2P_+]}{d\Omega'}}{\frac{d\sigma_m[^2P_-]}{d\Omega'}}, \tag{26}$$

should be equal to 3. Here, σ_m is the capture cross section into the particular state [63]. Thus, it came as a big surprise that the measured ratio R_m was found to significantly deviate for the C^{4+} beams [40], clearly calling for an explanation. A new two-electron process termed the Pauli exchange interaction was subsequently proposed [40], where a target electron, which is spin-aligned with the two projectile electrons in the $1s2s\,^3S$ state, experiences a (slightly) different potential than an anti-aligned electron. If the interaction potentials are different, the outcomes (^4P vs. ^2P formation) could also be different. This process does not necessarily involve excitation of the projectile $1s$ electron and could possibly be described within the independent electron model, but one would need spin-dependent potentials - a difficult task - not yet attempted [64].

However, more recently, it has been shown that capture to higher-lying $(1s2lnl'\,^{2,4}L_J)$ states with $n > 2$ should also be very probable. Therefore, selective cascade feeding of the $1s2s2p\,^4P$ can also be expected to occur [34, 41, 60]. All the higher-lying doublet states Auger decay strongly to the K-shell, while the higher-lying quartet Auger decay much more weakly due to spin selection rules. The $E1$ transition rates, however, are the strongest for quartet to quartet and/or doublet to doublet transitions. This results in a selective cascade feeding mechanism directly enhancing just the ^4P populations.

Utilizing our method, presented in Sec. 4.2, for separating the contributions for the $1s2s\,^3S$ metastable and the $1s^2\,^1S$ ground states, the ratio R_m can be directly obtained in terms of the measured parameters using Eqs. (14) and (15):

$$R_{m,exp} = \frac{\frac{1}{G_\tau\,\xi[^4P]}\left(\frac{dY_2[^4P]}{dY_2[^2D]} - \frac{dY_1[^4P]}{dY_1[^2D]}\right)}{\frac{1}{\xi[^2P_+]}\left(\frac{dY_2[^2P_+]}{dY_2[^2D]} - \frac{dY_1[^2P_+]}{dY_1[^2D]}\right) + \frac{1}{\xi[^2P_-]}\left(\frac{dY_2[^2P_-]}{dY_2[^2D]} - \frac{dY_1[^2P_-]}{dY_1[^2D]}\right)}. \tag{27}$$

As can be seen, R_m is expressed as the ratios of the peak areas of the same spectrum (the same measurement) in which the absolute normalization parameters κ have canceled, while there is no direct evidence of the metastable fractions. Clearly, this is a considerable advantage since both the absolute normalization and metastable fraction determination are generally more tedious to obtain because special attention is required regarding some additional experimental details. Thus, except for the theoretically determined P state the Auger yields $\xi[^2P_\pm]$ appearing in all the ratios, only the solid angle correction factor G_τ and the Auger yield $\xi[^4P]$ are needed for $R_{m,exp}$. Moreover, since the angular distributions of the Auger decays from the ^4P and ^2P states can be expected to be quite similar (both are the P states), it follows that $R_{m,exp}$ practically corresponds to the ratio $\sigma_m[^4P]/\sigma_m[^2P]$ of the total production cross sections.

In Fig. 10a, we present various results for R_m from the literature as well as from our recent measurements for collisions with helium targets. It is clearly seen that the results for B^{3+}+H_2 (filled rectangles), F^{7+}+H_2 (filled triangles), as well as for our recent C^{4+}+He work (filled circles) show R_m to be close to the statistically expected value of two. This is much better than the originally reported measurements of Strohschein *et al.* [34] (open

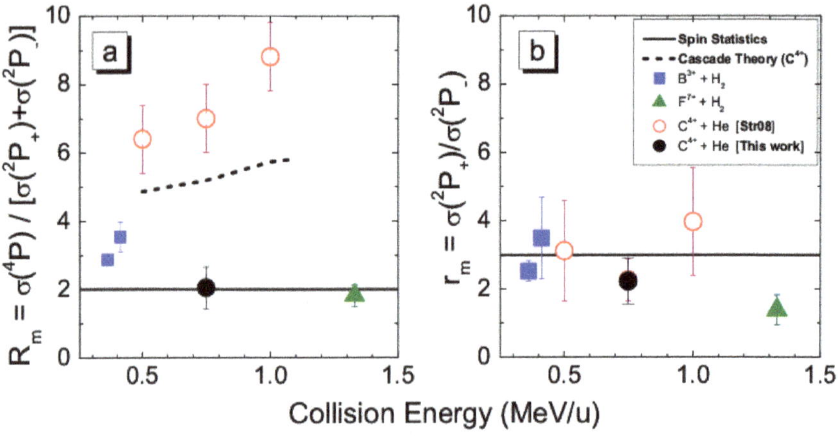

Fig. 10. Determination of the ratios R_m [a] and r_m [b] as a function of collision energy. Symbols: Experimental results. Lines: Theory. Filled rectangles: B^{3+}+H_2, filled triangles: F^{7+}+H_2 (reported by Benis *et al.* [63] from our older work). Open circles: C^{4+}+He (R_m from Strohschein *et al.* [34], with r_m values extracted from the same data and reported in Ref. [63]). Filled circles: C^{4+}+He (present). Dashed line: Theory including cascade feeding for $n \to \infty$+ Auger (Röhrbein *et al.* [60]).

circles) giving the values of R_m as large as 9. In Fig. 10b, we also include the results from the recent computation [63] of the ratio r_m via Eq. (26). This ratio, according to theory, is expected to be equal to 3 and, moreover, it should be independent of the solid angle correction factor G_τ. In this case, practically all the measurements, including those with very different R_m, are in much better agreement with theory. This, therefore, casts some doubt on the G_τ correction factors used by Strohschein *et al.* [34] which, in fact, were found recently [47] to be incorrectly computed.

In the context of this sub-section, it is pertinent to recall Fig. 2, where we display our recently obtained KLL spectra at 9 MeV for C^{4+}+He from two measurements having quite different metastable fractions. The high value metastable fraction was obtained after gas post-stripping the C^{3+} beam in the N_2 gas. The low value fraction was obtained by stripping the initial C^- beam in the N_2 gas inside the terminal of the tandem accelerator. The increased metastable fraction was readily manifested in the KLL spectra of Fig. 2 by the enhanced production of the ^4P state (squares) as opposed to the smaller production at the lower fraction (circles).

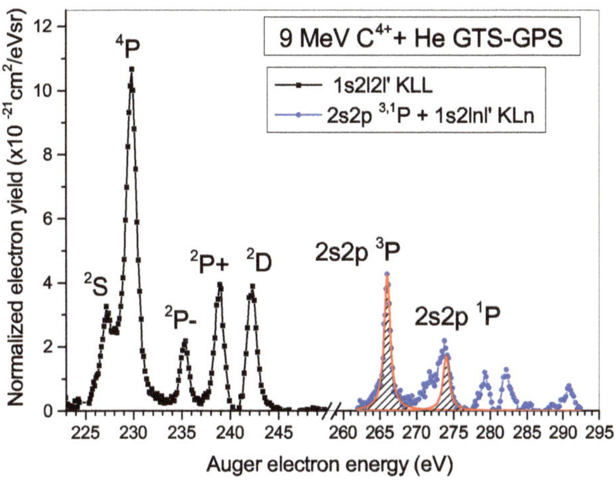

Fig. 11. The 9 MeV $C^{4+}(1s^2, 1s2s\,^3S)$ collisions with He. Auger lines: $C^{3+}(1s2l2l')$ KLL (squares), $C^{4+}(2s2p\,^{3,1}P)$ KLL (circles) and $C^{3+}(1s2s\,^3S)n\ell$ KLn ($n = 3 - 4$). The $1s \rightarrow 2p$ excitation lines are shown by the fits (shaded areas). The beam was produced by gas stripping in the accelerator terminal, GTS, and subsequent gas post stripping, GPS, of the analyzed C^{3+} ions.

5. $1s - 2p$ excitation of the metastable $1s2s\,^3S$ beam component

To obtain an independent cross-check on the $C^{4+}(1s2s\,^3S)$ metastable beam fraction (apart from the new method recently introduced and based upon the analysis of the Auger KLL lines in He-like beams with different metastable fractions [23]), we have also explored the production of various Auger lines of slightly higher electron energy such as the He-like $2s2p\,^{3,1}P$ KLL produced by $1s \rightarrow 2p$ excitation shown in Fig. 11. These lines are also of particular importance in the detailed study of fundamental mechanisms of excitation [42], i.e. the electron-nucleus, electron-electron and electron-electron excitation with spin exchange [65].

The presently unidentified lines in Fig. 11 are most probably due to $(1s2s\,^3S)n\ell$ Li-like states [66] produced by direct $n\ell$ capture with $n = 3$ and 4 to the metastable $1s2s\,^3S$ beam component. Their existence would be further proof that these higher lying states are produced by capture as required by the proposed cascade mechanisms leading to the reported enhancement of the ^4P population [34, 41, 60].

6. Summary, conclusions and future directions

We have presented the basics of the ZAPS technique used at the new experimental station dedicated to atomic collisions physics at the Demokritos TANDEM. This apparatus is now in operation and can perform high resolution Auger electron spectroscopy of projectile ions excited during their collision with gas targets which allows for the state selective determination of differential cross sections. Details of the experimental setup were described with particular application to studies of single electron capture to He-like ions in the pre-excited $1s2s\,^3S$ long-lived state. A terminal gas stripper along with both gas and foil post-strippers has also been installed in the TANDEM to provide more intense He-like beams with variable amounts of the $1s2s\,^3S$ beam fraction.

With the above tools, we have explored capture and excitation using the $C^{4+}(1s^2, 1s2s\,^3S)$ mixed state beams in collisions with various targets in the 0.5-1.5 MeV/u collision energy region. Using our new technique, already presented in detail in Ref. [63], whose results are also briefly presented here, we have been able to determine the capture to metastable ratios R_m and r_m, for various collision systems, including our own recent work for 9 MeV C^{4+}+He. Our results are supported by the accurate determination of the effective solid angle G_τ (crucial for the calculation of R_m) which has also been measured experimentally, and found to be in fair agreement with our MC-type SIMION calculations. The reported values of R_m are mostly close to the expected spin statistics value of 2, but in glaring disagreement both with the experimental results on 6-12 MeV C^{4+}+He from Ref. [34] and the theoretical calculations [60]. Theory, however, clearly demonstrates the selective strong cascade feeding of the ^4P state for these collision systems, in support of the proposed cascade enhancement mechanism [41]. In contrast, the newly reported values of r_m are found to be mostly in good agreement with the expected spin statistics value of 3 [61], even for the R_m disagreeing measurements of Ref. [34].

Thus, even though calculations have shown an enhancement of the ^4P state due to selective cascade feeding in both fluorine [41] and carbon [60] collisions with light targets, there is a rather strong disagreement between the calculated and measured values of R_m, especially for carbon in the 9-12 MeV range. It is still not clear why this cascade enhancement is not observable in the reported measurements. This remains a puzzle to be resolved.

It is evident that more results with careful measurements and theoretical calculations, including cascades, for some other low-Z_p He-like ions will be further needed to convincingly establish the existence of cascade feeding of the ^4P state, as well as to answer the question why this cascade feeding is not clearly identifiable in the measurements of R_m, as presented in here.

In the near future, we plan to finish our measurements and analysis of 6-18 MeV C^{4+} ions in collisions with various gas targets, including H_2, Ne and Ar in an effort to investigate all the possible TE contributions from both the ground state and metastable states to the Auger KLL states. Accurate theoretical calculations are also needed, particularly for NTE from the metastable state (never investigated before). In addition, we plan to finish our investigation of the related excitation channel briefly presented here. Again, accurate theoretical calculations are needed here, as well. Clearly, additional theoretical support would be most welcome in future collaborations.

Acknowledgments

During the period (2012-2015) we acknowledge support by the project APAPES co-financed by the European Union (European Social Fund - ESF) and Greek national funds through the Operational Program "Education and Lifelong Learning" of the National Strategic Reference Framework (NSRF) Research Funding Program: THALES. Investing in knowledge society through the European Social Fund, grant number MIS 377289. For the period of (2017-2020), we acknowledge support by the project "CALIBRA/EYIE" (MIS 5002799) which is implemented under the Action "Reinforcement of the Research and Innovation Infrastructures", funded by the Operational Program "Competitiveness, Entrepreneurship and Innovation" (NSRF 2014-2020) and co-financed by Greece and the European Union (European Regional Development Fund).

On a personal level, it is a pleasure to acknowledge the help of various colleagues during the development of the entire experimental program through 2015, including A. Dimitriou, Be. Sulik, I. Valastyan, O. Sise, G. Martinez, as well as the TANDEM accelerator staff of Tassos Lagoyannis, Michalis Axiotis and Miltos Andrianis. In addition, our thanks also extend to the staff of the J.R. Macdonald tandem accelerator laboratory of Kansas State University and in particular to K. Carnes for his invaluable help on the implementation of the recirculating gas terminal stripper and other components of the APAPES beam line optics and also to K. Bahner, former technical staff of the now defunct EN tandem accelerator in Aarhus Denmark, for his detailed advise on the implementation of the Steinmetz circuitry used to power the recirculating gas turbomolecular pump. We would also like to thank D. Manura of Scientific Instrument Services for his detailed help with SIMION Lua programming. Last, but not least, we thank the then director of the Institute of Nuclear and Particle Physics at Demokritos, S. Harissopulos, for his enthusiastic encouragement and support through out the entire APAPES program period. Finally, TJMZ also acknowledges the support of ELKE travel grant #4669 of the University of Crete that allowed for his participation at the 25th International Symposium on Ion-Atom Collisions held in Palm Cove, Queensland, Australia on July 23-25, 2017.

References

[1] T. J. M. Zouros and D. H. Lee, Zero Degree Auger Electron Spectroscopy of Projectile Ions, in S. M. Shafroth and J. C. Austin (eds.), *Accelerator-Based Atomic Physics Techniques and Applications*, Chap. 13. American Institute of Physics Conference Series, Woodbury, NY, pp. 426-479 (1997).

[2] N. Stolterfoht, R. D. Dubois and R. D. Rivarola, Electron emission in heavy ion-atom collisions, *Springer Series on Atoms and Plasmas*, Berlin (1997).

[3] Dž. Belkić, Review of theories on ionization in fast ion-atom collisions with prospects for applications to hadron therapy, *J. Math. Chem.* **47**, 1366-1419 (2010).

[4] APAPES Atomic Physics with Accelerators: Projectile Electron Spectroscopy, http://apapes.physics.uoc.gr (2015).

[5] S. Hagmann, T. Stöhlker, J. Ullrich, C. Kozhuharov, R. Moshammer, H. Kollmus, R. Mann and M. Nofal, A forward electron spectrometer for the ESR, *Nucl. Instrum. Methods Phys. Res. B* **205**, 207-209 (2003).

[6] W. Mampe, K. Schreckenbach, P. Jeuch, B. Maier, F. Braumandl, J. Larysz and T. von Egidy, The double focusing iron-core electron-spectrometer "*BILL*" for high resolution (n, e^-) measurements at the high flux reactor in Grenoble, *Nucl. Instrum. Methods* **154**, 127-149 (1978).

[7] Y. Litvinov, F. Attallah, K. Beckert, F. Bosch, D. Boutin, M. Falch, B. Franzke, H. Geissel, M. Hausmann, T. Kerscher, O. Klepper, H.-J. Kluge, C. Kozhuharov, K. Löbner, G. Münzenberg, F. Nolden, Y. Novikov, Z. Patyk, T. Radon, C. Scheidenberger, J. Stadlmann, M. Steck, M. Trzhaskovskaya and H. Wollnik, Observation of a dramatic hindrance of the nuclear decay of isomeric states for fully ionized atoms, *Physics Letters B* **573**, Supplement C, 80-85 (2003).

[8] H. Kankaanpää, P. Butlerb, P. Greenlees, J. Bastin, R. Herzberg, R. Humphreys, G. Jones, P. Jones, R. Julin, A. Keenan, H. Kettunen, M. Leino, L. Miettinen, T. Page, P. Rahkila, C. Scholey and J. Uusitalo,

In-beam electron spectrometer used in conjunction with a gas-filled recoil separator, *Nucl. Instrum. Methods Phys. Res. A* **534**, 503-510 (2004).
[9] R. L. Graham, G. T. Ewan and J. S. Geiger, A one-meter-radius iron-free double-focusing $\pi\sqrt{2}$ spectrometer for β-ray spectroscopy with a precision of 1:10^5, *Nucl. Instrum. & Methods* **9**, 245-286 (1960).
[10] K. Schreckenbach, H. Faust, F. V. Feilitzsch, A. Hahn, K. Hawerkamp and J. Vuilleumier, Absolute measurement of the beta spectrum from ^{235}U fission as a basis for reactor antineutrino experiments, *Phys. Lett.* **99B**, 251-256 (1981).
[11] N. Stolterfoht, High resolution Auger spectroscopy in energetic ion atom collisions, *Physics Reports* **146**, 315-424 (1987).
[12] R. Bruch, K. T. Chung, W. L. Luken and J. C. Culberson, Recalibration of the *KLL* Auger spectrum of carbon, *Phys. Rev. A* **31**, 310-315 (1985).
[13] E. P. Benis and T. J. M. Zouros, Improving the energy resolution of a hemispherical spectrograph using a paracentric entry at a non-zero potential, *Nucl. Instrum. Methods Phys. Res. A* **440**, 462-465 (2000).
[14] T. J. M. Zouros and E. P. Benis, Erratum to "The hemispherical deflector analyser revisited. I. Motion in the ideal $1/r$ potential, generalized entry conditions, Kepler orbits and spectrometer basic equation, *J. Electron. Spectrosc. Relat. Phenom.* **125**, 221-248 (2002).", *J. Electron Spectrosc. and Relat. Phenom.* **142**, 175-176 (2005).
[15] E. P. Benis and T. J. M. Zouros, The hemispherical deflector analyser revisited. II. Electron-optical properties, *J. Electron Spectrosc. and Relat. Phenom.* **163**, 28-39 (2008).
[16] O. Sise, M. Ulu, M. Dogan, G. Martinez and T. J. M. Zouros, Fringing field optimization of hemispherical deflector analyzers using BEM and FDM, *J. Electron Spectrosc. and Relat. Phenom.* **177**, 42-51 (2010).
[17] M. Dogan, M. Ulu, G. G. Gennarakis and T. J. M. Zouros, Experimental energy resolution of a paracentric hemispherical deflector analyzer for different entry positions and bias, *Rev. Sci. Instrum.* **84**, 043105 (2013).
[18] E. P. Benis, K. Zaharakis, M. M. Voultsidou, T. J. M. Zouros, M. Stöckli, P. Richard and S. Hagmann, A new hemispherical analyser with 2-D PSD and focusing lens for use in 0° electron spectroscopy, *Nucl. Instrum. Methods Phys. Res. B* **146**, 120-125 (1998).
[19] E. P. Benis, T. J. M. Zouros and P. Richard, High resolution RTE measurements at zero degrees using a Hemispherical analyser with lens and 2-D PSD, *Nucl. Instrum. Methods Phys. Res. B* **154**, 276-280 (1999).
[20] E. P. Benis, T. J. M. Zouros, H. Aliabadi and P. Richard, Hemispherical analyser with 2-D PSD for 0° Auger projectile spectroscopy, *Phys. Scr.* **T80**, 529-531 (1999).
[21] O. Sise, G. Martinez, I. Madesis, A. Laoutaris, A. Dimitriou, M. Fernandez-Martin and T. J. M. Zouros, The voltage optimization of a four-element lens used on a hemispherical spectrograph with virtual entry for highest energy resolution, *J. Electron Spectrosc. and Relat. Phenom.* **211**, 19-31 (2016).
[22] SIMION®Ion and Electron Optics Simulator, *SIS* Scientific Instrument Services, SIMION version 8.1.2.20, Ringoes, NJ, see http://www.simion.com (2014).

[23] E. P. Benis and T. J. M. Zouros, Determination of the $1s2\ell2\ell'$ state production ratios $^4P^o/^2P$, $^2D/^2P$ and $^2P_+/^2P_-$ from fast $(1s^2, 1s2s\,^3S)$ mixed-state He-like ion beams in collisions with H_2 targets, *J. Phys. B* **49**, 235202 (2016).
[24] H.-D. Betz, Charge states and charge-changing cross sections of fast Heavy ions penetrating through gaseous and solid media, *Rev. Mod. Phys.* **44**, 465-539 (1972).
[25] R. Sayer, Semi-empirical formulas for heavy-ion stripping data, *Rev. Phys. Appl. (Paris)* **12**, 1543-1546 (1977).
[26] A. Laoutaris, I. Madesis, A. Dimitriou, A. Lagoyannis, M. Axiotis, E. P. Benis and T. J. M. Zouros, Use of gas and foil strippers for the production of He-like ionic beams in both pure ground state $(1s^2)$ and mixed states $(1s^2, 1s2s)$ for zero-degree Auger projectile electron spectroscopy, *J. Phys: Conf. Ser.* **635**, 052062 (2015).
[27] G. W. F. Drake, Theory of relativistic magnetic dipole transitions: Lifetime of the metastable 2 ^{3}S state of the Heliumlike ions, *Phys. Rev. A* **3**, 908-915 (1971).
[28] M. Terasawa, T. J. Gray, S. Hagmann, J. Hall, J. Newcomb, P. Pepmiller and P. Richard, Electron capture by and electron excitation of two-electron fluorine ions incident on helium, *Phys. Rev. A* **27**, 2868-2875 (1983).
[29] M. Zamkov, H. Aliabadi, E. P. Benis, P. Richard, H. Tawara and T. J. M. Zouros, Energy dependence of the metastable fraction in $B^{3+}(1s^2\,^1S, 1s2s\,^3S)$ beams produced in collisions with thin-foil and gas targets, *Phys. Rev. A* **64**, 052702 (2001).
[30] M. Zamkov, H. Aliabadi, E. P. Benis, P. Richard, H. Tawara and T. J. M. Zouros, Stripping energy dependence of $B^{3+}(1s2s^3S)$ beam metastable fraction, *AIP* **576**, 149-152 (2001).
[31] M. Zamkov, E. P. Benis, P. Richard and T. J. M. Zouros, Fraction of metastable $1s2s\,^3S$ ions in fast He-like beams (Z = 5-9) produced in collisions with carbon foils, *Phys. Rev. A* **65**, 062706 (2002).
[32] E. P. Benis, M. Zamkov, P. Richard and T. J. M. Zouros, Techniques for the determination of the $1s2s\,^3S$ metastable fraction in two-electron beams, *Phys. Rev. A* **65**, 064701 (2002).
[33] E. P. Benis, M. Zamkov, P. Richard and T. J. M. Zouros, Comparison of two experimental techniques for the determination of the $\mathrm{1s2s}^3S$ metastable beam in energetic B^{3+} ions, *Nucl. Instrum. Methods Phys. Res. B* **205**, 517-521 (2003).
[34] D. Strohschein, D. Röhrbein, T. Kirchner, S. Fritzsche, J. Baran and J. A. Tanis, Nonstatistical enhancement of the $1s2s2p\,^4P$ state in electron transfer in 0.5-1.0 MeV/u $C^{4,5+}$ + He and Ne collisions, *Phys. Rev. A* **77**, 022706 (2008).
[35] M. Druetta, S. Martin and J. Desesquelles, VUV spectroscopy of charge exchange processes in collisions of multiply charged ions, *Nucl. Instrum. Methods Phys. Res. B* **23**, 268-273 (1987).
[36] S. Bliman, M. Cornille and K. Katsonis, Dielectronic recombination versus charge exchange: Electron capture by metastable ions, *Phys. Rev. A* **50**, 3134-3141 (1994).

[37] E. P. Benis, T. J. M. Zouros, T. W. Gorczyca, M. Zamkov and P. Richard, Isoelectronic study of triply excited Li-like states, *J. Phys. B* **36**, L341-L348 (2003).

[38] M. Zamkov, H. Aliabadi, E. P. Benis, P. Richard, H. Tawara and T. J. M. Zouros, Absolute cross sections and decay rates for the triply excited $B^{2+}(2s2p^2\,{}^2D)$ resonance in electron–metastable-ion collisions, *Phys. Rev. A* **65**, 032705 (2002).

[39] E. P. Benis, T. J. M. Zouros, T. W. Gorczyca, A. D. González and P. Richard, Elastic resonant and non-resonant differential scattering of quasi-free electrons from $B^{4+}(1s)$ and $B^{3+}(1s^2)$ ions, *Phys. Rev. A* **69**, 052718 (2004).

[40] J. A. Tanis, A. L. Landers, D. J. Pole, A. S. Alnaser, S. Hossain and T. Kirchner, Evidence for Pauli exchange leading to excited-state enhancement in electron transfer, *Phys. Rev. Lett.* **92**, 133201 (2004).

[41] T. J. M. Zouros, B. Sulik, L. Gulyas and K. Tokesi, Selective enhancement of $1s2s2p^4P_J$ metastable states populated by cascades in single-electron transfer collisions of $F^{7+}(1s^2/1s2s\,{}^3S)$ ions with He and H_2 targets, *Phys. Rev. A* **77**, 050701 (2008).

[42] D. H. Lee, P. Richard, J. M. Sanders, T. J. M. Zouros, J. L. Shinpaugh and S. L. Varghese, Electron capture and excitation studied by state-resolved *KLL* Auger measurement in 0.25-2 MeV/u $F^{7+}(1s^2\,{}^1S, 1s2s\,{}^3S)$ + H_2/He collisions, *Nucl. Instrum. Methods Phys. Res. B* **56/57**, 99-103 (1991).

[43] D. H. Lee, P. Richard, J. M. Sanders, T. J. M. Zouros, J. L. Shinpaugh and S. L. Varghese, *KLL* resonant transfer and excitation to $F^{6+}(1s2l2l')$ intermediate states, *Phys. Rev. A* **44**, 1636-1643 (1991).

[44] T. J. M. Zouros, Resonant transfer and excitation associated with Auger electron emission, in W. G. Graham, W. Fritsch, Y. Hahn and J. Tanis (eds.), *Recombination of Atomic Ions*, Vol. 296, NATO Advanced Study Institute Series B: Physics. Plenum Publishing Corporation, New York, pp. 271-300 (1992).

[45] S. Doukas, I. Madesis, A. Dimitriou, A. Laoutaris, T. J. M. Zouros and E. P. Benis, Determination of the solid angle and response function of a hemispherical spectrograph with injection lens for Auger electrons emitted from long lived projectile states, *Rev. Sci. Instrum.* **86**, 043111 (2015).

[46] E. P. Benis, S. Doukas, T. J. M. Zouros, P. Indelicato, F. Parente, C. Martins, J. P. Santos and J. P. Marques, Evaluation of the effective solid angle of a hemispherical deflector analyser with injection lens for metastable Auger projectile states, *Nucl. Instrum. Methods Phys. Res. B* **365**, 457-461 (2015).

[47] E. P. Benis, I. Madesis, A. Laoutaris, S. Nanos and T. J. M. Zouros, Experimental determination of the effective solid angle of long-lived projectile states in zero-degree Auger projectile spectroscopy, *J. Electron Spectrosc. and Relat. Phenom.* **222**, Supplement C, 31-39 (2018).

[48] D. H. Lee, Electron–electron interactions in ion-atom collisions studied by projectile state-resolved Auger electron spectroscopy, Ph.D. dissertation, Kansas State University (1990), (unpublished).

[49] E. P. Benis, A novel high-efficiency paracentric hemispherical spectrograph for zero-degree Auger projectile spectroscopy, Ph.D. dissertation, Dept. of Physics, University of Crete (2001), (unpublished).

[50] J. Doerfert, E. Träbert, A. Wolf, D. Schwalm and O. Uwira, Precision measurement of the electric dipole intercombination rate in C^{2+}, *Phys. Rev. Lett.* **78**, 4355-4358 (1997).

[51] E. Träbert, A. Wolf and G. Gwinner, Measurement of EUV intercombination transition rates in Be-like ions at a heavy-ion storage ring, *Phys. Lett. A* **295**, 44-49 (2002).

[52] A. Müller, S. Schippers, R. A. Phaneuf, A. L. D. Kilcoyne, H. Bräuning, A. S. Schlachter, M. Lu and B. M. McLaughlin, Fine-structure resolved photoionization of metastable Be-like ions C III, N IV, and O V, *J. Phys: Conf. Ser.* **58**, 383-386 (2007).

[53] N. Stolterfoht, P. D. Miller, H. F. Krause, Y. Yamazaki, J. K. Swenson, R. Bruch, P. F. Dittner, P. L. Pepmiller and S. Datz, Surgery of fast, highly charged ions studied by zero-degree Auger spectroscopy, *Nucl. Instrum. Methods Phys. Res. B* **24/25**, 168-172 (1987).

[54] S. W. J. Scully, A. Aguilar, E. D. Emmons, R. A. Phaneuf, M. Halka, D. Leitner, J. C. Levin, M. S. Lubell, R. Püttner, A. S. Schlachter, A. M. Covington, S. Schippers, A. Müller and B. M. McLaughlin, K-shell photoionization of Be-like carbon ions: experiment and theory for C^{2+}, *J. Phys. B* **38**, 1967-1975 (2005).

[55] M. M. A. Shorman, M. F. Gharaibeh, J. M. Bizau, D. Cubaynes, S. Guilbaud, N. E. Hassan, C. Miron, C. Nicolas, E. Robert, I. Sakho, C. Blancard and B. M. McLaughlin, K-shell photoionization of Be-like and Li-like ions of atomic nitrogen: experiment and theory, *J. Phys. B* **46**, 195701 (2013).

[56] A. Müller, S. Schippers, R. A. Phaneuf, S. W. J. Scully, A. Aguilar, C. Cisneros, M. F. Gharaibeh, A. S. Schlachter and B. M. McLaughlin, K-shell photoionization of Be-like boron (B^{+}) ions: experiment and theory, *J. Phys. B* **47**, 135201 (2014).

[57] B. M. McLaughlin, J.-M. Bizau, D. Cubaynes, S. Guilbaud, S. Douix, M. M. A. Shorman, M. O. A. E. Ghazaly, I. Sakho and M. F. Gharaibeh, K-shell photoionization of O^{4+} and O^{5+} ions: experiment and theory, *MNRAS* **465**, 4690-4702 (2017).

[58] M. H. Chen, B. Crasemann and H. Mark, Deexcitation of light Li-like ions in the $1s2s2p$ state, *Phys. Rev. A* **27**, 544-547 (1983).

[59] D. Schneider, R. Bruch, W. Butscher and W. H. E. Schwarz, Prompt and time-delayed electron decay-in-flight spectra of gas-excited carbon ions, *Phys. Rev. A* **24**, 1223-1236 (1981).

[60] D. Röhrbein, T. Kirchner and S. Fritzsche, Role of cascade and Auger effects in the enhanced population of the $C^{3+}(1s2s2p\,^4P)$ states following single-electron capture in $C^{4+}(1s2s\,^3S)$-He collisions, *Phys. Rev. A* **81**, 042701 (2010).

[61] E. P. Benis, T. J. M. Zouros, T. W. Gorczyca, A. D. González and P. Richard, Erratum: Elastic resonant and nonresonant differential scattering of quasifree electrons from $B^{4+}(1s)$ and $B^{3+}(1s^2)$ ions [*Phys. Rev. A* **69**, 052718 (2004)], *Phys. Rev. A* **73**, 029901 (2006).
[62] D. H. Lee, T. J. M. Zouros, J. M. Sanders, P. Richard, J. M. Anthony, Y. D. Wang and J. H. McGuire, K-shell ionization of O^{4+} and C^{2+} ions in fast collisions with H_2 and He targets, *Phys. Rev. A* **46**, 1374-1387 (1992).
[63] E. P. Benis, S. Doukas and T. J. M. Zouros, Evidence for the non-statistical population of the $1s2s2p\,^4P$ metastable state by electron capture in 4 Mev collisions of $B^{3+}(1s2s\,^3S)$ with H_2 targets, *Nucl. Instrum. Methods Phys. Res. B* **369**, 83-86 (2016).
[64] T. Kirchner (2016), private communications.
[65] T. J. M. Zouros, D. H. Lee and P. Richard, Projectile 1s→2p excitation due to electron-electron interaction in collisions of F^{6+} and O^{5+} ions with He and H_2 targets, *Phys. Rev. Lett.* **62**, 2261-2264 (1989).
[66] J. P. Santos, F. Parente, M. C. Martins, P. Indelicato, E. P. Benis, T. J. M. Zouros and J. Marques, Radiative transition rates of $1s2s(^3S)3p$ levels for Li-like ions with $5 \leq Z \leq 10$, *Nucl. Instrum. Methods Phys. Res. B* **408**, Supplement C, 100-102 (2017).

Chapter 2

Soft X-ray emission following charge exchange at solar wind velocities

R.T. Zhang* and C.C. Havener

Physics Division, Oak Ridge National Laboratory,
Oak Ridge, Tennessee 37831-6368, USA

A comprehensive review of charge exchange between slow highly charged ions and atoms is presented. Also addressed is a brief history of charge exchange studies, charge exchange mechanisms and soft X-ray spectroscopy following charge exchange. Emphasis is given to high-resolution soft X-ray measurements of highly charged ion-atom charge exchange at solar wind velocities. Experimental results for C^{6+}-He and O^{8+}-Kr are selected to illustrate the importance of high-resolution soft X-ray measurements. The discussions include X-ray emissions in single electron capture with cascades, X-ray emission following autoionizing double capture, and a possible cascade X-ray emission of doubly excited states. Future plans are presented for high-resolution measurements of electron capture by solar wind ions (C^{6+},N^{7+},O^{8+},O^{7+},...) from atomic hydrogen.

Contents

1. Introduction

Collisions between neutral atoms and highly charged ions include various processes, such as impact ionization, charge exchange (CX), transfer ionization, etc. Among these, the most probable is charge exchange or electron

*Corresponding author. Email: zhangrtimp@hotmail.com

capture. In this process, one or several electrons from one of the collision partners are transferred to the other [1]. Charge exchange plays a dominant role in understanding a number of other subfields of physics, such as astrophysics, plasma physics, and heavy ion physics [2, 3]. Much of the diagnostics of radiative environments relies on the availability of CX cross section data and spectroscopy of highly charged ions after CX. For example, X-ray emission following CX is widely used to diagnose astrophysical and laboratory plasma behavior [4]. Apart from its importance in fundamental research, it also has a strong impact on many applications, such as the X-ray laser technique, nuclear detection technique, and production of microelectronic devices [5, 6]. It is clear that CX between highly charged ions and atoms is a very fundamental area of research. Nowadays, the need for high-quality CX data and a better understanding of the CX mechanisms is as important as it has ever been.

Charge exchange was first predicted by Flamm and Schumann in 1916 [7] and experimentally observed by Henderson [8] and Rutherford [9]. Thereafter, CX studies were stimulated by the discovery of fission, because of its importance in the calculation of stopping powers of fast charged fragments during the slowing-down process [10, 11]. It was found that CX cross sections were dependent on target thickness, projectile species, and collision energies [12]. Early explanations include the classical two-step mechanisms proposed by Thomas [13], which were confirmed with a dedicated experiment 70 years later [14]. Oppenheimer [15], Brinkman and Kramers [16] described CX using the wave function overlaps within quantum theory. Continued interest in CX was promoted by the availability of accelerators, which made possible the absolute measurements of state-selective total cross sections as a function of certain variable parameters, such as q and v_p. Here, q and v_p are the charge state and the velocity of the projectile ion, respectively [17]. Typical state-selective measurements are performed with a photon emission spectrometer [18–20], translational energy gain spectrometer [21, 22], traditional electron spectrometer [23, 24]. Various theoretical methods, which often yield good agreement with measurements on charge exchange, have been reviewed in Refs. [25–30].

Generally speaking, in slow collisions ($\mathrm{v}_p \ll \mathrm{v}_e$, where v_e is the mean orbital velocity of the active target electron), the quasi-molecular picture proposed by Fano and Lichten [31] is valid. This is the case because the electron moves fast enough to adjust to the motion of nuclei from both collision partners. For fast collisions ($\mathrm{v}_p \gg \mathrm{v}_e$), the impulse approximation can be adopted, because the target electron is viewed as quasi-static in inverse kinematics. This has been used, for example, to model radiative electron capture (REC) [20] and resonant transfer excitation (RTE) [32]. For intermediate collisions ($v_p \approx v_e$), a "velocity matching" model is used, assuming that electron capture occurs via an overlap of the initial and final state wave functions that are shifted in momentum space by the projectile velocity [33, 34]. These and other related studies have become the basis of some of the CX theories and have improved the understanding of the CX process.

In the present work on CX, we are focusing on slow collisions with a projectile energy range from several hundred eV/u to several keV/u. The vast majority of studies published are related to single electron CX [35–38]. Some useful empirical laws have been obtained. For instance, the magnitude of cross sections involving an atomic hydrogen target can be estimated by $q \times 10^{-15}$ cm^2 [39–41]. The most probable shell in which capture occurs can be estimated by an empirical equation $n = q^{0.75}$ for an H target [42], and scaled by the square root of the binding energy ratios between H and other atoms [43]. Even though the total and dominant n capture cross sections can be qualitatively predicted, the state selectivity at different orbital angular momentum l is more elusive. They not only depend on the target species, but also on the collision energies.

To have a better understanding of the slow highly-charged ion-atom single CX process, various theoretical models were developed. These include the methods know as the classical trajectory Monte Carlo simulations (CTMC) [42, 44, 45], the atomic orbital close-couplingg (AOCC) [1, 46], the hidden crossing theory with adiabatic molecular states (advanced adiabatic approach) (HCT-AAA) [40], the molecular orbital close-coupling (MOCC) [47], the two-center basis generator method (TCBGM) [48], etc. In some cases, approximate calculations like the multichannel Landau-Zener (MCLZ) model are successful [49]. At very low energies, straight line or classical trajectories are not suitable, therefore, the quantum molecular orbital close-coupling (QMOCC) methods are more appropriate. Liu *et al.* [50] used a quantum molecular orbital close coupling method to compute the cross sections of C^{5+}-H single charge exchange process. Other simpler velocity dependent approximations like the classical over barrier (COB) model can be used to estimate the principal quantum number n and the average value of the angular momentum quantum number l of the captured electron [51, 52].

From the experimental perspective, the powerful cold target recoil ion momentum spectroscopy (COLTRIMS) techniques have advanced the CX measurement to the point where state-selective cross sections and impact parameter information have become accessible [53, 54]. For example, Kamber *et al.* [55] found that the population of n=3 state of C^{4+} is dominant in 0.5–50 keV/u C^{5+} single charge exchange from He. Fléchard *et al.* [56] determined that the electron is captured into levels n=3, 4, 5 and 6 of Ne^{9+} in the energy range from 50 keV to 150 keV for Ne^{10+}-He single capture process. Moreover, through measuring the transversal recoil ion momenta (or, indirectly, impact parameters), Abdallah *et al.* [57] suggested that dynamical coupling effects significantly affect state-selectivity during electron capture in slow Ar^{8+}-He collisions, which has been confirmed experimentally [58]. With a high-resolution COLTRIMS apparatus, Fisher *et al.* [59] have resolved state-selective differential cross sections of $4s$, $4p$, $4d$+$4f$ subshell capture in Ne^{8+}-He and Ne^{7+}-He single CX. However, these nl subshell state selectivity measurements are limited by recoil ion momentum resolution and target species. Thus, only a few systems are applicable with the COLTRIMS detecting technique.

In slow CX collisions, electrons are captured into excited states of projectile ions producing a certain n distribution with the de-excitation times 10^{-14}s and 10^{-10}s for autoionization and X-ray emission, respectively. These de-excitation times are much longer than the collision interaction time of 10^{-16}s. Therefore, an alternative study of subshell capture is to measure the soft X-ray emission following single CX. Dijkkamp *et al.* [60, 61] and Hoekstra *et al.* [62, 63] performed X-ray emission cross section measurements with a crossed-beam technique. Through measuring the signal for a particular line emission (the integrated area of the peak), the emission cross sections can be calculated via the relation:

$$\sigma_{em}\lambda = \frac{4\pi}{\omega}\frac{q}{K(\lambda)Q}\frac{S(\lambda)}{NL} \tag{1}$$

where ω is the solid angle of observation, Q is the accumulated ion charge, $K(\lambda)$ is the wavelength dispersive (10–80 nm) quantum yield of the detection system, $S(\lambda)$ is the measured signal, N is the effective target density, and L is the observation length. The measured X-ray emission cross sections agree with the theoretical calculations used for plasma modeling and plasma diagnostics [61]. However, these measurements did not cover a broad spectral range and have large systematic uncertainty.

Emission spectroscopy is important in many astrophysical environments. Electrons are captured from cometary neutrals into excited states of solar wind ions, which decay radiatively by emitting X-rays or extreme ultraviolet (EUV) radiation. The accurately determined line emission cross sections (or line ratios) can be used to estimate the abundance of solar wind ions. The relevant studies have become more and more attractive since the 1990s and this is now a very competitive as well as fruitful research field. In 1996, Lisse *et al.* [64] observed X-ray and EUV emission from Comet C/Hyakutake. A charge exchange interaction between the comet and the solar wind ions was proposed to be the most likely mechanism for the observed X-ray emission. Subsequently, Beiersdorfer *et al.* [65] carried out a series of laboratory simulations of charge exchange X-ray emission. They confirmed the CX mechanism of X-ray emission by fitting the simulation to the astrophysical observation. Recently, Koutroumpa *et al.* [66] concluded that nearly all the X-ray flux in a certain observation direction is due to charge exchange. Moreover, Bodewits *et al.* [67] observed that 0.3 keV-1 keV cometary X-ray emission mainly results from CX, involving fully stripped as well as H-like ions of C, N, O, Ne as the projectiles and H_2O or its dissociation products OH, O, H as the targets. However, Mullen *et al.* [68] theoretically modeled CX solar wind X-ray data from the XMM-Newton RGS detector and showed that the X-ray intensity is mainly determined by solar wind charge exchange with H as a target instead of H_2O.

Astrophysical observations also show that there are some other cool gases which probably produce soft X-ray by charge exchange collisions. Examples of such X-ray sources are terrestrial magnetosheath, geocorona and heliosphere (where the solar wind interacts with incoming interstellar neutral gas). Cravens [69] predicted that the heliospheric X-ray emission is from charge exchange between solar wind ions and interstellar neutral atoms/molecules accounting for 25%-50% of the observed soft X-ray background intensities. Up to now, however, the modeling of the reported solar wind charge exchange X-ray emissions is still suffering from the lack of CX cross section and spectroscopic data of highly charged ions. Therefore, laboratory simulations are needed for a more reliable model of soft X-ray background, from which soft X-ray emission associated with hot interstellar gas and the galactic halo might be distinguished.

At Oak Ridge National Laboratory (ORNL), we are concerned with single CX X-ray emission between fully stripped ions C^{6+}, N^{7+}, O^{8+} and neutral atoms/molecules like H, He, H_2, Kr. These species are key components of astrophysics gas emitting systems like solar wind ions and comets. The influence of metastable states can be avoided with fully stripped ions. These systems have high radiative decay rates and discrete structure. The following equation indicates the process to be considered:

$$A^{q+} + B \to A^{(q-1)+}(nl) + B^{+} \to A^{(q-1)+}(1s) + B^{+} + \gamma. \quad (2)$$

Another important issue is the multi-electron capture process. Such a process can occur for highly charged solar wind ions colliding with multi-electron targets. This possibility has been ignored in astrophysical CX models [70]. Simply taking two-electron capture as an example, it is known that two electrons are mainly captured into the doubly excited states at low collision energies [71–74]. Two kinds of doubly excited states are observed. One is categorized as asymmetric doubly excited states that probably stem from correlated double capture. The other one is symmetric doubly excited states which is due to uncorrelated double capture. A widely held perception is that doubly excited states always decay by autoionization, because autoionization is much faster than radiative decay. Nevertheless, recent studies illustrated that radiation following autoionization is an important diagnostic final step for doubly excited states [75]. Moreover, radiative decay from doubly excited states is worth investigation. For example, measurements of X-ray spectra on earth satellites revealed the lines corresponding specifically to radiative decay of the doubly excited C^{4+}, N^{5+} and O^{6+} ions [76]. Especially for the asymmetric doubly excited states, radiative stabilization becomes a preferred form of stabilization, which is confirmed by the photon experiment of Martin *et al.* [77]. Ali *et al.* [78] studied charge exchange collisions of slow Ne^{10+} with He, Ne, and Ar targets with simultaneous X-ray measurement and COLTRIMS, which also provided evidence for significant contribution from radiative decays of doubly excited states.

In the following sections, we will describe the current progress on the high-resolution measurements of X-ray emission in CX process. Details on the experimental method are given in Section 2, the radiative cascade decay scheme of H-like multicharged ions is presented in Section 3, X-ray emission from singly excited states and doubly excited states is outlined in Section 4, the conclusion is summarized in Section 5 and our future plans are stated in Section 6.

2. ORNL experimental apparatus

A series of atomic collision measurements involving slow multicharged ions have been carried out using the 250 kV high voltage platform located in the Multicharged Ion Research Facility (MIRF) at ORNL [37, 79, 80]. This high voltage platform contains a permanent magnet electron cyclotron resonance (ECR) ion source and has several experimental beam lines for molecular-ion diagnostics, ion-surface below Coulomb barrier nuclear reactions, and ion-atom collisions[a]. The microwave injected into the source can be varied in the frequency range of 12.75-14.5 GHz. The typical energy range of this high voltage platform is from 20q to 270q keV. The ion-atom beam line is used to measure absolute cross sections for charge exchange between multicharged ions and H using a merged-beam technique. The ground state H beam is produced via photon detachment of H^- as it passes through a cw Nd: YAG laser (1.06 um) inner cavity where circulating power is close to 2 kW. Details about the merged beam apparatus can be found elsewhere [37, 79, 81, 82].

This ion-atom beam line also has the capability to measure state-selective X-ray emission with a traditional gas cell and a high-resolution X-ray quantum micro-calorimeter (XQC) from the University of Wisconsin and Goddard Space Flight Center [83]. In this way, soft X-ray emission following charge exchange between multicharged ions and atoms can be measured. Soft X-ray emission measurements following charge exchange involving C^{6+}, O^{8+}, Ne^{8+} (projectiles) and He, H_2, Kr (targets) have been completed for collision energies from 1 keV/u to 30 keV/u [75, 84–86]. The sketch of the experimental setup can be seen in Fig. 1. Briefly, taking the O^{8+}-Kr charge exchange measurement as an example, $^{18}O^{8+}$ ions were produced by an ECR ion source, extracted at 17.7 kV, analyzed for momentum by a 90° dipole magnet on the platform, and then accelerated or decelerated off the platform to the desired ion energies of 445, 889, 1780, and 8180 eV/u. Two spherical deflectors, several sets of einzel lenses, and adjustable slits are used to focus and collimate the primary beam to a beam size of a few mm in diameter in a Kr gas cell. The Kr target gas was introduced into the gas cell volume via a leak valve with the total pressure being monitored via a bare Bayard-Alpert ion gauge and a quadrupole residual gas analyzer (SRS RGA 100). The background pressure in the cell was around 10^{-10} Torr. The ion beam passed through the gas cell where a small fraction of ions capture one or more electrons from the Kr atom as a target. The electrons are captured into excited states. For single electron capture subsequent direct radiative decay or cascade to the ground state leads to the soft X-ray emission.

The working principle and details of the XQC detector can be found elsewhere [83, 87]. It can be simply described as follows: the incident X-

[a]We measured the nuclear reactions for alpha particles on ^{13}C below the Coulomb barrier. The cross sections were extremely small and we were able to measure them due to large currents and a solid implanted target.

ray is absorbed in a temperature sensitive layer, where the energy of the X-ray results in a temperature rise of the absorber. Through a thermal link, the temperature rise is dumped into a heat sink. By measuring the amount of heat, the energy of X-ray is measured. There are several advantages of XQC, e.g. sensitivity to small amounts of deposited energy, high energy resolution, broad energy range for photon spectrum, and independence of the charge transport properties of absorbers.

During the measurement, the XQC detector was operated at 50 mK and positioned 23 cm above the beam, with a solid angle of 2.3×10^{-3} sr for measuring the prompt X-ray emitted at 90° with respect to the beam direction. The X-ray backgrounds from the ion beam (without gas) and dark counts were periodically measured and found to be insignificant. The target gas pressures were in the range of 2×10^{-7} Torr. Due to the thermal recovery time of the XQC detector between events, the pressure in the gas cell was adjusted to restrict the X-ray count rate of the detector to less than 1 Hz per pixel in order to limit pulse pileup. Further, the XQC detector was protected from thermal infrared (IR) radiation by several thin aluminum filters that were periodically defrosted. The transit time, Δt, of ions traveling across the field of view of the detector is in the range 16 ns – 68 ns ($1/\Delta t \sim 1.5$ - 6.3×10^7 s^{-1}). This is larger than the lifetimes of the populated excited states and covers all the possible transitions.

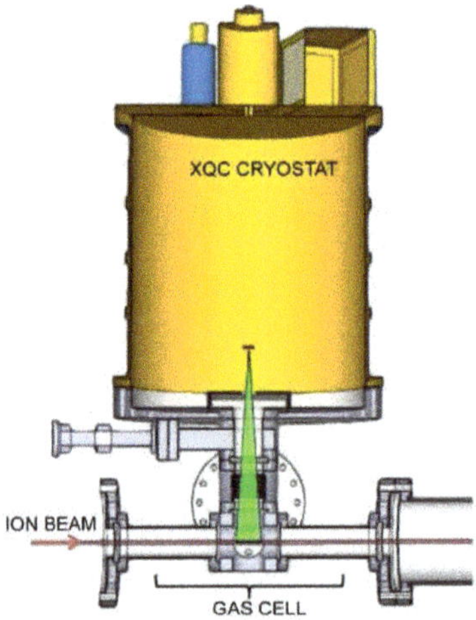

Fig. 1. Schematic view of the CX experiment with the XQC detector and gas cell at MIRF.

Ideally X-ray line intensity should be measured at 54° (so-called magic angle) [88], where a 4π averaged X-ray line intensity can be observed. Anisotropy of X-ray emission caused by electron cloud alignment or polarization of the de-excited electron can never have an influence on line intensity measured at the magic angle. However, we found that the calculated polarizations are small for the O^{8+}-H system [89]. Even if the X-rays are fully polarized, the error in cross sections due to anisotropy can never be larger than 30% (in fact, it is typically less than 15%) [48, 60]. Differences between the line ratios with the X-ray spectrum intensities detected at polar angle of 90° and 54° are not expected to be significant. The emission is isotropic with respect to the azimuthal angle. Figure 2 shows a

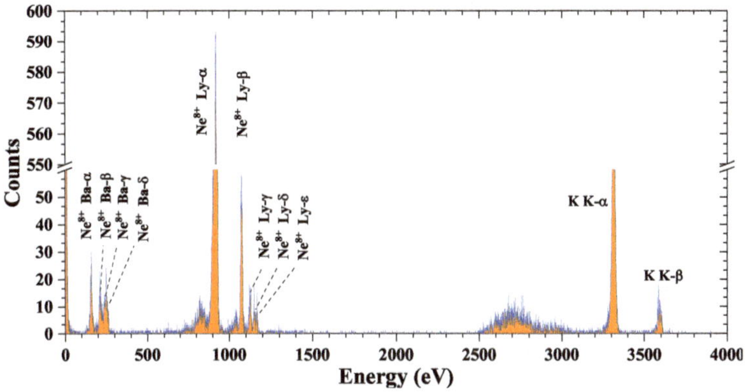

Fig. 2. X-ray spectrum in 929 km/s Ne^{9+}-Kr charge exchange process. The K-α and K-β are emission lines from the K calibration source with the energies 3313.8 eV and 3589.6 eV, respectively. The Ba-α to Ba-δ lines are identified in the figure with their specific transitions.

typical X-ray spectrum registered by the XQC detector for Ne^{9+} on Kr at a collision velocity of 929 km/s. The Ba-α, Ba-β, Ba-γ, and Ba-δ lines are all resolved. It is known from the O^{8+}-Kr experiment that the full-width at half-maximum (FWHM) line resolution of the XQC detector is better than 10 eV [75]. This is obtained through a multipeak fitting procedure for the Ly-α, Ly-β, Ly-γ, Ly-δ, and Ly-ε emission lines. When calculating the line ratios, the measured line intensities (obtained by fitting), were corrected for the energy-dependent efficiencies of the filters. Taking the O^{8+}-Kr CX collision as an example, the employed net filter efficiencies are 0.4997, 0.6057, 0.5334, 0.6453 and 0.6510 for the Ly-α, Ly-β, Ly-γ, Ly-δ and Ly-ε emissions, respectively. More details can be found elsewhere [75, 86].

3. Radiative cascade model

In the case of the single electron capture process involving multicharged ions, state-selective cross sections are a function of the quantum number (n, l) of the captured electron. These states can either cascade via dipole-allowed transitions to lower states or decay directly to the ground state resulting in soft X-ray emission. Since the ionization energy of Kr (14.0 eV) is very close to that of H (13.6 eV), state selectivity in single CX between O^{8+} and Kr is expected to be similar to O^{8+} and H. For single CX between O^{8+} and H, theoretical calculations indicate that capture occurs primarily to the n=5 states of O^{7+} for energies above 1 keV/u. At lower energies, there is some discrepancy among calculations for the relative dominance of the n=5, 6 channels, but the n=7 and the n=4 contributions to the total cross section are negligible (less than 5% for energies smaller than 6 keV/u) and there is no capture to $n \leq 3$ [91].

Figure 3 presents the observed cascade X-ray line emission from single CX between O^{8+} and Kr. These X-ray lines result from $6p(6s)$ to $2s(2p)$, $5p(5s)$ to $2s(2p)$, $4p(4s)$ to $2s(2p)$ transitions, where s and p are not resolved. For these low energy photons, the low efficiency of the present XQC filters results in poor statistics.

For single CX between O^{8+} and H, theoretical line ratios were determined from the calculated state-selective cross sections, $\sigma_{n,l}$ with n=4, 5, and 6 along with branching ratios. Expressions for O^{7+} line emission cross sections are given by:

$$\begin{aligned}\sigma(2p \to 1s) = \sigma(6h) &+ 0.956\sigma(6g) + 0.863\sigma(6f) + 0.627\sigma(6d) \\ &+ 0.038\sigma(6p) + 0.409\sigma(6s) + \sigma(5g) \\ &+ 0.908\sigma(5f) + 0.664\sigma(5d) + 0.056\sigma(5p) \\ &+ 0.454\sigma(5s) + \sigma(4d) + 0.746\sigma(4d) \\ &+ 0.042\sigma(4p) + 0.585\sigma(4s)\end{aligned} \quad (3)$$

$$\begin{aligned}\sigma(3p \to 1s) = 0.038\sigma(6g) &+ 0.105\sigma(6f) + 0.198\sigma(6d) \\ &+ 0.019\sigma(6p) + 0.240\sigma(6s) + 0.081\sigma(5f) \\ &+ 0.208\sigma(5d) + 0.007\sigma(5p) + 0.281\sigma(5s) \\ &+ 0.224\sigma(4d) + 0.366\sigma(4s)\end{aligned} \quad (4)$$

$$\begin{aligned}\sigma(4p \to 1s) = 0.015\sigma(6f) &+ 0.086\sigma(6d) + 0.008\sigma(6p) \\ &+ 0.161\sigma(6s) + 0.087\sigma(5d) + 0.190\sigma(5s) \\ &+ 0.839\sigma(4p)\end{aligned} \quad (5)$$

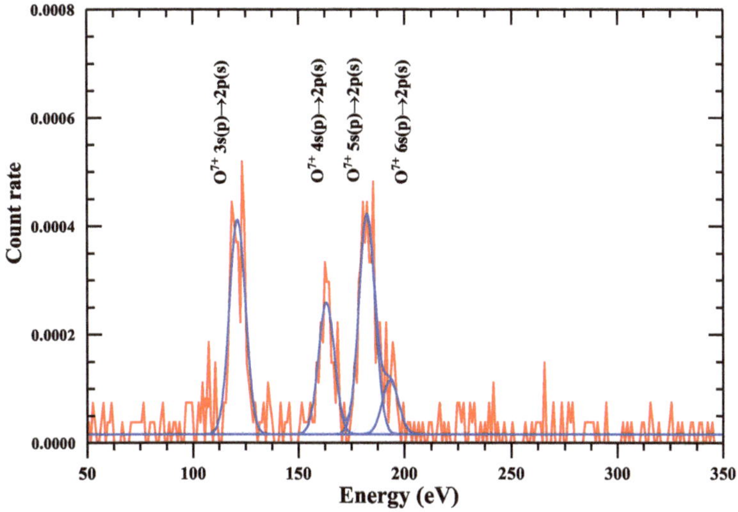

Fig. 3. Cascade X-ray emission in the 293 km/s O^{8+}-Kr collisions. The red line is the experimental measurement and the blue line denotes the Gaussian fitting.

$$\sigma(5p \to 1s) = 0.044\sigma(6d) + 0.117\sigma(6s) + 0.818\sigma(5p) \tag{6}$$

$$\sigma(6p \to 1s) = 0.780\sigma(6p). \tag{7}$$

The branching ratios were determined from Einstein's A-values for radiative transitions from n, l to n', l' and $A_{nl \to n'l'}$ were statistically averaged over total angular momentum j.

4. Results and discussion

The two systems C^{6+}-He and O^{8+}-Kr are used here to explain some physics aspects of soft X-ray emission in slow, fully-stripped ion charge exchange with atoms. For single charge exchange, X-ray spectra are from hydrogenic type transitions including Lyman series transitions. The transition energies are calculated from the energy level values provided by NIST. For example, as shown in Fig. 4, the Ly-α, Ly-β, Ly-γ, Ly-δ and Ly-ϵ emission lines are clearly observed for the 293 km/s O^{8+}-Kr single charge exchange, for which the corresponding spectra are from $2p \to 1s$, $3p \to 1s$, $4p \to 1s$, $5p \to 1s$ and $6p \to 1s$ transitions of the excited ion O^{7+}, respectively. These spectra are identified through multipeak Gaussian fitting, where the FWHM of each peak is fixed according to the observed energy resolution of the XQC detector [75]. Line ratios are defined by normalizing the X-ray spectra to the dominant emission line (usually the Ly-α in fully stripped ion-atom CX), and by incorporating the detection efficiencies.

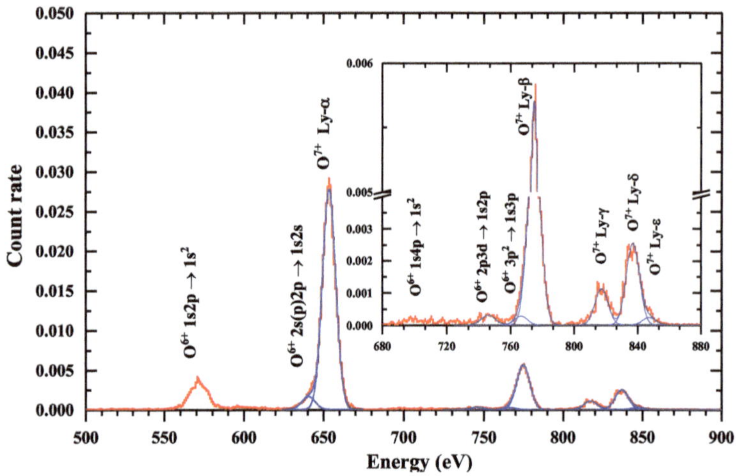

Fig. 4. X-ray spectrum for the O^{8+}-Kr collisions (the inset spectrum range is from 680 eV to 880 eV). The red line is the experimental measurement and the blue line denotes the Gaussian fitting.

For the case of the C^{6+}-He collisions, the line ratios of the Ly-β to the Ly-α lines are shown in Fig. 5. It is known that n=3 and n=4 are populated in single CX with n=3 capture being dominant [92], while the n=2 capture is negligible. The Ly-α, Ly-β and Ly-γ X-ray emission are observed [85]. The Ly-α is due to the cascade feeding from the higher populated states. Previous studies have suggested that contributions from autoionizing double capture are negligible for the present system [93]. Theoretical line ratios are constructed from the calculated l-subshell capture cross sections and by considering the cascade branching ratios. It can be seen in Fig. 5 that the line ratios calculated from the AOCC and TCBGM methods are in good agreement with the experimental data for velocities above 1500 km/s. The MOCC and TCBGM calculations predict the decreasing Ly-β to Ly-α ratio toward lower energies and are overall in satisfactory agreement with the measurement. By examining the effects of a time-dependent screening potential on the nl state population distributions, the TCBGM calculations show that the non-response potential is better described by the present system [48]. Koutroumpa [94] used a solar wind model where capture to the n=4 shell was assumed dominant. As seen in Fig. 5, this results in significant deviation from the measurement.

By comparing the C^{6+} + He measurements to a compilation of theoretical emission cross sections for CX between C^{6+} and H [67], the calculated ratios involving H (not shown) are from 0.1 to 0.16, which are significantly smaller than that of a He target for the collision velocity less than 1200 km/s. This indicates the prominent difference between the C^{6+}-H and C^{6+}-He electron capture processes. One qualitative explanation is that the electron is more prone to be captured into dominant excited states with a higher n for CX with H than He because of the smaller binding energy of H.

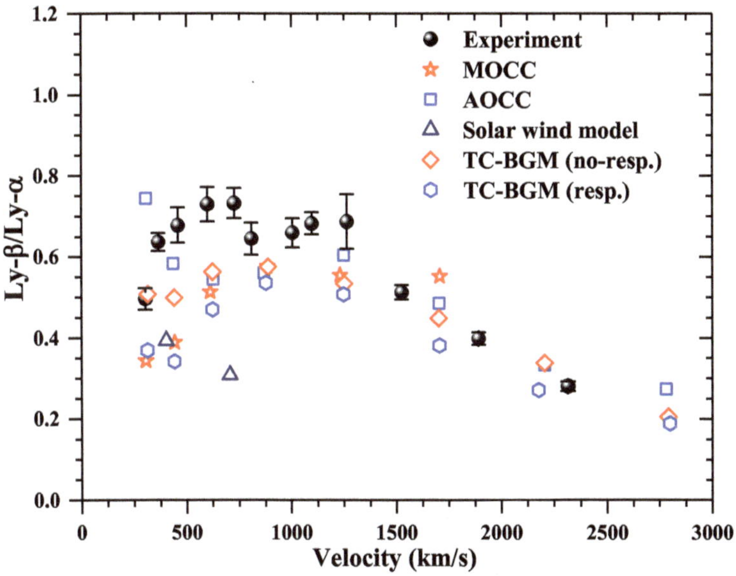

Fig. 5. The Ly-β/Ly-α line ratios for X-ray emission in C^{6+}-He collisions. The experimental data are from Defay *et al.* [85]. The corresponding theoretical values are calculated by Koutroumpa (solar wind model) [94], Fritsch and Lin (AOCC) [95], Kimura and Olson (MOCC) [96] and Leung *et al.* (TCBGM) [48].

For the case of CX in the O^{8+}-Kr collisions, it is known that n=4, n=5 and n=6 states are populated in single electron transfer. Herein, the contributions from the n=3 and n=7 capture are negligible, whereas the n=5 capture is dominant. Besides the Lyman series, there is some X-ray emission from cascades, as shown in Fig. 3. The n=4, n=5, n=6 states can also cascade to the n=2 state with subsequent Ly-α X-ray emission. Therefore, the line ratios from cascade contributions have to be taken into consideration through the branching ratios shown by Eqs. (3-7). By comparing the measured line ratios to theoretical calculations for the O^{8+}-H collisions, it can be seen that the ratios between the Ly-δ and Ly-α lines are in excellent agreement with the MOCC calculations. Note that these MOCC calculations have been used in modeling the X-ray emission from the Chandra comet by Bodewits *et al.* [67].

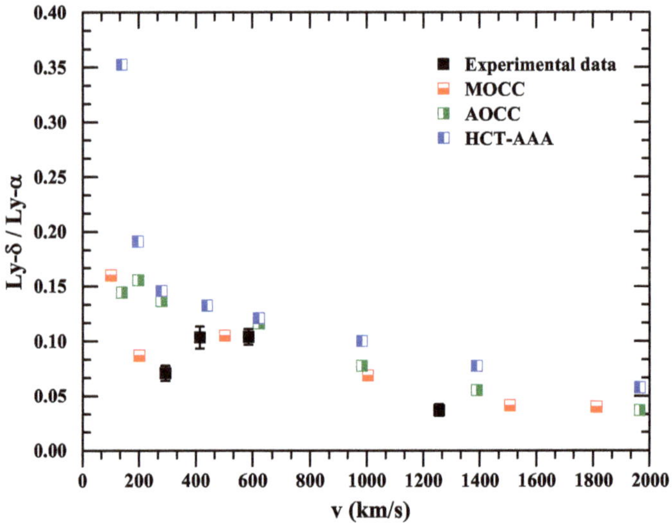

Fig. 6. The Ly-δ/Ly-α line ratios for X-ray emission in the O^{8+}+Kr collisions. The experimental data are measured by Seely *et al.* [75]. The theoretical ratios are calculated from the O^{8+}+H emission cross sections by Shipsey *et al.* (MOCC) [97], Fritsch *et al.* (AOCC) [98], and Richter *et al.* (HCT-AAA) [40], respectively.

For the non-dominant n=4 capture, the measured line ratios (Ly-γ)/(Ly-α) are shown in Table 1. First, the line ratios are smaller than those for (Ly-δ)/(Ly-α) from Fig. 6 in the case of the $n = 5$ dominant capture. Second, the theoretical calculations for a H target (not shown) did not reproduce the experimental measurements [75]. Line ratios from this non-dominant n=4 capture are complicated due to radiative cascade feeding from n=5 and n=6 cascade decay to n=4 as well as from direct capture. For the non-dominant n=6 capture, the measured line ratios (Ly-ϵ)/(Ly-α) are shown in Table 1. Further discussion is found elsewhere [75].

Table 1. Experimentally measured line ratios at energies 445, 889, 1780, and 8180 eV/u for CX in the $^{18}O^{8+}$ + Kr collisions (see also Fig. 6).

E (eV/u)	v (km/s)	Ly-γ/Ly-α	Ly-δ/Ly-α	Ly-ε/Ly-α
445	293	0.032 ± 0.008	0.071 ± 0.014	0.0065 ± 0.0038
889	414	0.039 ± 0.012	0.103 ± 0.020	0.0050 ± 0.0076
1780	586	0.035 ± 0.008	0.104 ± 0.015	0.0048 ± 0.0061
8180	1256	0.022 ± 0.004	0.037 ± 0.011	0.0048 ± 0.0028

Errors are determined from the standard deviations in the areas of the peaks; the measured data are from [75].

Electron capture can be strongly dependent on the binding energies of the considered target. The first ionization energies of multi-electron targets, such as CH_4(12.6 eV), H_2(15.43 eV), H_2O(12.6 eV) and CO_2(13.8 eV) are all close to that of H. Therefore, the X-ray emission distributions should be very similar to that of H for single CX in contrast to a He target [43]. However, any multielectron target can produce multi-electron capture. Chetiou *et al.* [99] have proposed that doubly excited states can autoionize into low singly excited states followed by the Ly-α or Ly-β X-ray emission in the 1.24 keV/u O^{8+} collisions with He and H_2. These can be described as follows:

$$\begin{aligned} O^{8+} + \mathrm{He}(1s^2\ {}^1S) &\to O^{6+}(3lnl\ n \sim 3-4) + \mathrm{He}^{2+} \\ &\to O^{7+}(2p) + \mathrm{He}^{2+} + e \\ &\to O^{7+}(1s) + \gamma + \mathrm{He}^{2+} + e \end{aligned} \tag{8}$$

$$\begin{aligned} O^{8+} + H_2 &\to O^{6+}(4l4l') + H^+ + H^+ \\ &\to O^{7+}(3p) + H^+ + H^+ + e \\ &\to O^{7+}(1s) + \gamma + H^+ + H^+ + e. \end{aligned} \tag{9}$$

Similar autoionizing double capture (ADC) channels have been found for the O^{8+} collisions with Kr [75] by comparing the ratios of the Ly-ϵ to Ly-α and Ly-β to Ly-α lines to the theoretical calculations from the O^{8+}-H CX.

Autoionizing double capture has also been found for various CX collisional systems, such as C^{6+}-H_2, O^{8+}-He, Ne^{10+}-H_2O, etc. [86, 100]. Ali *et al.* [101] performed a triple coincident measurement of the emitted X-ray scattered projectile, and recoil ions in collisions of Ne^{10+} with He, Ne, Ar, CO, and CO_2. These measurements also provide evidence of multi-electron capture.

Besides ADC, doubly excited states can also radiatively decay. It is shown in Fig. 4 that there are some minor peaks from doubly excited state radiation in cascade. For example, $O^{6+}(3p^2)$ first decays into $O^{6+}(1s3p)$ which itself decays to $O^{6+}(1s^2)$. The latter is buried in the Lyman-α emission. The associated fraction is estimated to be smaller than 6.3% [67]. Showing more detail in the following equation, O^{6+} $(2p3d)$ decays into $O^{6+}(1s2p)$ which further decays to $O^{6+}(1s^2)$ where the $O^{6+}(1s2p)$ $\rightarrow O^{6+}(1s^2)$ peak probably mixes with the nearby intercombination transition [75]:

$$\begin{aligned} O^{8+} + \mathrm{Kr}(4p^6) &\rightarrow O^{6+}(2p3d) + \mathrm{Kr}^{2+}(4p^5) \\ &\rightarrow O^{6+}(1s2p) + \mathrm{Kr}^{2+}(4p^5) + \gamma_1 \\ &\rightarrow O^{6+}(1s^2) + \gamma_1 + \gamma_2 + \mathrm{Kr}^{2+}(4p^5). \end{aligned} \tag{10}$$

The above discussions on single and double capture cover only a limited portion of the literature in the community of ion atom collisions. However, the l resolved spectroscopy data are very rare and the knowledge of relaxation of the multiply excited projectile states is still scarce. Much remains to be understood on several related topics. These include alignment and orientation of the states populated in capture as well as the evolution of the X-ray emission cross section data from slow CX to the intermediate and high velocity regime. High resolution measurements with a microcalorimeter detector are a promising method to carry out these studies.

5. Conclusions and future plans

State-selective soft X-ray emission measurements following charge exchange have been performed for C^{6+} and O^{8+} impinging on He and Kr at solar wind velocities. The normalized line ratios are measured and found to be sensitive to collision velocities. By comparing to various theoretical models, it is found that the line ratios are in excellent agreement with the MOCC calculations for capture to the dominant n shell for the O^{8+}-H charge exchange process. Also, fairly good agreement of measurements with the theory is recorded using the MOCC method for CX in the C^{6+}-He collisions. For non-dominant capture, cascade feeding from higher levels and direct electron capture competitively come into play. For the Kr target, an autoionizing double capture mechanism substantially contributes to line emissions from low n and enhances the intensity of those emissions. This indicates that the effects of multi-electron capture should be considered for X-ray emission. More experimental data are needed for further testing of theoretical methods on charge exchange state selectivity and for modeling X-ray emission in astrophysics.

From the above presentation, it is seen that laboratory measurements are capable of resolving CX emission lines. These can be used to verify the CX theories and to model soft X-ray emission in astrophysics. Previous measurements of CX relevant to the solar wind ions have been carried out by three experimental groups: Chutjian's group at the Jet Propulsion Laboratory using a solid germanium detector with a resolution of only about 100 eV [102], Beiersdorfer's group at Lawrence Livermore National Laboratory using an electron beam ion trap with a solid state and a microcalorimeter detector, but restricted to low collision energies (less than 20 eV) [103], and Ali's group combining COLTRIMS and X-ray detection with the energy resolution of 200 eV [78]. Most of these measurements were using multi-electron atoms or molecules (He, CO, H_2O) as targets, where multi-electron capture comes into play and modifies the line intensity. Until now, there are only limited UV spectroscopic measurements involving H atoms produced by dissociating H_2 in a microwave discharge [104, 105]. Further X-ray spectroscopic measurements with H are needed.

At ORNL, we are working to perform a series of benchmark measurements involving ground state of H using a microcalorimeter detector and a modified merged-beam technique. The merged-beam technique provides good collision energy resolution. Moreover, the microcalorimeter positioned a few cm from the merged beams increases the solid angle for detecting X-ray emission. In order to reduce the background noise, the ion and H beam will be merged in a 24 cm long tube with the inner diameter of few mm. The tube will be cooled down to 4 K by an exchange-gas vibration isolation crycooler system. The vacuum inside the tube is expected to be around 10^{-14} Torr. The ratio between signal and noise is estimated to be around 10. The measured spectra of soft X-ray emission from dipole allowed transitions, intercombination transitions and radiative cascades will not only test basic CX theory, but also benchmark the current astrophysical models.

At much lower energies (less than 10 eV/u), the cross sections can scale as the reciprocal of collision velocity, as predicted by the Langevin model [106, 107]. It will be interesting to study the state selectivity and clarify the mechanism for angular momentum state population. Moreover, by measuring X-ray emission following CX between slow H-like ions and ground state H atoms, it may bring the CX studies to the level of spin multiplicity and explore the Pauli exclusion principle during the collisions.

Acknowledgments

We thank the Editors for the invitation to contribute with the present work. This research is supported in part by NASA Astrophysics Research and Analysis Grant No. NNH13ZDA001N. We thank D.G. Seely and D. Wulf for helpful conversations and their help in data analysis for preparation of this manuscript. In addition, much of the experimental data presented was obtained in collaboration with D. McCammon's group at the University of Wisconsin, Madison. Our special appreciation is given to them.

References

[1] C. D. Lin, ed., *Review of fundamental processes and applications of atoms and ions.* Word Scientific, Singapore (1993).
[2] J. Eichler, *Lectures on ion atom collisions: from nonrelativistic to relativistic velocities.* Elsevier, Amsterdam (2005).
[3] R. K. Janev, L. P. Presnyakov, and V. P. Shevelko, *Physics of highly charged ions.* Springer Press, Berlin (1985).
[4] H. S. W. Massey, E. W. McDaniel, and B. Bederson, *Applied Atomic Collision Physics Vol 1.* Academic Press, San Diego (1982).
[5] F. Aumayr, G. Betz, and H. Winter, Proceedings of the 7. international conference on the physics of highly charged ions (HCI-94), *Nucl. Instrum. Meth. Phys. Res. B.* **98**, 1–615 (1995).
[6] M. Nastasi, J. K. Hirvonen, and J. W. Mayer, *Ion-solid interations: Fundamentals and applications.* Cambridge University Press, Cambridge (1996).
[7] L. Flamn and R. Schumann, Die geschwindigkeitsabnahme der α-strahlen in materie, *Ann Phsik.* **50**, 655–699 (1916).
[8] G. H. Henderson, Changes in the charges of an alpha-particle passing through matter, *Proc. Roy. Soc. A.* **102**, 496–506 (1923).
[9] E. Rutherford, The capture and loss of electrons by alpha particles, *Philos Mag.* **47**, 277–303 (1924).
[10] N. O. Lassen, Total charges of fission fragments as functions of the pressure in the stopping gas, *Mat. Fys. Medd. Dan. Vid. Selsk. 26.* **12**, 1–49 (1951).
[11] N. O. Lassen, The total charges of fission fragments in gaseous and solid stopping media, *Mat. Fys. Medd. Dan. Vid. Selsk. 26.* **5**, 1–28 (1951).

[12] S. K. Allison, Experimental results on charge-changing collisions of hydrogen and helium atoms and ions at kinetic energies above 0.2 keV, *Rev. Mod. Phys.* **30**, 1137–1168 (1958).
[13] L. H. Thomas, On the capture of electrons by swiftly moving electrified particles, *Proc. Roy. Soc.* **114**, 561–576 (1927).
[14] V. Mergel, R. Dörner, M. Achler, K. Khayyat, S. Lencinas, J. Euler, O. Jagutzki, S. Nöttgens, M. Unverzagt, L. Spielberger, W. Wu, R. Ali, J. Ullrich, H. Cederquist, A. Salin, J. C. Wood, R. E. Olson, C. L. Cocke, and H. Schmidt-Böcking, Intra-atomic electron-electron scattering in p-He collisions (Thomas process) investigated by cold target recoil ion momentum spectroscopy, *Phys. Rev. Lett.* **79**, 387–390 (1997).
[15] J. R. Oppenheimer, On the quantum theory of the capture of electrons, *Phys. Rev.* **31**, 349–356 (1928).
[16] H. C. Brinkman and H. A. Kramers, Zur theorie der einfangung von elektronen durch alpha-teichen, *Proc. Roy. Acad. Amsterdam.* **33**, 973–984 (1930).
[17] H.-D. Betz, Charge states and charge-changing cross sections of fast heavy ions penetrating through gaseous and solid media, *Rev. Mod. Phys.* **44**, 465–539 (1972).
[18] R. K. Janev and H. Winter, State-selective electron capture in atom-highly charged ion collisions, *Phys. Rep.* **117**, 265–387 (1985).
[19] R. Hoekstra, D. Ćirić, A. N. Zinoviev, Y. S. Gordeev, F. J. de Heer, and R. Morgenstern, Emission cross sections for fully stripped carbon colliding with atomic hydrogen, *Z. Phys. D: At. Mol. Clust.* **8**, 57–61 (1988).
[20] J. Eichler and T. Stöhlker, Radiative electron capture in relativisitic ion-atom collisions and the photoelectric effect in hydrogen-like high-Z systems, *Phys. Rep.* **439**, 1–99 (2007).
[21] F. G. Wilkie, R. W. McCullough, and H. B. Gilbody, State-selective electron capture by slow C^{3+} and N^{3+} ions in H and H_2, *J. Phys. B: At. Mol. Opt.* **19**, 239–251 (1986).
[22] T. K. McLaughlin, R. W. McCullough, and H. B. Gilbody, State-selective electron capture by slow C^{4+} in H_2 and O_2, *J. Phys. B: At. Mol. Opt.* **25**, 1257–1264 (1992).
[23] N. Stolterfoht, K. Sommer, J. K. Swenson, C. C. Havener, and F. W. Meyer, Electron-correlation effects in double-electron-capture collisions of 60-keV C^{6+} with He, *Phys. Rev. A.* **42**, 5396–5405 (1990).
[24] M. Mack, J. H. Nijland, P. v. d. Straten, A. Niehaus, and R. Morgenstern, Correlation in double electron capture in collisions of fully stripped ions on He and H_2, *Phys. Rev. A.* **39**, 3846–3854 (1989).
[25] B. H. Bransden and M. R. C. McDowell, *Charge Exchange and the Theory of Ion-Atom Collisions*, 2nd edn. Oxford University, London (1992).
[26] D. P. Dewangan and J. Eichler, Charge exchange in energetic ion atom collisions, *Phys. Rep.* **247**, 59–219 (1994).
[27] Dž. Belkić, R. Gayet, and A. Salin, Electron capture in high-energy ion-atom collisions, *Phys. Rep.* **56**, 279–369 (1979).
[28] Dž. Belkić, I. Mančev, and J. Hanssen, Four-body methods for high-energy ion-atom collisions, *Rev. Mod. Phys.* **80**, 249–314 (2008).

[29] Dž. Belkić, *Principle of Quantum Scattering Theory.* The Institute of Physics Publishing, Bristol (2003).

[30] Dž. Belkić, *Quantum Theory of High-Energy Ion-Atom Collisions.* Taylor & Francis, London (2008).

[31] U. Fano and W. Lichten, Interpretation of Ar^+-Ar collisions at 50 keV, *Phys. Rev. Lett.* **14**, 627–629 (1965).

[32] J. A. Tanis, E. M. Bernstein, W. G. Graham, M. P. Stockli, M. Clark, R. H. McFarland, T. J. Morgan, K. H. Berkner, A. S. Schlachter, and J. W. Stearns, Resonant electron transfer and excitation in two, three, and four-electron $^{20}Ca^{q+}$ and $^{23}V^{q+}$ ions colliding with helium, *Phys. Rev. Lett.* **53**, 2551–2554 (1984).

[33] N. Andersen and K. Bartschat, eds., *Polarization, Alignment, and Orientation in atomic collisions.* Springer-Verlag, New York (2001).

[34] E. Y. Sidky and H.-J. T. Simonsen, Velocity-matching model for electron capture in keV atomic collisions, *Phys. Rev. A.* **54**, 1417–1429 (1996).

[35] R. A. Phaneuf, Electron capture in very slow C^{q+}+H collisions, *Phys. Rev. A.* **24**, 1138–1141 (1981).

[36] F. W. Meyer, A. M. Howald, C. C. Havener, and R. A. Phaneuf, Low-energy total-electron-capture cross sections for fully stripped and H-like projectiles incident on H and H_2, *Phys. Rev. A.* **32**, 3310–3318 (1985).

[37] C. C. Havener, A. Möller, P. A. Z. van Emmichoven, and R. A. Phaneuf, Low-energy electron capture by C^{3+} from hydrogen using merged beams, *Phys. Rev. A.* **51**, 2982–2988 (1995).

[38] J. S. Thompson, A. M. Covington, M. Pieksma, J. L. Shinpaugh, P. C. Stancil, and C. C. Havener, Low-energy electron capture by Cl^{7+} from D using merged beams, *Phys. Rev. A.* **63**, 012717 (2000).

[39] A. Salop and R. E. Olson, Charge exchange between H(1s) and fully stripped heavy ions at low-keV impact energies, *Phys. Rev. A.* **13**, 1312–1320 (1976).

[40] K. Richter and E. A. Solovév, Application of the advanced adiabatic approach to charge exchange in slow collisions between H and O^{8+}, *Phys. Rev. A.* **48**, 432–441 (1993).

[41] G. Drake, *Springer handbook of atomic, molecular, and optical physics.* Springer Press, New York (2006).

[42] R. E. Olson, n, l distributions in A^{q+}+H electron capture collisions, *Phys. Rev. A.* **24**, 1726–1733 (1981).

[43] S. Otranto, R. E. Olson, and P. Beiersdorfer, X-ray emission cross sections following charge exchange by multiply charged ions of astrophysical interest, *Phys. Rev. A.* **73**, 022723 (2006).

[44] S. Otranto and R. E. Olson, X-ray emission cross sections following Ar^{18+} charge exchange collisions on neutral argon: The role of the multiple electron capture, *Phys. Rev. A.* **83**, 032710 (2011).

[45] J. Simćić, D. R. Schultz, R. J. Mawhorter, I. Čadež, J. B. Greenwood, A. Chutjian, C. M. Lisse, and S. J. Smith, Measurement and calculation of absolute single-and multiple-charge-exchange cross sections for Fe^{q+} ions impacting CO and CO_2, *Phys. Rev. A.* **81**, 062715 (2010).

[46] W. Fritsch and C. D. Lin, Theory of ion-atom collisions-coupled channel methods, *Phys. Rep.* **202**, 1–97 (1991).
[47] R. J. Allan, A. S. Dickinson, and R. McCarroll, Molecular treatment of charge exchange in H^++Li collisions, *J. Phys. B: At. Mol. Opt.* **16**, 467–480 (1983).
[48] A. C. K. Leung and T. Kirchner, Analysis of x-ray emission spectra in charge-exchange collisions of C^{6+} with He and H_2, *Phys. Rev. A.* **93**, 052710 (2016).
[49] C. C. Havener, R. Rejoub, C. R. Vane, H. F. Krause, D. W. Savin, M. Schnell, J. G. Wang, and P. C. Stancil, Electron capture by Ne^{4+} ions from atomic hydrogen, *Phys. Rev. A.* **71**, 034702 (2005).
[50] C. H. Liu, J. G. Wang, and R. K. Janev, Spin-resolved state-selective electron capture in C^{5+}-H collisions, *Journal of Physics B: At. Mol. Opt.* **49**, 175202 (2016).
[51] H. Ryufuku, K. Sasaki, and T. Watanabe, Osillatory behavior of charge transfer cross sections as a function of the charge of projectiles in low energy collisions, *Phys. Rev. A.* **21**, 745–750 (1980).
[52] J. Burgdörfer, R. Morgenstern, and A. Niehaus, Angular momentum distribution in the classical over-barrier model for electron capture into highly charged slow projectiles, *J. Phys. B: At. Mol. Phys.* **19**, L507–L513 (1986).
[53] J. Ullrich, R. Moshammer, A. Dorn, R. Dörner, L. P. H. Schmidt, and H. Schmidt-Böcking, Recoil-ion and electron momentum spectroscopy: reaction-microscopes, *Rep. Prog. Phys.* **66**, 1463–1545 (2003).
[54] R. Dörner, V. Mergel, O. Jagutzki, L. Spielberger, J. Ullrich, R. Moshammer, and H. Schmidt-Böcking, Cold target recoil ion momentum spectroscopy: A 'momentum microscope' to view atomic collision dynamics, *Phys. Rep.* **330**, 95–192 (2000).
[55] E. Y. Kamber, M. A. Abdallah, C. L. Cocke, and M. Stöckli, Longitudinal and transverse recoil electron capture processes in 0.5-50 keV/u C^{5+}-He collisions, *Phys. Scr.* **1999**(T80B), 367–369 (1999).
[56] X. Fléchard, C. Harel, H. Jouin, B. Pons, L. Adoui, F. Fremont, A. Cassimi, and D. Hennecart, Single and double-electron capture in low energy Ne^{10+} and He collisions, *J. Phys. B: At. Mol. Opt. Phys.* **34**, 2759–2779 (2001).
[57] M. A. Abdallah, W. Wolff, H. E. Wolf, E. Sidky, E. Y. Kamber, M. Stöckli, C. D. Lin, and C. L. Cocke, Cold-target recoil-ion-momentum spectroscopy study of single electron capture from He by slow Ar^{8+} ions, *Phys. Rev. A.* **57**, 4373–4378 (1998).
[58] R. T. Zhang, X. L. Zhu, X. Y. Li, L. Liu, S. F. Zhang, W. T. Feng, D. L. Guo, Y. Gao, D. M. Zhao, J. G. Wang, and X. Ma, Single-electron capture in 3-keV/u Ar^{8+} on He collisions, *Phys. Rev. A.* **95**, 042702 (2017).
[59] D. Fisher, B. Feuerstein, R. D. DuBois, R. Moshammer, J. R. C. López-Urrutia, I. Draganić, H. Lórch, A. N. Perumal, and J. Ullrich, State-resolved measurements of single-electron capture in slow Ne^{7+}- and Ne^{8+}-helium collisions, *J. Phys. B: At. Mol. Opt. Phys.* **35**, 1369–1377 (2002).
[60] D. Dijkkamp, Y. S. Gordeev, A. Brazuk, A. G. Drentje, and F. J. de Heer, Selective single-electron capture into (n,l) subshells in slow collisions of C^{6+},

N^{6+}, O^{6+} and Ne^{6+} with He, H_2 and Ar, *J. Phys. B: At. Mol. Phys.* **18**, 737–756 (1985).

[61] D. Dijkkamp, D. Ćirić, and F. J. de Heer, Total capture and line-emission cross sections for C^{6+}, N^{7+}, O^{8+}-H in the energy range of 3-7.5 keV/u, *Phys. Rev. Lett.* **54**, 1004–1007 (1985).

[62] J. E. Hansen, ed., *Atomic spectra and oscillator strengths for astrophysics and fusion research.* North-Holland, Amsterdam (1990).

[63] J. P. M. Beijers, R. Hoekstra, A. R. Schlatmann, R. Morgenstern, and F. J. de Heer, State-selective electron capture in slow collisions of C^{6+} and O^{6+} with He, *J. Phys. B: At. Mol. Opt. Phys.* **25**, 463–474 (1992).

[64] C. M. Lisse, K. Dennerl, J. Englhauser, M. Harden, F. E. Marshall, M. J. Mumma, R. Petre, J. P. Pye, M. J. Ricketts, J. Schmitt, J. Trömper, and R. G. West, Discovery of x-ray and extreme ultraviolet emission from comet C/Hyakutake 1996 B2, *Science.* **274**, 205–209 (1996).

[65] P. Beiersdorfer, K. R. Boyce, G. V. Brown, H. Chen, S. M. Kahn, R. L. Kelley, M. May, R. E. Olson, F. S. Porter, C. K. Stahle, and W. A. Tillotson, Laboratory simulation of charge exchange produced x-ray emission from comets, *Science.* **300**, 1558–1559 (2003).

[66] D. Koutroumpa, R. Lallement, J. C. Raymond, and V. Kharchenko, The solar wind charge transfer x-ray emission in 1/4 keV energy range: inferences on local bubble hot gas at low Z, *Astrophys. J. Lett.* **696**, 1517–1525 (2009).

[67] D. Bodewits, D. J. Christian, M. Torney, M. Dryer, C. M. Lisse, K. Dennerl, T. H. Zurbuchen, S. J. Wolk, A. G. G. M. Tielens, and R. Hoekstra, Spectral analysis of the chandra comet survey, *A & A.* **469**, 1183–1195 (2007).

[68] P. D. Mullen, R. S. Cumbee, D. Lyons, L. Gu, J. Kaastra, R. L. Shelton, and P. C. Stancil, Line ratios for solar wind charge exchange with comets, *Astrophys. J.* **844**, 7–15 (2017).

[69] T. E. Cravens, Heliospheric x-ray emission associated with charge transfer of the solar wind with the interstellar neutrals, *Astrophys. J. Lett.* **532**, L153–L156 (2000).

[70] V. Kharchenko, M. Rigazio, A. Dalgarno, and V. A. Krasnopolsky, Charge abundances of the solar wind ions inferred from cometary x-ray spectra, *Astrophys. J.* **585**, L73–L75 (2003).

[71] N. Stolterfoht, C. C. Havener, R. A. Phaneuf, J. K. Swenson, S. M. Shafroth, and F. W. Meyer, Evidence for correlated double-electron capture in low-energy collisions of O^{6+} with He, *Phys. Rev. Lett.* **57**, 74–77 (1986).

[72] Z. Chen and C. D. Lin, Double electron capture and the angular distribution of ejected electrons in Ne^{8+}-He collisions, *Phys. Rev. A.* **48**, 1298–1307 (1993).

[73] J.-Y. Chesnel, H. Merabet, X. Husson, F. Frémont, D. Lecler, H. Jouin, C. Harel, and N. Stolterfoht, Energy dependence of cross sections for double-electron capture in 48-132-keV C^{6+}+He collisions, *Phys. Rev. A.* **53**, 2337–2344 (1996).

[74] R. T. Zhang, W. T. Feng, X. L. Zhu, S. F. Zhang, D. L. Guo, Y. Gao, D. B. Qian, S. Xu, S. C. Yan, P. Zhang, Z. K. Huang, H. B. Wang, B. Hai, D. M. Zhao, and X. Ma, Two-electron transfer and ionization mechanism in 80-keV/u Ne^{8+} on He collisions, *Phys. Rev. A.* **93**, 032709 (2016).
[75] D. G. Seely, V. M. Andrianarijaona, D. Wulf, K. Morgan, D. McCammon, M. Fogle, P. C. Stancil, R. T. Zhang, and C. C. Havener, Line ratios for soft x-ray emission following charge exchange between O^{8+} and Kr, *Phys. Rev. A.* **95**, 052704 (2017).
[76] N. O. Vasetskaya and V. K. Ivanov, Radiative decays of doubly excited states of oxygen, carbon, and nitrogen ions, *Optika i spektroskopiya.* **105**, 726–731 (2008).
[77] S. Martin, A. Salmoun, Y. Ouerdane, M. Druetta, J. Désesquelles, and A. Delon, Double rydberg states of high angular momentum (l=6-8) produced in Ar VIII by Ar^{9+}-Cs collisions, *Phys. Rev. Lett.* **62**, 2112–2115 (1989).
[78] R. Ali, P. Beiersdorfer, C. L. Harris, and P. A. Neill, Charge exchange x-ray spectra: Evidence for significant contributions from radiative decays of doubly excited states, *Phys. Rev. A.* **93**, 012711 (2016).
[79] C. C. Havener, M. S. Huq, H. F. Krause, P. A. Schulz, and R. A. Phaneuf, Merged-beams measurements of electron-capture cross sections for O^{5+}+H at electron-volt energies, *Phys. Rev. A.* **39**, 1725–1740 (1989).
[80] M. A. Abdallah, C. R. Vane, C. C. Havener, D. R. Schultz, H. F. Krause, N. Jones, and S. Datz, Multiple ionization in fast ion-atom collisions: Simultaneous measurement of recoil momentum and projectile energy loss, *Phys. Rev. Lett.* **85**, 278–281 (2000).
[81] S. M. Shafroth and J. C. Austin, eds., *Accelerator-based atomic physics techniques and applications.* AIP press, New York (1997).
[82] R. A. Phaneuf, C. C. Havener, G. H. Dunn, and A. Möller, Merged-beams experiments in atomic and molecular physics, *Rep. Prog. Phys.* **62**, 1143–1180 (1999).
[83] D. McCammon, K. Barger, D. E. Brandl, R. P. Brekosky, S. G. Crowder, J. D. Gygax, R. L. Kelley, C. A. Kilbourne, M. A. Lindeman, F. S. Porter, L. E. Rocks, and A. E. Szymkowiak, The x-ray quantum calorimeter sounding rocket experiment: Improvements for the next flight, *J. Low Temp. Phys.* **151**, 715–720 (2008).
[84] V. M. Andrianarijaona, D. Wulf, D. McCammon, D. G. Seely, and C. C. Havener, Radiance line ratios Ly-β/Ly-α, Ly-γ/Ly-α, Ly-δ/Ly-α, and Ly-ϵ/Ly-α for soft x-ray emissions following charge exchange between C^{6+} and Kr, *Nucl. Instr. Meth. Phys. Res. B.* **345**, 69–73 (2015).
[85] X. Defay, K. Morgan, D. McCammon, D. Wulf, V. M. Andrianarijaona, M. Fogle, D. G. Seely, I. N. Draganić, and C. C. Havener, X-ray emission measurements following charge exchange between C^{6+} and He, *Phys. Rev. A.* **88**, 052702 (2013).
[86] M. Fogle, D. Wulf, K. Morgan, D. McCammon, D. G. Seely, I. N. Draganić, and C. C. Havener, X-ray-emission measurements following charge exchange between C^{6+} and H_2, *Phys. Rev. A.* **89**, 042705 (2014).

[87] D. A. Wollman, K. D. Irwin, G. C. Hilton, L. L. Dulcie, D. E. Newbury, and J. M. Martinis, High-resolution, energy-dispersive microcalorimeter spectrometer for x-ray microanalysis, *J. Microsc.* **188**, 196–223 (1997).

[88] B. L. Moiseiwitsch and S. J. Smith, Electron impact excitation of atoms, *Rev. Mod. Phys.* **40**, 238–253 (1968).

[89] C. D. Lin and J. H. Macek, Theory of anisotropy transfer and calculations of alignment of np states populated in electron capture by highly charged ion, *Phys. Rev. A.* **35**, 5005–5011 (1987).

[90] J. B. Greenwood, I. D. Williams, S. J. Smith, and A. Chutjian, Experimental investigation of the processes determining x-ray emission intensities from charge-exchange collisions, *Phys. Rev. A.* **63**, 062707 (2001).

[91] L. Liu, J. G. Wang, and R. K. Janev, Dynamics of O^{8+}+H electron capture in Debye plasmas, *Phys. Rev. A.* **79**, 052702 (2009).

[92] M. Barat, M. N. Gaboriaud, L. Guillemot, P. Roncin, H. Laurent, and S. Andriamonje, Coincident energy gain spectroscopy of elelctron catpure in multiply charged ions colliding with He, H_2 and heavy rare-gas targets, *J. Phys. B: At. Mol. Opt.* **20**, 5771–5783 (1987).

[93] C. Harel, H. Jouin, and B. Pons, Double capture in C^{6+} with He collisions at low impact energies, *J. Phys. B: At. Mol. Phys.* **24**, L425–L430 (1991).

[94] D. Koutroumpa. *Etude des émissions gazeuses UV/X et des phénoménes déchange de charge dans ĺéliosphére: application á des systémes astrophysiques analogues.* PhD thesis, ÍUniversite Pierre et Marie Curié Verrieres-le-Buisson, Paris, France (2007).

[95] W. Fritsch and C. D. Lin, Atomic-orbital expansion study for the (quasi-) two-electron collision system O^{6+}+He and C^{6+}+He, *J. Phys. B: At. Mol. Opt.* **19**, 2683–2694 (1986).

[96] M. Kimura and R. E. Olson, Electron capture to (nl) states in collisions of C^{4+} and C^{6+} with He, *J. Phys. B: At. Mol. Opt. Phys.* **17**, L713–L719 (1984).

[97] E. J. Shipsey, T. A. Green, and J. C. Browne, Modified method of perturbed stationary states. V. Electron-capture cross sections for the reaction O^{8+}+H → O^{7+}+H^+, *Phys. Rev. A.* **27**, 821–832 (1983).

[98] W. Fritsch and C. D. Lin, Atomic-orbital-expansion studies of electron transfer in bare nucleus Z(Z=2,4,8)-hydrogen-atom collisions, *Phys. Rev. A.* **29**, 3039–3051 (1984).

[99] A. Chetioui, F. Martín, M. F. Politis, J. P. Rozet, A. Touati, L. Blumenfeld, D. Vernhet, K. Wohrer, C. Stephan, M. Barat, M. N. Gaboriaud, H. Laurent, and P. Roncin, Doubly excited states populated in collisions of O^{8+} ions with He and H_2, *J. Phys. B: At. Mol. Opt. Phys.* **23**, 3659–3675 (1990).

[100] S. Otranto and R. E. Olson, Charge exchange and x-ray emission cross sections for multiply charged ions colliding with H_2O, *Phys. Rev. A.* **77**, 022709 (2008).

[101] R. Ali, P. A. Neill, P. Beiersdorfer, C. L. Harris, M. J. Rakvić, J. G. Wang, D. R. Schultz, and P. C. Stancil, On the significance of the contribution of multi-electron capture processes to cometary x-ray emission, *Astrophys. J.* **629**, L125–L128 (2005).

[102] J. B. Greenwood, A. Chutjian, and S. J. Smith, Measurements of absolute, single charge-exchange cross section of H^+, He^+ and He^{2+} with H_2O and CO_2, *Astrophys. J.* **529**, 605–609 (2000).

[103] P. Beiersdorfer, Laboratory x-ray astrophysics, *Annu. Rev. Astron. Astrophys.* **41**, 343–390 (2003).

[104] R. Hoekstra, F. J. de. Heer, and R. Morgenstern, State-selective electron capture in collisions of He^{2+} with H, *J. Phys. B: At. Mol. Phys.* **24**, 4025–4048 (1991).

[105] J. Slevin and W. Stirling, Radio frequency atomic hydrogen beam source, *Rev. Sci. Instrum.* **52**, 1780–1782 (1981).

[106] N. Shimakura and M. Kimura, Electron capture in collisions of N^{5+} ions with H atoms from the meV to keV energy regions, *Phys. Rev. A.* **44**, 1659–1667 (1991).

[107] M. Pieksma, M. Gargaud, R. McCarroll, and C. C. Havener, Electron-capture cross section at near-thermal collision energies for Si^{4+}+H, *Phys. Rev. A.* **54**, R13–R16 (1996).

Chapter 3

Ionization of large biomolecules by fast heavy ions

L.C. Tribedi*

Tata Institute of Fundamental Research, Homi Bhabha Road, Colaba, Mumbai, 400005 India

In this chapter, we review our recent measurements on ionization and fragmentation of large biomolecules, particularly the DNA and RNA nucleobases. There have been several important experimental investigations on double differential, single differential and total cross section of electron emission in ionization of some of the nucleobases impacted by fast protons, carbon ions and other highly charged ions. These experiments are carried out using time-of-flight recoil-ion mass spectroscopy, recoil-ion momentum spectroscopy and ejected continuum electron spectroscopy. A few recent results from different research groups are also addressed. The existing data sets for nucleobases are used to obtain certain approximate scaling formulae in terms of charge state and energy of the ions. A brief account of the experimental techniques is given, including the normalization procedure, the target-preparation, electron and recoil ion detection techniques. With the advent of the recent theoretical calculations for DNA and RNA nucleobases, it is possible to compare the existing experimental data for ionization with different models, such as, the "continuum distorted wave - eikonal initial state", the "first-order Born with correct boundary conditions" and the "classical trajectory Monte Carlo - classical over the barrier" approximations. Some of the examples of these comparisons are presently discussed. The measured quantities to be analyzed are fragmentation pattern, fractional yields of fragments, energy and angular distributions of electron emission, total cross sections as a function of energy and charge states, the forward-backward angular asymmetry of electron emission and total cross section scaling rules. In addition, a comparative study has been made between the results obtained for biomolecules with those for simple atoms and molecules. The scope of this chapter is limited to mainly experimental investigations. The progress in the theoretical investigation has not been addressed except through occasional references.

*Email: lokesh@tifr.res.in

Contents

1. Introduction

The studies on ionization and breakup of large biomolecules and water are of immense importance in radiobiological applications. In particular, modeling of the energy loss of such ions in biological matter demands the cross sections for these basic collisional processes near the Bragg peak region. In fast ion-atom collisions of relevance to these applications, various physical events can take place. These include e.g. soft collisional ionization, two-center interaction, electron capture to both bound and continuum states, binary-encounters responsible for electron emission and the characteristic Auger electron emissions due to inner shell vacancy production. Such processes constitute the well-known mechanisms that characterize the double differential distribution of electron emission in a typical collision with a simple atom. Such collisions provide a basis towards the understanding of few-body collisional dynamics involving molecules or clusters. Although experimental studies with elaborated equipments and data analysis techniques have been pursued over several decades, many of the investigations are renewed in the last decade due to its importance in fundamental ion-molecule interactions as well as for its applications. On the theoretical side, finding a suitable wave function and calculations of cross sections of different processes for many-body molecules is one of the most challenging tasks. However, there has also been a rapid progress on the theoretical framework, which was inspired by a series of new experimental investigations. The study of complex molecules of specific interests, such as biologically relevant molecules and their interactions with swift ions have now been the prime research goal of many investigators across different fields of science: physics, biology, chemistry and medicine. Such truly interdisciplinary research is likely to open new avenues to fundamental understanding as well as to applications in e.g. radiation biology, particularly for hadron therapy. The investigation of collisions with protons and carbon ions, is particularly crucial in modeling the cell damage, since these ions at about 200-400 MeV/u energies are commonly used in treatments to remove deep-seated tumorous tissue in the human body. Unlike the case of gamma-ray induced radiation treatments, in heavy-ion therapy, the energy loss of swift ions during their passage through tissue exhibits a maximum at a certain distance which is dependent on the impact energy. Therefore, the study of energetic ions impinging on nucleobases over a wide energy range, across the Bragg-peak region is of primary importance. Since it is well-known that the low energy electrons created in such collisions are primarily responsible for single and double strand breaks, a detailed knowledge of the ion-induced electron emission mechanisms is of great importance for radiobiologal investigations within hadron therapy.

The low-energy proton and electron ionization and fragmentation of uracil was investigated by Märk *et al.* [1–3]. They showed that the essential feature of the fragmentation pattern of uracil is similar to that under the impact of electrons and protons. These authors derived the branching ratios of the capture and ionization processes for keV energy proton collisions. Measurements of ionization and fragmentation of nucleobases for highly charged Xe ions by de Vries *et al.* [4] demonstrated the dependence of kinetic energies of the fragments on the initial charge state of the intermediate parent ions. Brédy *et al.* [5] investigated the fragmentation of singly and doubly charged adenine induced by electron capture processes in collisions with 6 keV F^{2+} ions as projectiles. There are several other measurements on the fragmentation of varieties of biomolecules by different groups [6–12]. Most of these studies on biomolecular ionization and fragmentation dynamics are based on gas-phase collisions with isolated biomolecules. However, for radiobiological applications, it would be desirable to place the biomolecular targets in some more realistic biochemical environment. To clarify this aspect, Schlathölter *et al.* [8] studied the interactions of keV energy ions not only with the isolated nucleobases, but also with the nucleobase clusters using time-of-flight (ToF) spectrometry. They found significant changes in the molecular fragmentation patterns in the case of small clusters of DNA-bases compared to those for a single nucleobase.

The electron spectroscopic investigations that can provide the details of the energy and angular distributions of the double differential cross sections (DDCS) of electron emission in such collision processes are very scarce [13–15]. One such measurement was reported by Moretto-Capelle *et al.* [13] for ionization of uracil in collisions with low energy protons (around 25-100 keV). However, this experiment was performed only for one emission angle, namely 35°. For higher energy protons (0.5-2.0 MeV), the experiments to measure full angular distribution were conducted by the group of Itoh *et al.* [14, 16, 17] for the adenine and uracil molecules. The details of energy and angular distributions of low energy electron emission for highly charged and high energy ions (3.5 MeV/u bare C-ions bombarding a single uracil molecule), were reported by Agnihotri *et al.* [15]. They provide detailed comparisons of their data with the recent theoretical calculations. Finally, a series of measurements on the total cross sections of ionization and fragmentation, their dependence on the projectile charge state (q_p) and the projectile velocity (v_p) and, hence, on the perturbation strength (q_p/v_p), have been reported by Tribedi *et al.* [18–20]. For these latter series of studies, the highly charged ions of C, O, F-ions are used. The influence of the charge transfer process for C-ions colliding with uracil was also investigated by Bacchus-Montabonel *et al.* [11]. Many experiments on ionization and fragmentation of water molecule have been reported in the last few decades and some of them can be found in Refs. [21–28].

The detailed theoretical techniques developed to deal with such large biomolecules and water molecule are beyond the scope of this chapter. However, one may find such model calculations in Refs. [12, 29–32] for biomolecules and in Refs. [33–36] for water. Some of the recent articles on theory pursued by Rivarola and colleagues can be found in Refs. [37, 39–44]. In brief, the classical trajectory Monte Carlo (CTMC) based model used by Bacchus-Montabonel *et al.* [11] was combined with a classical over-barrier (COB) model by Champion *et al.* [12, 29]. They calculated single and multiple ionization cross sections of RNA and DNA base molecules by several bare ions, such as fast protons, He^{2+} and C^{6+}. Additionally, two different quantum-mechanical models used in ion-atom collisions were adapted to biomolecular targets: the continuum distorted wave – eikonal initial state (CDW-EIS) [32, 37] and the boundary-corrected first Born (CB1) [31, 32] methods. In particular, the CDW-EIS model is an approximation of the continuum distorted wave (CDW) method which uses the full continuum wave functions within both the initial and final scattering states [38]. In order to compare with experimental data, these theoretical studies have provided the DDCS, single differential cross sections (SDCS) and total cross section (TCS) for highly charged ion-impact ionization of DNA bases. The complete neglect of differential overlap (CNDO) approximation, similar to Senger *et al.* [45], was also used. Both the mentioned models are briefly described, as well as applied in Refs. [15, 19]. In order to represent the 29 orbitals of uracil, the fundamental state of uracil was described by the restricted Hartree-Fock method with geometry optimization [37]. Such a molecular description was used to avoid the difficulty of implementing the more elaborated models based on the density functional theory [47]. In the CDW-EIS model for one active electron, each molecular orbital is distorted by a continuum projectile eikonal phase for the initial scattering states in the entrance channel. On the other hand, the continuum state of the electron from the final state in the exit channel, moving in the field of the residual target, is distorted by the projectile continuum wave function [48, 49], which is identical to that of the CDW method [38]. In both the CDW-EIS and CB1 approaches, the effective Coulomb wave functions are employed to represent the continuum of the electron in the field of the residual target, with an effective nuclear charge $\xi = \sqrt{(-2n_{ji}^2\epsilon_j)}$. Here, ϵ_j denotes the energy of the jth-orbital and n_{ji} is the principal quantum number of the corresponding i^{th}-atomic wave function associated with this orbital. A theoretical review on ion-atom collisions in the context of hadron therapy has been provided by Belkić [46].

2. General experimental techniques

Most of the experimental techniques to study the TCS involve detection of recoil-ions after ionization and fragmentation in a coincidence time-of-flight detection setup (see Fig. 1). A few measurements are also available which employ continuum electron spectroscopy to measure the DDCS and SDCS.

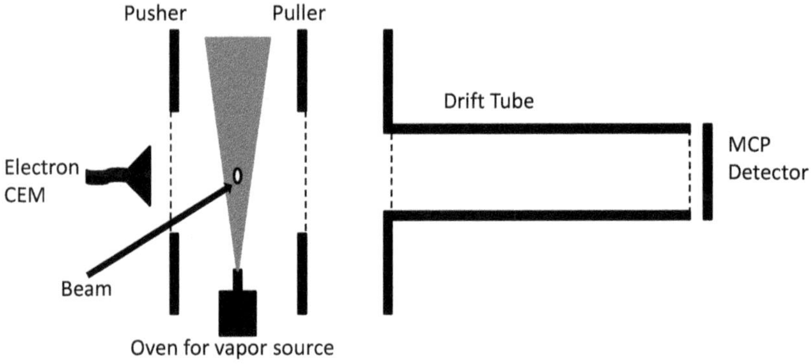

Fig. 1. Schematic diagram of the ToF recoil-ion detection setup including the oven as vapor source for uracil.

2.1. *Time-of-flight mass spectrometer*

A time-of-flight mass spectrometer (see Fig. 1) is commonly used in experiments where a large number of recoil ions with varying mass and charge are produced. It essentially analyzes the recoil ions according to their charge to mass ratio, q/m. We used a Wiley-Mclaren type ToF mass spectrometer. In one of the experiments, the gap between the pusher and puller plates was 30 mm and that between the puller plate and the flight tube was 21 mm, with an acceptance opening of 20 mm. The flight tube was 280 mm long with an inner diameter of 25 mm. The pusher and puller plate openings and the flight tube ends were covered with nickel wire mesh to maintain field uniformity in the respective regions. The extracted ions were detected using a channel electron multiplier (CEM) detector placed at the end of the flight tube. A 10 mm x 0.8 mm slit was placed in front of the electron CEM to reduce the excessive electron count rate at the CEM placed behind the pusher plate, so as to have E_A=700 V/cm and E_S=250 V/cm in the extraction region. The signal from the electron CEM was used as a starter (a trigger), while the signal from the ion CEM was used as the stopper to obtain the recoil-ion ToF spectra. Resolution of the ToF spectrometer was $m/\Delta m$ = 130, as determined by detecting the isotopes of xenon. The collection efficiency of the ToF mass spectrometer depends on the electric fields in the extraction and acceleration region, as well on the kinetic energy of the recoil ions for a given geometry. For gases, the field in the extraction region was about 120 V/cm and in the accelerating region about 320 V/cm. A simulation was done using the SIMION package, which showed that, for recoil energies of the order of few eV, the collection efficiency was 100% excluding the wire mesh attenuation which was assumed to be the same for all the recoiled ions. The fields applied in the extraction and acceleration regions were sufficient to collect all the ions. Detection efficiency of the particle detectors is an important factor in determining cross sections for recoil ion production. For example, in Ref [4], a reflectron time-of-flight spectrometer combined with a multi-sphere plate (MSP) was used to stop the fragmentation products while an electron was used as a start pulse. The details of these setups can be found in Refs. [50–52].

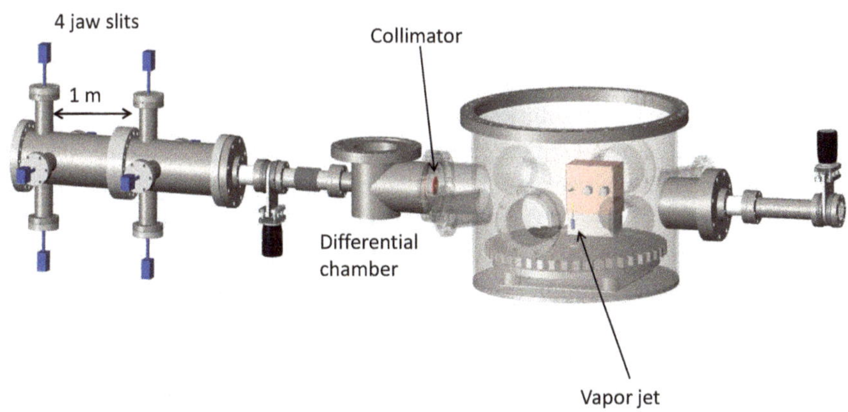

Fig. 2. Schematic diagram of the electron spectrometer based scattering chamber and the part of the beamline [88].

2.2. *Ejected electron spectroscopy*

Figure 2 displays the electron spectrometer-based scattering chamber along with the beamline components. The secondary electrons emitted in the collision process were energy analyzed by a hemispherical electrostatic energy analyzer. The inner and outer electrodes of the analyzer were made up of oxygen free high conductivity (OFHC) copper hemispheres with radii 2.5 and 3.5 cm, respectively. The electrodes were housed inside the OFHC copper housing. In order to reduce the secondary electron background, curved surfaces of the electrodes were coated with carbon soot. A small pre-acceleration voltage is usually applied to have a better collection efficiency for these electrons. This is done by applying a small positive potential (~ 6 volts) to the entrance and the exit aperture of the spectrometer. The entrance and exit aperture have dimensions of 4x5 mm^2 and 3x5 mm^2, respectively. The analytical relationship between the electron energy and applied voltages to the electrodes is given by $HV_1 = (V_c + \epsilon_e)(R_2\text{-}R_1)/R_1 + V_c$ and $HV_2 = (V_c + \epsilon_e)(R_2\text{-}R_1)/R_2 + V_c$. Here, $\{HV_1, R_1\}$ and $\{HV_2, R_2\}$ are the voltages and radii associated with the inner and outer hemispheres, respectively. Further, V_c is the pre-acceleration potential and ϵ_e is the energy of the ejected electron. These relationships were verified by simulating the electron trajectories inside the analyzer for the given geometry using the SIMION software. The analyzer energy resolution depends mostly on the exit slit width and the acceptance angle of the entrance slit. The resolution of the analyzer can be analytically written as $\Delta\epsilon_e/\epsilon_e = (x_1 + x_2)/(R_1 + R_2) + \alpha^2 + \beta^2$ [53], where x_1 and x_2 are the deviations from the object point and the image point respectively, while α is the acceptance angle of the analyzer. The energy resolution of the analyzer is measured using the quotient FWHM/ϵ_e, where FWHM denotes the full width at half maximum. For the FWHM, the value of $x_1 + x_2$ becomes equal to the larger slit width. The acceptance angle α is determined by the collimating slit geometry in front of the entrance slit of the analyzer. In some cases, when a gaseous target is used under static gas pressure, the chamber is flooded with the target gas. Under this circumstance, the collision volume is extended along the electron beam, and the acceptance angle of the analyzer varies between 3.4° to 3.6° depending on the emission angle. The corresponding uncertainty in electron emission angle is ~ 3.7°. Energy resolution of the analyzer is about 6% for which the contribution due to the acceptance angle is about 1% [see Refs. [54, 55]]. At each energy, the number of ejected electrons was detected for a specified amount of the incident projectile charge collected on a Faraday cup which was measured using a standard charge integrator. The spectrometer mounted on a turn table could be rotated manually or by a stepper motor control system using "Labview", between the electron emission angles θ_e =15° and θ_e =160° measured from the projectile trajectory. In order to check the accuracy of the angles and also to assess any change in the solid angles from the calculated angles due to any inclination of e.g. the turn table, we cross-checked the data in three ways. First, the DDCS was found to be the same for +90° and -90° (or 270°) within about 10%. This could be checked in every experiment. Secondly, the binary-encounter electrons with velocity $v_e = 2v_p cos\theta_e$ were detected in which case v_e is sensitive to θ_e. Finally, we measured the angular distributions of the elastically scattered electrons from O_2 and Ar. The Rutherford predicted angular distribution ~ $\mathrm{cosec}^4(\theta_e/2)$ was well reproduced each time.

The energy analyzed electrons were detected by a channel electron multiplier placed after the exit slit of the hemispherical analyzer. The cone of the CEM was kept at a positive potential of 100 V. The reason is that the detection efficiency of the CEM remains constant (~ 0.85–0.90) for electron energies between 100 and 500 eV, as obtained from the manual operation of the detector. A "LabView" based data acquisition system along with the "National Instruments" hardware were used for scanning the voltages on the hemispheres and to collect the data for different electron energies.

Itoh *et al.* [17] used a 45°-parallel-plate-electrostatic spectrometer mounted on a turntable for collision experiments involving biomolecules. Their spectrometer had a resolution of about 8%. They also used a pre-acceleration voltage (which was as large as 40 V) applied to the entrance/lower electrode plate.

2.3. *Normalization procedure*

2.3.1. *Static gas pressure*

The double differential cross section can be calculated from the two measured electron counts N_e and N_b taken with and without target, respectively, at a given angle, by using the relation:

$$\frac{d^2\sigma}{d\Omega_e d\varepsilon_e} = \frac{\left(\frac{N_e(\epsilon_e,\theta_e)}{N_p\Delta\epsilon} - \frac{N_b(\epsilon_e,\theta_e)}{N_p'\Delta\epsilon}\right)}{\epsilon_{el} n_t (l\Omega)_{eff}}. \tag{1}$$

Here, $n_t = 9.659\times10^{15} P_c(mTorr)/T(K)$ is the number density of the target molecules at a static gas pressure (P_c) and temperature T(in °K). The quantities N_p and $N_p^{'}$ are the number of incident projectile ions in the presence and absence of the target gas, respectively (the latter number being the background counts). The resolution of the spectrometer is given by $\Delta\epsilon = 0.06(\epsilon_e + V_e)$. Here, ϵ_e denotes the efficiency of the CEM and ϵ is the ejected electron energy. The quantity $(l\Omega)_{eff}$ is the solid angle path length integral which is given by:

$$(l\Omega)_{eff} = \frac{w_1 w_2 h_2}{LRsin\theta} \tag{2}$$

where $\{w_1, w_2\}$ and h_2 are the widths and height of the rectangular apertures on the front and back of the entrance collimator mounted on the spectrometer. Here, L is the length of the collimator and R is the distance of the front of the collimator from the center of the interaction region.

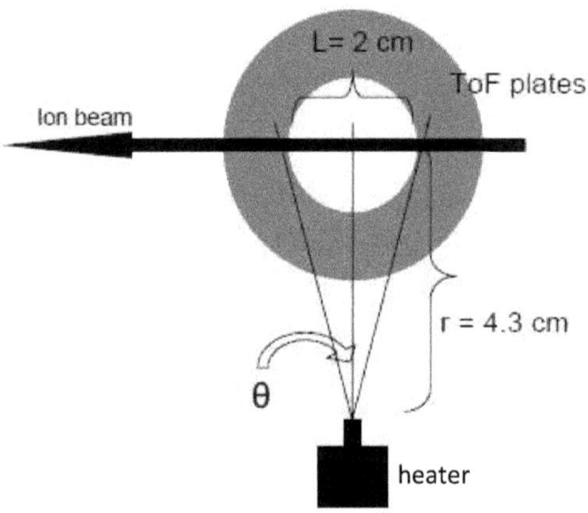

Fig. 3. Parameters for calculation of absolute cross sections in the case of recoil ion ToF spectrometer and an effusive jet from an oven-based vapor source.

2.3.2. *Effusive jet: vapor source*

In the case of the ToF-based experiments with a vapor jet target, the determination of recoil ion cross section depends on the collection efficiency of the ToF spectrometer which is typically made to be 100% using the SIMION simulation. The vapor jet target was also used for the electron spectroscopy experiments. In both kind of experiments, it is most important to determine the number density of the molecules in the effusive jet. This was estimated following the formalism of Scoles *et al.* [56] for expansion of an effusive molecular jet. The relevant geometry is shown in Figs. 3 and 4. In the case of an effusive jet, the solid-angle-path-length-integral has to be convoluted with the molecular density distribution within the jet whose variation as a function of spread angle at a given distance from the nozzle tip is given by:

$$j(\theta) = \alpha \cos\theta + \frac{2}{\pi} \cos\theta[(1-\alpha)R(q) + \frac{2}{3q}(1-2\alpha)\{1-(1-q^2)^{3/2}\}],\ q \le 1, \quad (3)$$

$$j(\theta) = \alpha \cos\theta + \frac{4}{3\pi q}(1-2\alpha)\cos\theta,\ q \ge 1, \quad (4)$$

with

$$R(q) = \cos^{-1}q - q(1-q^2)^{1/2} \quad , \quad q = \frac{\tan\theta}{\beta} \quad , \quad \beta = \frac{L}{2r}, \quad (5)$$

$$\alpha = \frac{1}{2} - \frac{1}{3\beta^2}\frac{1-2\beta^2+(2\beta^2-1)(1+\beta^2)^{1/2}}{(1+\beta^2)^{1/2}-\beta^2\sinh^{-1}(1/\beta)}, \quad (6)$$

where β is the aspect ratio of the jet nozzle. Then the effective solid-angle-path-length-integral can be written as:

$$(l\Omega)_{\text{eff}} = \int_z \Omega(\theta, z)j(z)dz. \quad (7)$$

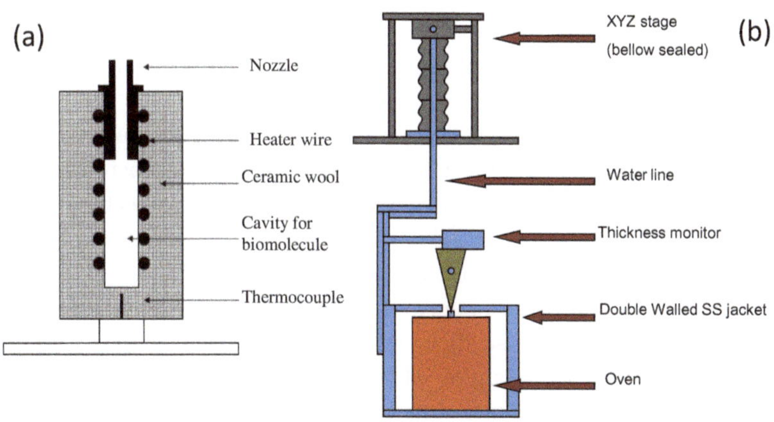

Fig. 4. Schematic diagram of: (a) oven and (b) assembly showing the oven, thickness monitor, water jacket and XYZ stage manipulator used in electron spectroscopy and ToF-based experiments.

The absolute normalization of the DDCS and the TCS data in the case of a vapor jet target is non-trivial since it may not be easy to estimate the vapor density, the exact solid-angle-path-length-integral, unlike the case of a flooded chamber with static gas pressure. In the present case, therefore, the DDCS data were normalized in two steps, which were then used for the TCS normalization. The procedure is based on the KLL-Auger electron cross sections. In the first step, the absolute values of the DDCS for a relevant gaseous target, were obtained from the first principles [54]. For this purpose, the beam intensity and the target thickness were determined accurately. In the case of normalization of the uracil data, the oxygen gas at static pressure was used and the KLL-Auger peak of oxygen was employed for normalization. In the Auger electron spectrum of uracil ($C_4H_4N_2O_2$), one can find that there are three K-LL Auger electron peaks arising from C (at 230 eV), N (330 eV) and O(480 eV) that are all well separated. It is relatively simpler to obtain the K-LL peak area for oxygen than for carbon and nitrogen atoms for most of the forward and backward angles, excepting 75° and 90° for which the K-LL peaks are not so pronounced due to strong background from Coulomb ionization. However, this does not contribute much to the error in the overall analysis (see below). The quality of the Auger peaks of C, N and O arising from the uracil target is shown in the inset of Fig. 5 for two different angles. The absolute DDCS for O_2 target was then integrated over the oxygen-KLL Auger peak region for each angle to provide the SDCS, $d\sigma/d\Omega$. Then, after integrating the SDCS (KLL) over angles, the absolute total Auger emission cross sections σ_{KLL} were obtained.

In the second step, the DDCS data corresponding to the jet target of uracil were plotted. These data had to be corrected for each angle since the solid-angle-path-length integral is different for each spectrometer angle. This correction was carried out using Scole's prescription as given by Eq. 3. We plotted the solid angle corrected DDCS as a function of emission angles for a given energy of ejected electron. The similar plots for different energies were studied, as well. In the next step, the relative SDCS spectrum $d\sigma/d\epsilon_e$ was plotted by integrating the DDCS over angles. Therefore, the spectrum which was obtained from the jet target has two parts: A) the continuum electron part and B) the KLL-Auger electron peaks (C, N, O etc). Both A and B are produced from the same target thickness, jet-profile, beam overlap and are associated with the same solid angle. The total yield Y_{KLL} of the oxygen KLL-Auger line was then obtained by integrating the SDCS spectrum $d\sigma/d\epsilon_e$ over the oxygen-Auger line (see Fig. 3 in Ref [15]). A linear baseline subtraction was done by suitably choosing the points on either sides of the peak. The yield is given by $Y_{KLL} = \sigma_{KLL} G\epsilon(E) N_p$ where G represents the target thickness convoluted with the jet-profile and the path-length integral and N_p is the number of the projectile ions. Here, we have assumed that the oxygen KLL Auger emission cross section σ_{KLL} is the same in the case of O_2 gas target and the uracil molecule, since it is an inner shell ionization process. In this way, the unknown quantity G was determined which was subsequently used to normalize the entire electron DDCS spectrum (i.e. A and B) obtained for uracil. Itoh *et al.* [14], in their experiments on 1 MeV proton collisions with adenine, measured the effective target thickness of adenine using the Rutherford forward scattering technique and a vapor deposition method. These authors compared the measured ionization cross sections with several other carbohydrate molecules and predicted that these cross sections scale fairly well in terms of the total number of valence electrons.

The error estimated in the present normalization procedure is ~ 20%. For most of the angles and most of the energies, the statistical error varies between 15-25% thus giving total error of about 25-30%. In a few cases, e.g. for extreme higher energies (around 560 eV) and for a few angles, the statistical error itself can be about 25-30%. In addition, a systematic error of about 10% arising from background subtraction method cannot be ruled out. In the case of gas targets, the error due to statistical fluctuations was around 2-5% and for high energies at extreme backward angles, the error was about 10%.

2.3.3. *Vapor source holder and thickness monitor*

The biomolecules (uracil or adenine) or fullerene were used in the form of an effusive vapor jet for the experiments. For this, commercially available powder is heated in a metal oven at high temperatures inside the vacuum chamber. Figure 4 displays the schematic diagram of the heater/oven, jet, water cooled jacket and thickness monitor, all of which are mounted together to a 3D translation stage for proper alignment of the target vapor with the ion beam. It was observed that at such high temperatures, the channel electron multiplier of the electron energy analyzer gives rise to some dark counts. In order to suppress these dark counts, we had housed the oven inside a double walled water cooled jacket. The density of evaporated material in the effusive jet is very sensitive to the temperature fluctuations

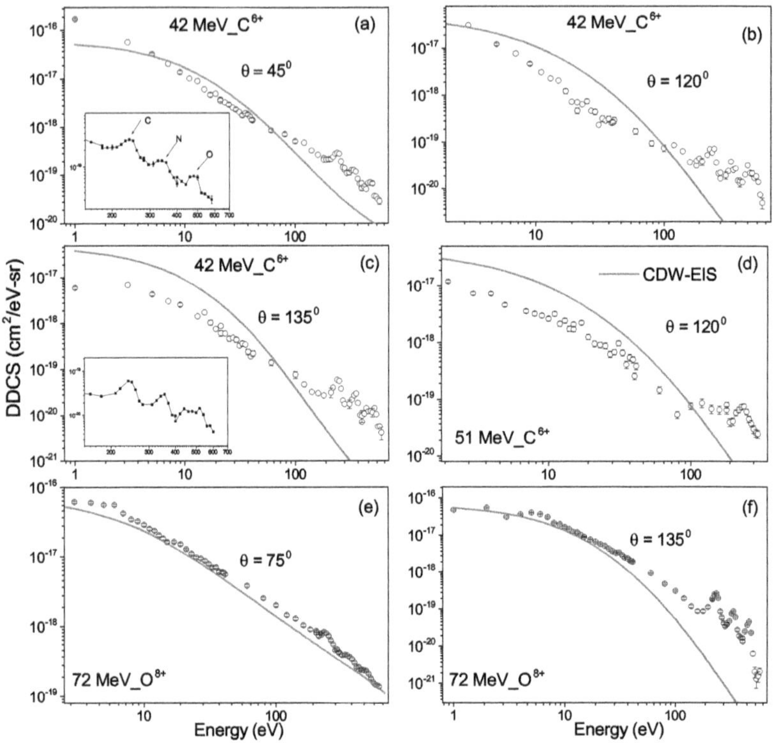

Fig. 5. Electron DDCS, for different emission angles, induced by bare C and O ions with different energies: (a),(b), (c) 42 MeV C^{6+} [15], (d) 51 MeV C^{6+} [95], (e) and (f) 72 MeV O^{8+} colliding with uracil vapor target (present). The solid lines are from the CDW-EIS calculations. Inset in (a) and (c): the K-LL Auger peaks of C, N and O.

of the metal oven. To monitor the vapor density fluctuations, we employed a quartz crystal thickness monitor (INFICON SQM 160 Multi-Film Rate/Thickness monitor) inside the vacuum chamber. The thickness monitor was mounted about 10 cm above the oven nozzle. These monitors use a Cr/Au 6 MHz crystal. The crystal oscillations were derived through an external oscillator. The deposition rate of the target sample was measured from the change in the oscillation frequency. The deposition rate was kept constant throughout the experiment and the variation of the rate (if at all) was smooth. Figure 6 shows the deposition rate of uracil for a time period of 8 hours. Two different sets of data corresponding to two separate experiments are displayed and in both cases the deposition rate of molecules is seen to be fairly constant.

2.3.4. *Accelerators used : 14 MV Pelletron and 14.5 GHz ECRIA*

The experiments involve high-energy C, O and F ion beams in the energy range of 2-5 MeV/u (v_p=7 to 15 a.u.). These projectiles have been extracted from a 14 MV tandem Pelletron accelerator at TIFR, Mumbai. The energy and charge state of the selected ion beams were determined by means of an analyzing magnet. To generate even higher charge states, the ions were made to pass through a post accelerator thin carbon foil stripper. A switching magnet, which has five outlet beam lines, was used to select the charge states of the ions produced in the foil stripper and to direct the beam to the specified beamline. In some cases, low-energy ion beams of carbon of selected charge states were also used for fragmentation studies. For this purpose a permanent magnet-based 14.5 GHz electron cyclotron resonance ion source with four beamlines has been installed and used by our research group [57]. The microwave supply could provide maximum power of about 300 W. The ECR plasma chamber was followed by an Einzel lens, Faraday cup and a double focusing dipole magnet. All of these components were mounted on a high voltage deck which could be raised to a maximum power of about 380 kV. The magnet chamber is followed by the modular accelerating column and a switching magnet. At present, there are four beam lines after the switching magnet.

3. Results and discussion

3.1. *Electron double differential distribution*

The lowest energy electrons (even below the ionization-threshold energy) are known to be primarily responsible for the damage of the single and double strands in DNA and RNA molecules [58]. The primary process which is responsible for such damage is dissociative electron attachment (DEA) which, as a resonance process, is important below 30 eV or so. Relatively higher energy electrons or the so called δ-electrons can also cause damages by fragmentation of the base molecules. Thus far, measurements on the damage of molecules of biological interest were mostly limited to the study of the TCS [5–8, 15, 19], particularly for heavy ions as projectiles. These

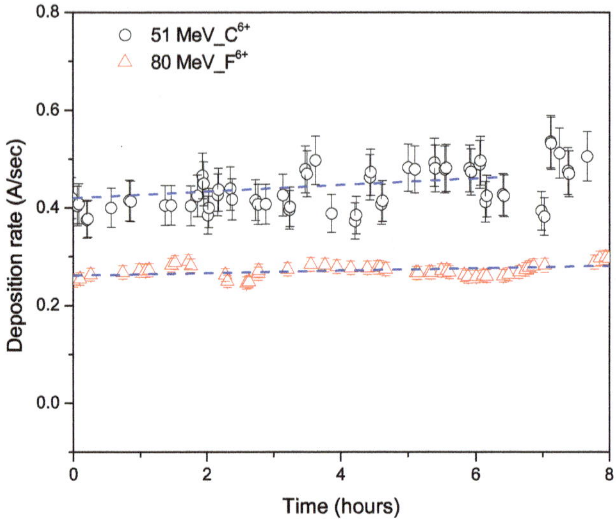

Fig. 6. The rate of deposition as a function of time. The deposition rate is shown to be approximately constant over several hours. The two plots correspond to two different experiments: during electron spectroscopy (circles) and ToF-based (triangles) experiment. The line is drawn to guide the eye.

studies, alongside other similar recent investigations of charge particle interaction with large biomolecules, water molecules and fullerenes [3, 59–65] have created a broad data base. However, the TCS cannot give the microscopic details, such as energy and angular distribution of emitted electrons. In collisions with ions, a large number of low energy electrons are produced with various energies along the ion tracks inside a body which can cause large scale damage [58, 66, 67]. Therefore, it is important to measure the energy as well as the angular distributions of such low energy electrons. In the past, the electron DDCS investigations for DNA- and RNA-base molecules have been reported only for a few cases [1, 13, 14], which are limited to mainly protons as projectiles. In addition, the DDCS measurements on large biomolecules, in collisions with highly charged heavy ions, such as C^{6+}, have been addressed by Agnihotri *et al.* [15]. As already mentioned, the dynamics of electron emissions in collisions involving such heavy ions are more complicated than those for protons or electrons as projectiles.

Therefore, heavy ions impinging on biomolecules need to be investigated in detail. In particular, highly charged ions, due to their high charge state (q_p), generate a considerable Coulomb perturbation strength q_p/v_p. This causes that the electron final state is significantly influenced by the Coulomb field of the recoil ion and the fast scattered projectile. Even in the initial state, the wave function of the bound electron is distorted by the Coulomb field of the projectile ions.

The degree of the two center effect and the post collision interaction depends on the perturbation strength q_p/v_p, as stated. The electron DDCS, ($d^2\sigma/d\epsilon_e d\Omega$), which is differential in the emission angle (θ) and the ejected electron energy (ϵ_e), can provide an experimental signature of different mechanisms of electron emission. These include a soft collisional mechanism [68–73], electron capture to continuum, electron loss in continuum cusp or binary encounters. The Young-type electron interference arising from electron scattering from a molecular double slit, is a relatively new feature which has implications in the basic quantum physics. The influence of interference on electron emission spectrum, such as observation of oscillatory structure and the related features are investigated using double differential electron emission from H_2 [74–78] and other diatomic molecules [55, 79–89].

It is obvious that the DDCS data give a more stringent test to theoretical models than the SDCS such as $d\sigma/d\Omega$ or $d\sigma/d\epsilon_e$ and the TCS. The high resolution recoil-ion momentum spectroscopy technique, which provides cross sections differential in longitudinal and transverse momentum of all three particles in the final state, are also being used to provide a stringent test of various theoretical models of ion-atom collisions and to discover many interesting features.

Agnihotri *et al.* [15] investigated the energy and angular distributions of low-energy electron emission from uracil in collisions with 42 MeV and 51 MeV bare C ions and 72 MeV bare oxygen ions. A part of these data (i.e. for 42 MeV C ions) have been published. Presently, we will give the examples of some of these published as well as unpublished experimental data. The examples of some other measurements will also be given. Thus, panels (a)-(c) on Fig. 5 display the measured DDCS for three different angles, 45^0, 120^0 and 135^0, for 42 MeV C^{6+} impact on uracil. The insets on panels (a) and (c) show the characteristic KLL Auger electron peaks for C, N and O. The DDCS falls by 3-4 orders of magnitudes between the electron energies 1 and 600 eV. In ion-atom collisions, such a fall is well

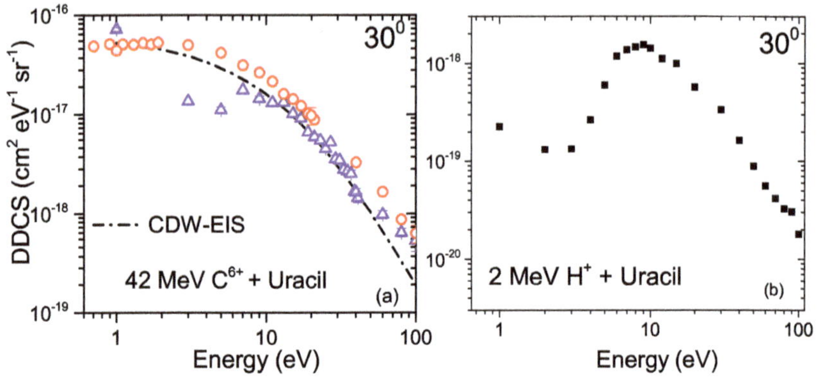

Fig. 7. Two examples of the DDCS spectra of uracil taken at 30 degree (one with C^{6+} and the other with H^+ at different impact energies). (a) Use of 42 MeV bare carbon ions as projectiles. The two symbols correspond to two different spectra taken at different times and in different conditions. The circles clearly indicate the smooth spectrum down to 0.6 eV for which the overall shape is in excellent agreement with theory. The earlier data under slightly different condition show a fall below 5 eV and then a rise at 1 eV. (b) Use of 2.0 MeV protons as projectiles, showing a gradual fall below 8 eV according to Itoh *et al.* [17].

known. The fall is steeper in the case of the backward angles. The DDCS values are as large as nearly 10^{-17} cm^2 for the lowest energies (around 7 eV). The CDW-EIS calculations for uracil was carried out by the group of Rivarola *et al.* [15]. It can be seen that, in general, this theoretical model crosses over the DDCS data at around 60-100 eV. At higher energies, the CDW-EIS model underestimates the experimental data by a large factor. The best agreement is found around the forward angle as seen on panel (a) in Fig. 5. The results are not very different for the bare C ions of energy 51 MeV, shown on panel (d)for only for one angle. The panels (e) and (f) in Fig. 5 display the representative results (whose details will be published elsewhere) on the DDCS for uracil under the impact of 72 MeV bare oxygen ions at two angles (75° and 135°). The cross sections in panels (e) and (f) are larger compared to those displayed in panels (a)-(d) when compared with the data for similar angles. For example, on panel (e) of Fig. 5, at the lowest energy (around 1-10 eV), the DDCS values at 75° are as large as 5-7 x 10^{-17} cm^2. Another aspect is the comparison with the CDW-EIS model which, for this case, shows a better agreement. This aspect is being investigated from a detailed measurement of angular distribution and the data analysis, which is in progress, will be reported separately.

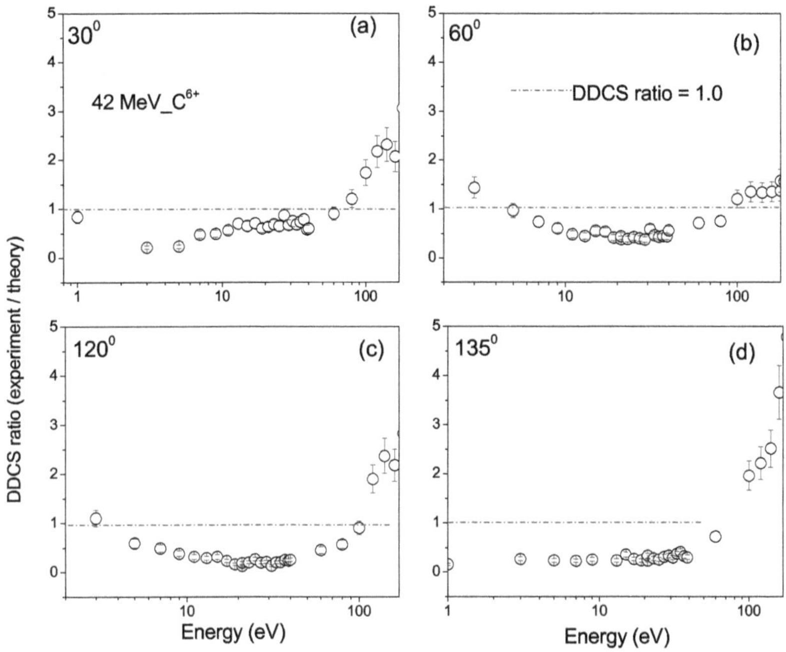

Fig. 8. Ratio of experimental-to-theoretical (CDW-EIS model) DDCS values for different electron emission angles in the case of 42 MeV bare C ions impact on uracil vapor target. The blue dash-dotted line corresponds to the DDCS ratio value of 1.0.

The lowest energy part of the spectrum (typically below 10 eV or so), often suffers a gradual fall which could be due to the deposition of the uracil (or adenine, as the case may be) on the chamber walls and spectrometer surfaces. Accumulated charges on these sublimated layers can create a static field, which is responsible for deflecting the low energy electrons. A special precaution is to be taken to avoid such a problem. The spectrum collected in the beginning part of the experiment with negligible deposition shows a flat distribution in the lowest energy part, as expected from the theory as well as from the ion-atom collision physics. At times, when the charging up is negligible, we initially collected only the lowest energy spectrum (1-21 eV) for all the electron emission angles. One such example is given on panel (a) of Fig. 7 where, under certain conditions, a structure appears in the DDCS spectrum below 6 eV. This is seen as the blue triangles showing a fall and then rise at 1 eV for 42 MeV C^{6+} ions colliding on uracil. Such a fall is not unusual in continuum electron spectroscopy and it may not have any physical origin. The same spectrum collected at some different times, with a much "cleaner" situation, shows a smooth behavior (red circles) and the shape is in excellent agreement with CDW-EIS model (dash-dotted line). The DDCS spectra measured by Itoh *et al.* [14, 16, 17] for proton collisions with adenine and uracil also show a larger fall in the low-energy part of the spectrum. Thus, panel (b) of Fig 7 displays the low-energy part of the DDCS spectrum for 2 MeV proton colliding on uracil [17], where a sharp fall below 10 eV is observed.

Comparisons of the measurements with theory is better represented by looking into the experimental-to-theoretical DDCS ratios as shown in Fig.8 for our data and the CDW-EIS model. This clearly reveals substantial differences between the model prediction and the measured data. The ratio varies between 0.2 to about 3.0 for the backward angles, whereas for the forward angles the ratio is slightly closer to 1, i.e. between 0.5 and 2.0.

3.2. *Angular distribution of DDCS: Uracil versus* O_2

The shape of the angular distributions of the electron DDCS depends on the impact parameter and, hence, on the ejected electron energy. In collisions at large impact parameters, most electrons are emitted with the lowest energies. At large impact parameters, the projectile "sees" the entire molecule, and since a very low momentum transfer is involved, the distribution is mostly governed by the molecular structure, the binding energy, the size of the molecule, etc. As the impact parameter reduces, the momentum transfer is large, and the projectile starts "seeing" the individual constituent atoms and the electrons. The resultant sharp distribution is thus related to the two-body or binary-collisions between the target electrons and the swift projectile nucleus. In the case of ion-atom collisions, the lowest energy electrons show relatively flat angular distributions, whereas a sharp binary-type peak is displayed at higher energy electrons. Figure 9 shows several results on the angular distributions of electrons with different energies emitted from uracil impacted by 42 MeV and 51 MeV bare C ions. The DDCS for low-energy electrons (e.g. 17 and 40 eV) clearly show stronger angular asymmetry between the extreme forward and backward angles, compared to the theoretical prediction. One can easily see that the CDW-EIS predicts more symmetric distribution. At higher energies, however, the peak shape due to binary collisions is somewhat more visible. These distributions can be compared to those for simpler molecules, such as O_2 which shows much more symmetric distributions. The asymmetry, in the case of simple atoms or molecules, primarily arises due to a two-center effect and post-collision interactions of the projectile ions with the emitted electron. However, if this was the only factor to determine the angular asymmetry then one would have similar angular asymmetry for O_2 or uracil.

For example, in the case of electrons with energy of 17 eV, the uracil data at the forward angles are larger (by a factor which is nearly 6) than the data for the extreme backward angles, whereas this surpassing factor is 2-2.5 for the O_2 data at electron energies between 10-40 eV. Similarly, for the same electron energies around 30-50 eV taken for these two targets, the forward-to-backward ratio is about 7 for uracil, while it is much lower for O_2. If the ionization mechanism is the same for these two molecules, one would expect approximately the same forward-to-backward ratio for both molecules. The theoretical calculation predicts the ratio to be about 2.5 for 17 eV electron emission for uracil, and this is much lower compared to the experimental data. The details of the theoretical calculations have already been published in our earlier article [15].

3.3. *Forward backward asymmetry: Uracil versus O_2*

Figure 10 displays the asymmetry, in the form of forward-to-backward DDCS ratio (FBDR) for uracil and O_2 as a function of ejected electron velocity, v_e (in atomic units). The ratios are derived by using the uracil DDCS data for two angles (30° and 135°) that are compared with the similar ratios derived for O_2. From this figure, one can see that at low velocities the asymmetry in the FBDR is much larger than that for oxygen. For example, at the lowest velocity (near $v_e \sim 1.0$), the FBDR for uracil is nearly 10 times larger than that for O_2. In the case of O_2, the asymmetry increases gradually and reaches the value nearly the same as that of uracil for $v_e \sim 4.0$. For uracil, the large value of the ratio (about 8) at low velocity gradually falls to a value of around 4 for 42 MeV of impact energy. High energy electrons are emitted from small impact parameters at which the projectile

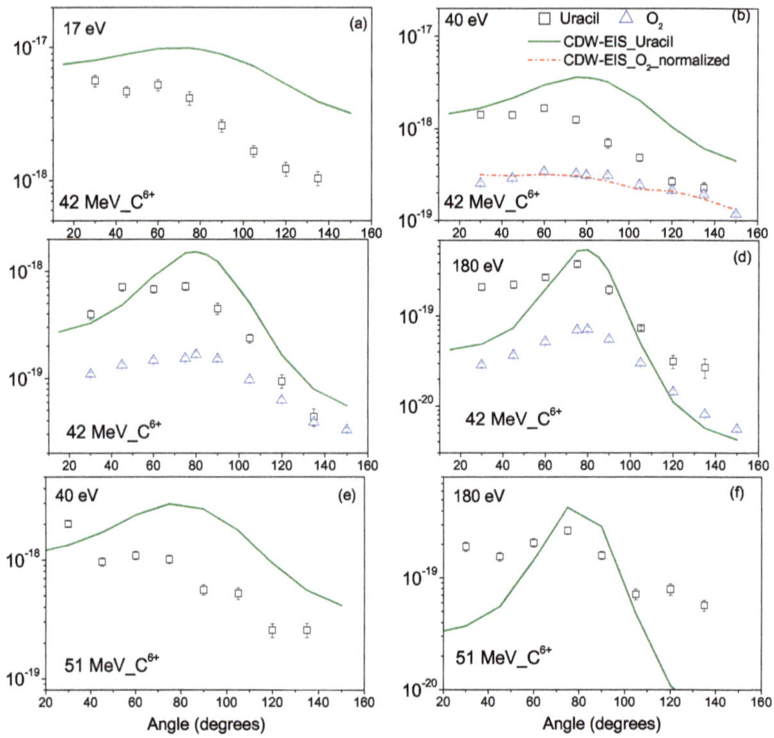

Fig. 9. Electron angular distributions for bare C ions with two different energies colliding with uracil vapor target (black open circles). The green solid line shows the results from the CDW-EIS calculations for the uracil target. Blue open triangles are the angular distributions in the case of the O_2 target. The red dash-dotted line on panel (b) shows the findings from the CDW-EIS model for O_2. The data presented in (b) and (d) are taken from Agnihotri *et al.* [15].

"sees" mainly the individual atoms. A similarly large value is also observed for 51 MeV ion beam energy (not shown here). To compare these data for low energy proton data, we have derived the asymmetry ratio from the DDCS data published by Itoh *et al.* [17]. The asymmetry ratio for electron energies above 15 eV is found to be around 2 or so and nearly independent of the velocity. The lower value of this latter ratio compared to that for O_2 is due to the high charge state for oxygen. The origin of this large forward enhancement is not clear. It may be due to the size and structure of the uracil molecule. The bigger size of the molecule can cause the screening

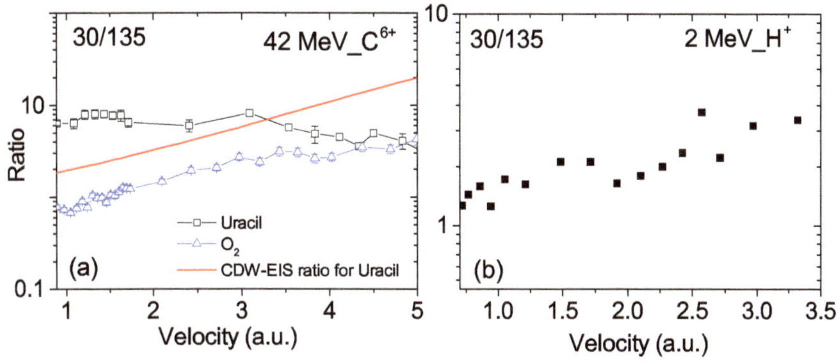

Fig. 10. Forward-to-backward DDCS ratio representing the angular asymmetry [17]: (a) uracil (black open circles) and O_2 (blue open triangles) for electron emission under the impact of 42 MeV C^{6+} and (b) 2 MeV proton impact on uracil target.

of the residual charge on the molecule, which will be different than that for smaller molecules. Such a screening on the parent molecule may reduce the backward scattering of the emitted electrons. These many-body effects or size effects or even multi-electron correlations should be included in a more comprehensive theoretical model describing the electron emission from uracil.

3.4. *Perturbation strength dependence of DDCS*

Next we test in Fig. 11 the dependence of the measured DDCS for electron emission on the perturbation strength. Therein, shown are the experimental data at 45° for different collisional systems at different values of Coulomb perturbation strength. Thus, for 72 MeV O^{8+} ions with q_p/v_p = 0.6, the DDCS data are the highest at all the electron energies. The data for 51 MeV C^{6+} with q_p/v_p = 0.5 are well below the O^{8+}-data. Finally, for 42 MeV C^{6+} ions with q_p/v_p = 0.45, the DDCS values are the lowest. Overall, with higher q/v_p, the DDCS values show an increasing trend at almost all the electron energies. This is approximately along the lines of the earlier observations of the TCS which increase with the q/v values at a rate given by $(q/v)^{1.5}$.

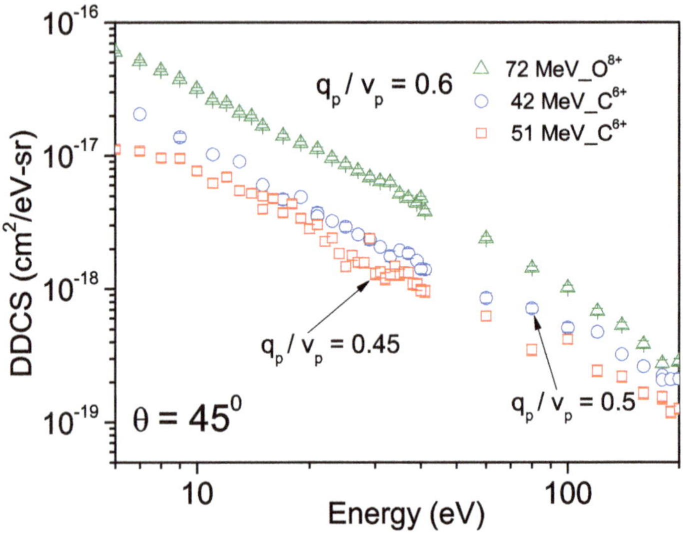

Fig. 11. The DDCS for electron emission induced by C^{6+} of two impact energies (42 MeV, 51 MeV) by O^{8+} at 72 MeV for a fixed emission angle 45°. These collision systems correspond to the perturbation strength (q_p/v_p) with the values 0.45, 0.5 and 0.60 as indicated in the figure.

4. Time-of-flight recoil-ion spectroscopy for ionization and fragmentation: TCS

In the present investigation, the total ionization cross sections, TCS, are measured via the detection of the recoil-ions produced in ionization and fragmentation. Some two-electron processes, e.g. transfer-ionization (TI) would also contribute within the ECRIA-based low-energy collisions. Here, we cannot estimate the separate contributions from such double-electron transitions. However, for this latter purpose, momentum resolved recoil ions recorded in coincidence with capture channels can be used (our work in progress). Energetic ion beams are obtained from two different accelerators (both are at TIFR, Mumbai): low-energy 14.5 GHz ECR-based ion accelerator [57] and 14 MV tandem Pelletron accelerator. The ECRIA, mounted on a 400 kV voltage deck, was used to produce low-energy (80 keV to 1200 MeV) He-like C and O ions. The Pelletron machine was used

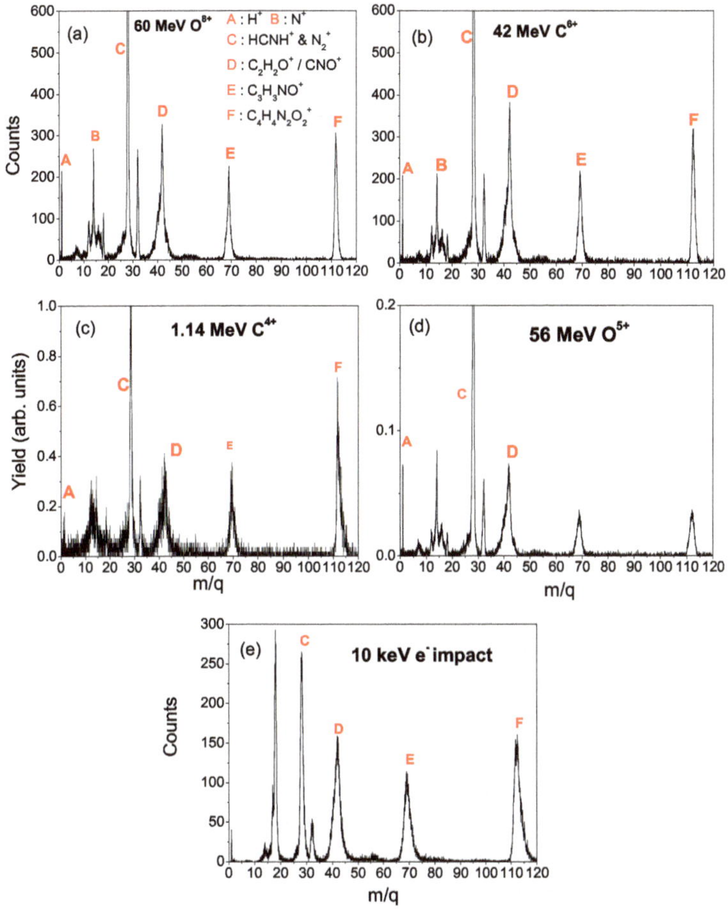

Fig. 12. The recoil ion spectra arising from different processes: single ionization, fragmentation, transfer ionization, etc. Different peaks at the masses 112, 69, 42, 28, 14, 1 are clearly visible for heavy ion impact and fast electron impact. The data in spectrum (b) are taken from our earlier article [19] and rescaled.

for He-like C, O and F ion-beams of energy 2-5 MeV/u corresponding to the velocity range of $v_p = 7$ to 15 a.u. In some cases, other charge states of C, O and F ions were also used. We investigated the charge state dependence of the recoil ion spectra.

Figure 12 displays the typical ToF spectrum of uracil breakup under the impact of 60 MeV O^{8+} (a), 42 MeV C^{6+} (b), 1.14 MeV C^{4+} (c), 56 MeV O^{5+} (d) and 10 keV electron (e). For the electron impact fragmentation experiment, an electron gun beam line was used. The recoil ions of several m/q values are detected, such as (F) 112 ($C_4H_4N_2O_2{}^+$ i.e. Ur^{1+}), (E) 69 ($C_3H_3NO^+$), (D) 42 ($C_2H_2O^+$), (C) 28 ($HCNH^+$), (B) 14 (N^+) and (A) 1 (H^+) are seen. The fragment $C_3H_3NO^+$ of mass 69 involves the rupture of two bonds in the parent cation. This fragment is formed by the loss of HCNO (mass 43) from the parent cation by the so-called Retro-Diels-Alder (RDA) reaction. Further fragmentation of $C_3H_3NO^+$ leads to the lower mass ions. For example, a loss of CNO^+ from $C_3H_3NO^+$ will result in $C_2H_2O^+$ (m=42), which will involve a complex nuclear rearrangement. Most favorable structures of the uracil fragments have been reported by Jochims *et al.* [59] for photoionization. Therein, the strongest peak was observed at $m/q = 42$. A similar fragmentation pattern has been observed in photon, ion and electron impact ionization of uracil [59, 60, 90–92]. In the present experiments, we have also studied the fragmentation spectrum in the case of 10 keV electron impact. For low-energy 1.14 MeV C^{4+} ions, the projectile velocity is ~ 1.0 a.u. which is nearly the same as that of the loosely bound electrons of uracil. Under such a circumstance, electron transfer and TI will be quite strong. In this case, one can see in Fig. 12 that the peak F (m=112) is the strongest among all the structures, except for peak 'C' which is a mixed fragmentation peak. No doubly ionized uracil (with m/q= 56) is observed. The electron impact fragmentation for three different energies (20 eV, 70 eV and 3 keV) has also been reported [93], [94], [95]. Further, the detailed partial cross sections have been presented by Tabet *et al.* [2] for direct ionization as well as for single capture by measuring the charge states of the final state. In the present measurement, because the final scattering state is not detected, there will be a substantial contribution from TI besides ionization at the low energies (around 20-100 keV/u for C ions).

4.1. *Total ionization cross sections*

The measured data from two experiments, namely the low-energy data at ECRIA and high-energy data were normalized for the deposition rate and the collection efficiencies. In fact, the experiments were repeated several times in order to check the consistency. The ToF spectra were recorded for C and O ions of different charge states with energies ranging from 100 keV to 1.6 MeV (from ECRIA) and from 15 MeV to 78 MeV (from Pelletron). The sum of the areas under all the (parent + fragment) ion peaks is used to derive the total ionization yield. The absolute TCS was then derived by adopting a novel normalization procedure, as discussed before. The derived TCS data for the uracil target are displayed on panel (a) of Fig. 13. The TCS values increase as a function of energies in the low-energy region. At higher energies. the TCS data show a sharp fall as seen in Fig. 13(a). Therein, also displayed are the results from the theoretical calculations whose details were published in Refs. [15, 19, 96]. The data are shown for He-like C-ions. It was observed that at higher energies, the experimental data fall (by a factor of 2) below the findings from the CDW-EIS model. For convenience, the results of the CDW-EIS calculations are divided by 2 in Fig. 13(a). Further, the results from the CTMC-COB calculations (no multiplication by any factor) underestimate the experimental C^{4+} data. There is a large deviation of the findings of the CTMC-COB model from the low-energy part of the experimental data. The Bragg peak is approximately covered by the energy range studied here. The CDW-EIS approximation, although viewed as one of the most successful models for ionization, does not reproduce the distribution at the Bragg peak; rather it overestimates the experimental data by a factor of about 5 or so. The TCS data for bare carbon ions are also measured at different energies, as displayed on the inset, alongside the different models. The CB1-COB model (not shown) considerably overestimates the experimental. In the low-energy part investigated by ECRIA, the beam energies are quite low at which electron capture is very strong. At such energies, the TI process is also reasonably strong and can give rise to electron emission. In addition, the double transfer-ionization (T2I) and triple transfer-ionization (T3I) processes can contribute to the ionization events which have now been demonstrated (to be published) using a newly built recoil ion momentum spectrometer (RIMS). To obtain a better agreement with the theory, the TI processes should be considered in theoretical models. Thus, as shown in Fig. 13(b), Montenegro *et al.* [97] have added the contribution from the TI process to the ionization yield using a scaling formulation. The proton impact TCS data for ionization of uracil, measured by Itoh *et al.* [17] seem to be in good agreement (within $\approx$ 20%) with the CTMC-COB model of Champion *et al.* [17, 98]. In Fig. 13, the results from the CB1 model fall a bit more significantly (by about 20%-40%), which seems reasonable, given a complication of the uracil molecule. Märk *et al.* [1–3] studied the absolute total and partial cross sections for ionization of nucleobases by proton impact in the Bragg peak region. The data measured at 80 keV proton on uracil by Tabet *et al.* [3] are much above (i.e. larger by a factor of 3 to 6) the results of the CB1, CDW-EIS or CTMC-COB models.

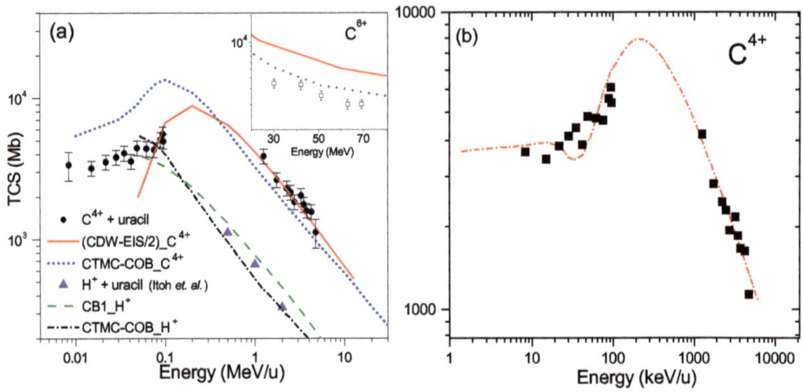

Fig. 13. (a) Measured TCS data of uracil under the impact of C^{4+} ions using a ToF spectrometer (black solid circles) [19]. The ion beam energy covers a wide range by using ECRIA and 14 MV Pelletron accelerator. The results of the CDW-EIS (divided by 2) and CTMC-COB models are displayed by red solid and blue dotted lines, respectively. The TCS data for the H^{1+} impact on uracil [17] are shown by solid violet triangles. The findings from the CB1 (green dashed line) and CTMC-COB (black dash dotted line) models for proton impact are also displayed. The black open triangle shows the TCS at 80 keV H^{1+} impact on uracil [3]. Inset: TCS for the C^{6+} impact. Most of the data in this figure are taken from Tribedi *et al.* [19] with addition of new data from Itoh *et al.* [17] and the new calculations. (b) The uracil TCS data [19] for C^{4+} along with the prediction of scaling prescription (line) given by Montenegro *et al.* [97].

4.2. *Scaling of total ionization cross sections*

Energy loss of projectiles in ion-molecule collisions, and thereby in collisions with biological matter, depends on the perturbation strength q_p/v_p. Charge state q_p is normally varied in order to vary the perturbation strength in a collision keeping the velocity v_p constant. For modeling in radiation therapy, particularly in hadron therapy, one has to deal with a large number of systems (varieties of target molecules), different charge states of projectiles, different impact energies, different atomic number of projectiles and also with the physical state of the targets (gas, solid, liquid). The *ab initio* calculations taking care of such a large number of parameters is a very challenging task. As seen above, the state-of-the art models show much deviation around the Bragg peak region. Such deviations attain a factor of two in the higher energy region where ionization is the main mechanism for energy loss. At the Bragg peak (which is also the TCS peak) and at lower ion-beam energies, the TI process is quite a strong mechanism along with ionization. Thus, with TI neglected, the calculations overestimate the experimental data by a large factor (about 5.0 or so). Because it is difficult to include the TI mechanism in an *ab initio* model, varieties of scaling formulae have been proposed.

4.2.1. *The q_p-dependence of TCS for uracil and H_2O*

A quadratic charge-state dependence of single ionization cross section of atoms or simple molecules in fast collisions is well-known from the first-order perturbative models and also from experimental studies. Deviations from the q_p^2-dependence have been observed with large molecules, such as C_{60} [64, 99] as well as coronene [100]. For fullerene molecules, C_{60}, one observes a linear q_p-dependence for ionization cross sections, which is shown to be related to the additional mechanisms for electron emission, such as collective plasmon resonance. In the case of large molecules, such as DNA and RNA bases, it is not obvious whether the q_p^2-dependence can be used for predicting the energy loss in the case of modeling in hadron therapy. To study the effect of the q_p-dependence of the ionization cross-sections, the ToF spectra were collected for different q_p values for three projectiles (C, O and F) [18, 20]. In these latter articles, the q_p-dependence of TCS for uracil was measured in collisions with the said projectiles of varieties of charge states in the energy range from ~ 2 MeV/u to 5 MeV/u. The experimental data for the TCS versus q_p were fitted to a power law (~ q_p^n) and the n values were found to be varying from a maximum value of 2 to a minimum of 1.1 with the average value ~ 1.5. In light of this circumstance, we have also scaled the TCS for uracil in the high velocity region for C,O and F ions with different q_p and v_p, as demonstrated in Ref. [20]. Therein, it was shown that all the data scaled using the $q^{1.5}$-dependence fall almost on the same line (or band) when plotted against energy in MeV/u. The line, proportional to $1/v_p^{1.5}$ and normalized at one point, agrees quite well with all the data as seen in Fig. 14. The heavy-ion data, therefore, follow the scaling formula $\sigma/q_p^{1.5} \sim 1/v_p^{1.5}$. The data for MeV/u energy proton impact ionization on uracil and adenine are also included in Fig. 14. Since adenine has a greater number of valence electrons (50), compared to uracil (42), the TCS for adenine was multiplied by (42/50) in order to compare with cross sections for uracil. However, the scaled adenine data lie above the uracil data. The lowest energy proton data for uracil (0.5 and 1 MeV) are seen to lie above this line and, therefore, do not follow this scaling rule which was applied for heavier ions above 2 MeV/u. It is seen that as the energy decreases, the deviation from the model increases. A similar scaling rule has also been recently found for data with ion impinging on water molecule for which it was found [27, 28] that $\sigma/q_p^{1.7} \sim 1/v_p^{1.5}$ (not shown here).

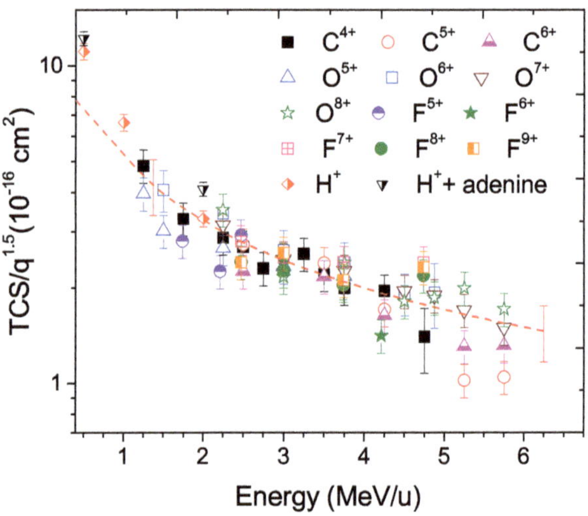

Fig. 14. The scaled TCS data of uracil under heavy ion impact (p and ions of C, O,F) via TCS/$q_p^{1.5}$ as a function of E/M (MeV/amu) using ToF spectrometer and electron spectrometer. The data for proton impact on uracil [17] and on adenine [16] refer to electron spectrometric measurements and are taken from Itoh *et al.* [17]. The ToF data for the ions of C,O and F are taken from Tribedi *et al.* [20]. The line represents the function TCS/$q_p^{1.5} \sim 1/v_p^{1.5}$ normalized at one point. Since adenine has a greater number of valence electrons (50) compared to uracil (42), the adenine TCS has been multiplied by (42/50) to facilitate comparison with the uracil cross sections.

4.2.2. *Target dependence*

Besides the scaling of the TCS data in terms of q_p and v_p, one can also check how the TCS for adenine and uracil total cross sections scale with the number of the loosely bound target electrons for a given projectile energy and charge state. Such a scaling was checked by Itoh *et al.* [14, 17] for proton beams on adenine and uracil, following the original scaling by Toburen *et al.* who demonstrated that ionization cross sections for different hydrocarbons can be scaled linearly in terms of number (n_v) of the loosely bound electrons in these molecules. They had established this linear behavior of the TCS as a function of n_v for different molecules up to n_v=30 i.e. for benzene target and proton beams. This scaling was extended [14, 17] up to n_v= 42 (uracil) and 50 (adenine) for three different energies of proton beams. It was found that the linear behavior was well established for several hydrocarbon molecules and including the two nucleobases (adenine and uracil), giving the following formulae $TCS = K(E_p)n_v \times 10^{-16}$ cm^2. Here, $K(E_p)$ is a fitting constant which depends on the proton energy, E_p and with the value 0.282 (0.5 MeV), 0.163(1.0 MeV) and 0.094 (2.0 MeV).

4.2.3. *A general scaling*

There have been many attempts to develop different scaling models in terms of q_p, v_p, the target atomic number or binding energy, as well as to allow for different mechanisms, such as ionization, electron capture or even double ionization [101–109]. In one such attempt, Montenegro *et al.* [97] have shown a very good agreement between a scaling model and the measured total cross sections for ionization of uracil in collisions with highly charged fast C ions as seen in Fig. 13(b). It was demonstrated that inclusion of ionization alone in the model did not reproduce the low-energy data. Both electron capture and ionization had to be included for the low-energy data to obtain a much better agreement. The present set of the TCS data for highly charged heavy ions like C, O and F colliding with uracil, along with those published by Itoh *et al.* [14, 16, 17] for proton beams should provide an important set of the input data for radiation therapy of tumors.

5. Conclusions and summary

In this chapter, we review the recent work on heavy-ion induced ionization and fragmentation of DNA and RNA base molecules with the main emphasis on our work on the uracil molecule. The results based on electron emission spectroscopy as well as on time-of-flight recoil ion mass spectroscopy are presented. The salient features of the experimental techniques are given. Recent data from other groups are also included. Comparisons with the theoretical models are presented for the double differential electron emission distributions with respect to ejected electron energy and emission angles. The angular distributions of electron emission reveal unusually large forward-backward angular asymmetry, especially relative to simpler molecules such as O_2. The theoretical model which reproduces the asymmetry in ion-collision data for simple atoms and molecules cannot explain similar data for large DNA and RNA base molecules. The large value of asymmetry has also been observed for proton induced ionization data of uracil and adenine. The total cross sections data are displayed for a wide energy range of projectiles (protons and ions of C, O, F) obtained from ECRIA and from different higher energy accelerators. The lower energy data could be explained only after adding the transfer-ionization contribution along with ionization in a model which was, however, based on a scaling rule via fitting [97]. In general, the energy and angular distributions are qualitatively reproduced by theory such as the CDW-EIS, CB1 and CTMC-COB models that show different degrees of success. These models provide relatively better agreement with the experimental data with proton beams at impact energies from 0.5 to 2 MeV. The TCS data for uracil and adenine along with other hydrocarbon molecules for protons of impact energy of a few MeV are shown to follow a simple scaling rule in terms of the number of loosely bound target electrons. The scaling properties are also investigated using different formulations in terms of energy and charge states of the projectile. At higher energies, the total cross sections for the uracil target seem to follow the $(q_p/v_p)^{1.5}$-dependence, whereas for water a similar scaling follows the rule $q_p^{1.7}/v_p^{1.4}$.

Acknowledgments

The author would like to express his sincere thanks to several experimental and theoretical collaborators. These experimental investigations on biomolecules and other large molecules used in this chapter, have been carried out in active collaboration with my Ph.D students: A. Agnihotri, A. Kelkar, S. Kasthurirangan, D. Misra, S. Nandi, S. Bhattachrjee and M.R. Chowdhury, who contributed to this work at different times. Its a pleasure to extend my sincere thanks to R.D. Rivarola, C. Champion and J. Monti for very active theoretical support and for many fruitful discussions/collaborations over the years. It is my pleasant duty to express my sincere thanks to my student M.R. Chowdhury for her sincere help in preparing this manuscript. Finally, I acknowledge the Pelletron accelerator staff and my colleagues in ECRIA team at TIFR for their technical support during the allocated 'beam times' at the accelerators.

References

[1] B. Coupier, B. Farizon, M. Farizon, M.J. Gaillard, F. Gobet, N.V. de Castro Faria, G. Jalbert, S. Ouaskit, M. Carre, B. Gstir, G. Hanel, S. Denifl, L. Feketeova, P. Scheier and T.D. Märk, Inelastic interactions of protons and electrons with biologically relevant molecules, *Eur. Phys. J. D*, **20**, 459-468 (2002).

[2] J. Tabet, S. Eden, S. Feil, H. Abdoul-Carime, B. Farizon, M. Farizon, S. Ouaskit and T.D. Märk, 20-150 keV proton-impact-induced ionization of uracil: fragmentation ratios and branching ratios for electron capture and direct ionization, *Phys. Rev. A*, **81**, 012711 (2010).

[3] J. Tabet, S. Eden, S. Feil, H. Abdoul-Carime, B. Farizon, M. Farizon, S. Ouaskit and T.D. Märk, Absolute total and partial cross sections for ionization of nucleobases by proton impact in the Bragg peak velocity range, *Phys. Rev. A*, **82**, 022703 (2010).

[4] J. deVries, R. Hoekstra, R. Morgenstern and T. Schlathölter, Charge driven fragmentation of nucleobases, *Phys. Rev. Lett.*, **91**, 053401 (2003).

[5] R. Brédy, J.Bernard, L. Chen, M.C. Buchet-Poulizac, S. Martin, Dissociation of charge selected biomolecules induced by electronic capture processes in collisions with positive ions, *Nucl. Instr. Meth. B*, **261**, 114-117 (2007).

[6] M. Imhoff, Z. Deng, M.A. Huels, Identification of ion fragments produced from thymine and deuterated thymine by low energy ion impact in films and electron impact in the gas phase, *Int. J. Mass Spec.*, **245**, 68-77 (2005).

[7] J. Bernard, R. Brédy, L. Chen, S. Martin, B. Wei, Relative partial cross-sections and target fragmentation in ion-adenine collisions at 56 keV, *Nucl. Instr. Meth. B*, **245**, 103-107 (2006).

[8] T. Schlathölter, F. Alvarado, S. Bari, A. Lecointre, R. Hoekstra, V. Bernigaud, B. Manil, J. Rangama and B. Huber, Ion-induced biomolecular radiation damage: from isolated nucleobases to nucleobase clusters, *Chem. Phys. Chem.*, **7**, 2339-2345 (2006).

[9] H. Deutsch, K. Becker, S. Matt, T.D. Märk, Theoretical determination of absolute electron-impact ionization cross sections of molecules, *Int. J. Mass Spec.*, **197**, 37-69 (2000).

[10] Y.K. Kim and M.E. Rudd, Binary-encounter-dipole model for electron-impact ionization, *Phy. Rev. A*, **50**, 3954-3967 (1994).

[11] M.C. Bacchus-Montabonel, M. Labuda, Y.S. Tergiman, J.E. Sienkiewicz, Theoretical treatment of charge-transfer processes induced by collision of C^{q+} ions with uracil, *Phys. Rev. A*, **72**, 052706 (2005).

[12] I. Abbas, C. Champion, B. Zarour, B. Lasri and J. Hanssen, Single and

multiple cross sections for ionizing processes of biological molecules by protons and $I^{\pm}$-particle impact: a classical Monte Carlo approach, *Phys. Med. Biol.*, **53**, N41-N51 (2008).
[13] P. Moretto-Capelle and A. Le Padellec, Electron spectroscopy in proton collisions with dry gas-phase uracil base, *Phys. Rev. A* , **74**, 062705 (2006).
[14] Y. Iriki, Y. Kikuchi, M. Imai, A. Itoh, Absolute doubly differential cross sections for ionization of adenine by 1.0-MeV protons, *Phys. Rev. A*, **84**, 032704 (2011).
[15] A. Agnihotri, S. Nandi, S. Kasthurirangan, A. Kumar, M. Galassi, R. Rivarola, C. Champion and L. Tribedi, Doubly differential distribution of electron emission in ionization of uracil in collisions with 3.5 MeV/u bare C ions, *Phys. Rev. A*, **87**, 032716 (2013).
[16] Y. Iriki, Y. Kikuchi, M. Imai and A. Itoh, Proton-impact ionization cross sections of adenine measured at 0.5 and 2.0 MeV by electron spectroscopy, *Phys. Rev. A*, **84**, 052719 (2011).
[17] A. Itoh, Y. Iriki, M. Imai, C. Champion and R. D. Rivarola, Cross sections for ionization of uracil by MeV energy proton impact, *Phys. Rev. A*, **88**, 052711 (2013).
[18] L.C. Tribedi, A.N. Agnihotri, M.E. Galassi, R.D. Rivarola, C. Champion, Ionization of uracil in collisions with fast bare ions, *Euro. Phys. Jour. D*, **66**, 303 (2012).
[19] A.N. Agnihotri, S. Kasthurirangan, S. Nandi, A. Kumar, M. Galassi, R. Rivarola, O. Fojón, C. Champion, J. Hanssen, H. Lekadir, P.F. Weck and L.C. Tribedi, Ionization of uracil in collisions with highly charged carbon and oxygen ions of energy 100 keV to 78 MeV, *Phys. Rev. A*, **85**, 032711 (2012).
[20] A. Agnihotri, S. Kasthurirangan, S. Nandi, A. Kumar, C. Champion, H. Lekadir, J. Hanssen, P. Weck, M. Galassi, R. Rivarola, O. Fojón and L.C. Tribedi, Absolute total ionization cross sections of uracil ($C_4H_4N_2O_2$) in collisions with MeV energy highly charged carbon, oxygen and fluorine ions, *J. Phys B: At, Mol and Opt Phys*, **46**, 185201 (2013).
[21] D. Ohsawa, H. Kawauchi, M. Hirabayashi, Y. Okada, T. Honma, A. Higashi, S. Amano, Y. Hashimoto, F. Soga and Y. Sato, An apparatus for measuring the energy and angular distribution of secondary electrons emitted from water vapor by fast heavy-ion impact, *Nucl. Instrum. Methods Phys. Res. Sec. B*, **227**, 431-449 (2004).
[22] D. Ohsawa, Y. Sato, Y. Okada, V.P. Shevelko and F. Soga, 6.0 – 10.0 – MeV/u He^{2+} ion-induced electron emission from water vapor, *Phys. Rev. A*, **72**, 062710 (2005).
[23] D. Ohsawa, Y. Sato, Y. Okada, V.P. Shevelko and F. Soga, Unexpectedly large cross sections of high-energy electrons ejected from water vapor by 6.0-10.0 MeV/u He^{2+} ions, *Phys. Lett. A*, **342**, 168-174 (2005).
[24] F. Gobet, S. Eden, B. Coupier, J. Tabet, B. Farizon, M. Farizon, M.J. Gaillard, M. Carre, S. Ouaskit, T.D. Märk and P. Scheier, Ionization of water by 20-150 keV protons: separation of direct ionization and electron capture processes, *Phys. Rev. A*, **70**, 062716 (2004).
[25] S. Nandi, S. Biswas, A. Khan, J.M. Monti, C.A. Tachino, R.D. Rivarola,

D. Misra and L. C. Tribedi, Double-differential cross sections for ionization of H_2O by fast bare O ions: comparison with continuum distorted wave eikonal initial state calculations in prior and post forms, *Phys. Rev. A*, **87**, 052710 (2013).

[26] S. Bhattacharjee, S. Biswas, C. Bagdia, M.R. Chowdhury, S. Nandi, D. Misra, J.M. Monti, C.A. Tachino, R.D. Rivarola, C. Champion and L.C. Tribedi, Double differential distribution of electron emission in the ionization of water molecules by fast bare oxygen ions, *J. Phys. B: At. Mol. Opt. Phys.* ,**49**, 065202 (2016).

[27] S. Bhattacharjee, S. Biswas, J.M. Monti, R.D. Rivarola and L. C. Tribedi, Double-differential cross section for ionization of H_2O molecules by 4 MeV/u C^{6+} and Si^{13+} ions, *Phys. Rev. A*, **96**, 052707 (2017).

[28] S. Bhattacharjee, S. Biswas, J.M. Monti, R.D. Rivarola and L. C. Tribedi Euro Phys J D (2017- in press).

[29] H. Lekadir, I. Abbas, C. Champion, O. Fojón, R.D. Rivarola and J. Hanssen, Single electron loss cross sections of DNA and RNA bases impacted by energetic multicharged ions: a classical Monte Carlo approximation, *Phys. Rev. A*, **79**, 062710 (2009).

[30] H. Lekadir, I. Abbas, C. Champion and J. Hanssen, Total cross sections for ionizing processes induced by proton impact on molecules of biological interest: a classical trajectory Monte Carlo approach, *Nucl. Instrum. Methods Phys. Res. Sec. B* , **267**, 1011-1014 (2009).

[31] C. Dal Cappello, P.A. Hervieux, I. Charpentier and F. Ruiz-Lopez, Ionization of the cytosine molecule by protons: ab initio calculation of differential and total cross sections, *Phys. Rev. A*, **78**, 042702 (2008).

[32] C. Champion, H. Lekadir, M.E. Galassi, O. Fojón, R.D. Rivarola and J. Hanssen, Theoretical predictions for ionization cross sections of DNA nucleobases impacted by light ions, *Phys. Med. Biol.*, **55**, 6053-6067 (2010).

[33] C. Champion, O. Boudrioua, C. Dal Cappello, Y. Sato, and D. Ohsawa, Theoretical and experimental investigations of electron emission in He^{2+} + H_2O collisions, *Phys. Rev. A*, **75**, 032724 (2007).

[34] O. Boudrioua, C. Champion, C. Dal Cappello and Y.V. Popov, Ab initio calculation of differential and total cross sections for the ionization of water vapor by protons, *Phys. Rev. A*, **75**, 022720 (2007).

[35] C. Dal Cappello, C. Champion, O. Boudrioua, H. Lekadir, Y. Sato and D. Ohsawa, Theoretical and experimental investigations of electron emission in C^{6+}+H2O collisions, *Nucl. Instrum. Meth. Phys. Res. Sec. B*, **267**, 781-790 (2009).

[36] C. Champion and C. Dal Cappello, Theoretical investigations of electron emission after water vapour ionization by light ion impact, *Nucl. Instrum. Meth. Phys. Res. Sec. B*, 267, 881-884 (2009).

[37] M.E. Galassi, C. Champion, P.F. Weck, R.D. Rivarola, O. Fojón and J. Hanssen, Quantum-mechanical predictions of DNA and RNA ionization by energetic proton beams, *Phys. Med. Biol.*, **57**, 2081-2099 (2012).

[38] Dž. Belkić, Quantum theory of ionization in fast collisions between ions and atomic systems, *J. Phys.B*, **11**, 3529-3552 (1978).
[39] C. Champion, P.F. Weck, H. Lekadir, M.E. Galassi, O. A. Fojón, P. Abufager, R.D. Rivarola and J. Hanssen, Proton-induced single electron capture on DNA/RNA bases, *Phys. Med. Biol.*, **57**, 3039-3049 (2012).
[40] C.A. Tachino, J.M. Monti, O.A. Fojón, C. Champion and R.D. Rivarola, Ionization of water molecules by ion beams: on the relevance of dynamic screening and the influence of the description of the initial state, *J. Phys. B*, **47**, 035203 (2014).
[41] C. Champion, M.A. Quinto, J.M. Monti, M.E. Galassi, P.F. Weck, O.A. Fojón, J. Hanssen and R.D. Rivarola, Water versus DNA: new insights into proton track-structure modelling in radiobiology and radiotherapy, *Phys. Med. Biol.*, **60**, 7805-7828 (2015).
[42] M.A. Quinto, J.M. Monti, P.D. Montenegro, O.A. Fojón, C. Champion and R.D. Rivarola, Single ionization and capture cross sections from biological molecules by bare projectile impact, *Eur. Phys. J. D*, **71**, 35 (2017).
[43] M.A. Quinto, J.M. Monti, P.F. Weck, O.A. Fojón, J. Hanssen, R.D. Rivarola, P. Senot and C. Champion, Monte Carlo simulation of proton track structure in biological matter, *Eur. Phys. J. D*, **71**, 130 (2017).
[44] C.A. Tachino, J.M. Monti, O.A. Fojón, C. Champion and R.D. Rivarola, Double differential cross sections for ionization of water molecules by ion impact, *Phys. Scr. T156*, 014040 (2013).
[45] B. Senger, Calculated molecular double-differential cross-sections for ionisation under proton impact, *Z. Phys. D : At. Mol. Clusters*, **9**, 79-89 (1988).
[46] Dž. Belkić, Review of theories on ionization in fast ion-atom collisions with prospects for applications to hadron therapy, *J. Math. Chem.*, **47**, 1366-1419 (2010).
[47] O.V. Shishkin, L. Gorb, A.V. Luzanov, M. Elstner, S. Suhai and J. Leszczynski, Structure and conformational flexibility of uracil: a comprehensive study of performance of the MP2, B3LYP and SCC-DFTB methods, *J. Mol. Struct. (Theochem)* , **625**, 295-303 (2003).
[48] P.D. Fainstein, V.H. Ponce and R.D. Rivarola, A theoretical model for ionisation in ion-atom collisions: application for the impact of multicharged projectiles on helium, *J. Phys. B*, **21**, 287-299 (1988).
[49] M.E. Galassi, R.D. Rivarola and P.D. Fainstein, Multicenter character in single-electron emission from H_2 molecules by ion impact, *Phys. Rev. A*, **70** 032721 (2004).
[50] J. de Vries, R. Hoekstra, R. Morgenstern, and T. Schlatholter, C^{q+} -induced excitation and fragmentation of uracil: effects of the projectile electronic structure, *J. Phys. B*, **35**, 4373-4381 (2002).
[51] O. Hadjar, R. Hoekstra, R. Morgenstern, and T. Schlatholter, Projectile atomic-number effect on ion-induced fragmentation and ionization of fullerenes, *Phys. Rev. A*, **63**, 033201 (2001).
[52] J. de Vries, R. Hoekstra, R. Morgenstern, and T. Schlatholter, Multiple ionization and fragmentation of the DNA base thymine by interaction with C^{q+} ions, *Eur. Phys. J. D* **24** 161-164 (2003).

[53] K.D. Sevier, Low Energy Electron Spectrometry, *Wiley Interscience, New York*, 1972.

[54] D. Misra, K.V. Thulasiram, W. Fernandes, A.H. Kelkar, U. Kadhane, A. Kumar, Y. Singh, L. Gulyás and L.C. Tribedi, Double differential distributions of electron emission in ion-atom and electron-atom collisions using an electron spectrometer, *Nucl. Instru. Meth. B*, **267**, 157-162 (2009).

[55] M.R. Chowdhury, C.R. Stia, C.A. Tachino, O.A. Fojón, R.D. Rivarola and L.C. Tribedi, Ionization of N_2 in collisions with fast electrons: evidence of an interference effect, *Phys. Rev. A*, **94**, 052703 (2016).

[56] *Atomic and Molecular Beam Methods* Vol. 1 (New York, Oxford, 1988).

[57] A.N. Agnihotri, A.H. Kelkar, S. Kasthurirangan, K.V. Thulasiram, C.A. Desai, W.A. Fernandez and L.C. Tribedi, An ECR ion source-based low-energy ion accelerator: development and performance, *Phys. Scr.*, **T144**, 014038 (2011).

[58] B. Boudaiffa, P. Cloutier, D. Hunting, M.A. Huels and L. Sanche, Resonant formation of DNA strand breaks by low-energy (3 to 20 eV) electrons, *Science*, **287**, 1658-1660 (2000).

[59] H.W. Jochims, M. Schwell, H. Baumgartel and S. Leach, Photoion mass spectrometry of adenine, thymine and uracil in the 6-22 eV photon energy range, *Chem. Phys.*, **314** 263-282 (2005).

[60] M. Imhoff, Z. Deng, M.A. Huels, Ionizing fragmentation of uracil and 5-bromouracil by electron impact in gas phase and hyperthermal Ar^+ ion irradiation in condensed phase, *Int. J. Mass Spec.*, **262**, 154-160 (2007).

[61] H.J. Lüdde, T. Spranger, M. Horbatsch and T. Kirchner, Nonperturbative, quantum-mechanical approach to ion collisions from molecular targets, *Phys. Rev. A*, **80** 060702(R) (2009).

[62] W. Friedland, M. Dingfelder, P. Kundrat, and P. Jacob, Track structures, DNA targets and radiation effects in the biophysical Monte Carlo simulation code PARTRAC, *Mutation Research*, **711**, 28-40 (2011).

[63] E. Surdutovich, G. Setzler, W.E. Kauppila, S.J. Rehse and T.S. Stein, Measurements of total cross sections for positron scattering by uracil molecules, *Phys. Rev. A*, **77**, 054701 (2008).

[64] A.H. Kelkar, U. Kadhane, D. Misra, L. Gulyás and, L.C. Tribedi, Single and multiple ionization of C_{60} fullerenes and collective effects in collisions with highly charged C, F, and Si ions with energy 3 MeV/u, *Phys. Rev. A*, **82**, 043201 (2010).

[65] U. Kadhane, A. Kelkar, D. Misra, A. Kumar and L. C. Tribedi, Effect of giant plasmon excitations in single and double ionization of C_{60} in fast heavy-ion collisions, *Phys. Rev. A*, 75, 041201(R) (2007).

[66] L. Sanche, Nanoscopic aspects of radiobiological damage: fragmentation induced by secondary low-energy electrons, *Mass. Spectrom. Rev.*, **21**, 349-369 (2002).

[67] L. Sanche, Low energy electron driven damage in biomolecules, *Eur. Phys. J. D* 35, 367-390 (2005).

[68] N. Stolterfoht, H. Platten, G. Schiwietz, D. Schneider, L. Gulyás, P.D. Fainstein and A. Salin, Two-center electron emission in collisions of fast,

highly charged ions with He: experiment and theory, *Phys. Rev. A*, **52**, 3796-3802 (1995).
[69] J.O.P. Pedersen, P. Hvelplund, A. Petersen and P.D. Fainstein, Two-centre effects in ejected-electron spectra from ionization of helium atoms in collisions with fast, fully stripped ions, *J. Phys. B*, 24, 4001-4016 (1991).
[70] S. Suárez, C. Garibotti, W. Meckbach and G. Bernardi, Experimental evidence of the asymmetry of the soft electron peak in ion-atom ionization, *Phys. Rev. Lett.*, **70**, 418-421 (1993).
[71] L.C. Tribedi, P. Richards, W. DeHaven, L. Gulyás, M.W. Gealy and M.E. Rudd, Double-differential distributions of electron emission in a pure three-body collision: ionization of atomic hydrogen by highly charged ions, *J. Phys. B*, **31**, L369-L378 (1998).
[72] P.D. Fainstein, L. Gulyás, F. Martin and A. Salin, Angular asymmetry of low-energy electron emission in ion-atom collisions, *Phys. Rev. A*, **53**, 3243-3246 (1996).
[73] L.C. Tribedi, P. Richard, Y.D. Wang, C.D. Lin and R.E. Olson, Doubly differential final-state momentum distributions of the ionization products in collision of bare ions with hydrogen, *Phys. Rev. Lett.*, **77**, 3767-3770 (1996).
[74] H.D. Cohen and U. Fano, Interference in the photo-ionization of molecules, *Phys. Rev.*, **150**, 30-33 (1966).
[75] N. Stolterfoht, B. Sulik, V. Hoffmann, B. Skogvall, J.Y. Chesnel, J. Rangama, F. Fremont, D. Hennecart, A. Cassimi, X. Husson, A.L. Landers, J.A. Tanis, M.E. Galassi and R.D. Rivarola, Evidence for interference effects in electron emission from H_2 colliding with 60MeV/u Kr^{34+} ions, *Phys. Rev. Lett.*, **87**, 023201 (2001).
[76] D. Misra, U. Kadhane, Y.P. Singh, L.C. Tribedi, P.D. Fainstein and P. Richard, Interference effect in electron emission in heavy ion collisions with H_2 detected by comparison with the measured electron spectrum from atomic hydrogen, *Phys. Rev. Lett.* , **92**, 153201 (2004).
[77] D. Misra, A. Kelkar, U. Kadhane, A. Kumar, Y.P. Singh, L.C. Tribedi, P.D. Fainstein, Angular distribution of low-energy electron emission in collisions of 6-MeV/u bare carbon ions with molecular hydrogen: two-center mechanism and interference effect, *Phys. Rev. A*, **75**, 052712 (2007).
[78] S. Chatterjee, D. Misra, A.H. Kelkar, L.C. Tribedi, C.R. Stia, O.A. Fojón and R.D. Rivarola, Fast electron impact ionization of molecular hydrogen: energy and angular distribution of double and single differential cross sections and Young-type interference, *J. Phys. B: At. Mol. Opt. Phys.*, **42**, 065201 (2009).
[79] J.L. Baran, S. Das, F. Járai – Szábo, K. Póra, L. Nagy and J.A. Tanis, Suppression of primary electron interferences in the ionization of N_2 by $1-5$–MeV/*mathrmu* protons, *Phys. Rev. A*, **78**, 012710 (2008).
[80] M. Winkworth, P.D. Fainstein, M.E. Galassi, J.L. Baran, B.S. Dassanayake, S. Das, A. Kayani and J. A. Tanis, Interferences in electron emission from O_2 by 30MeV $O^{5,8+}$ impact, *Nucl. Instrum. Methods Phys. Res., Sect. B* , **267**, 373-376 (2009).

[81] H. Chaluvadi, N.O. Zehra, M. Dogan, C. Ning, J. Colgan and D. Madison, Observation of two-center interference effects for electron impact ionization of N_2, *J. Phys. B*, **48**, 155203 (2015).

[82] M. Ilchen, L. Glaser, F. Scholz, P. Walter, S. Deinert, A. Rothkirch, J. Seltmann, J. Viefhaus, P. Decleva, B. Langer, A. Knie, A. Ehresmann, M. Braune, G. Hartmann, A. Meissner, L. Tribedi, M. Al-Khaldi, O. Al-Dossary and U. Becker, Angular momentum sensitive two-center interference, *Phys. Rev. Lett.*, **112**, 023001 (2014).

[83] S. Nandi, S. Biswas, C.A. Tachino, R.D. Rivarola and L.C. Tribedi, Double differential electron emission from N_2 under impact of fast C^{6+} ions and Young-type interference, *Eur. Phys. J. D* , **69**, 192 (2015).

[84] D. Rolles, M. Braune, S. Cvejanović, O. Geβner, R. Hentges, S. Korica, B. Langer, T. Lischke, G. Prümper, A. Reinköster, J. Viefhaus, B. Zimmermann, V. McKoy and U. Becker, Isotope-induced partial localization of core electrons in the homonuclear molecule N_2, *Nature*, **437**, 711-715 (2005).

[85] B. Zimmermann, D. Rolles, B. Langer, R. Hentges, M. Braune, S. Cvejanović, O. Gebner, F. Heiser, S. Korica, T. Lischke, A. Reinkoster, J. Viefhaus, R. Dorner, V. Mckoy and U. Becker, Localization and loss of coherence in molecular double-slit experiments, *Nat. Phys.*, **4**, 649-655 (2008).

[86] K. Ueda, X.J. Liu, G. Prumper, T. Lischke, T. Tanaka, M. Hoshino, H. Tanaka, I. Minkov, V. Kimberg, F. Gelmukhanov, Role of the recoil effect in two-center interference in X-ray photoionization, *Chem. Phys.*, **329**, 329-337 (2006).

[87] S.E. Canton, E. Plésiat, J.D. Bozek, B.S. Rude, P. Decleva and F. Martin, Direct observation of Young's double-slit interferences in vibrationally resolved photoionization of diatomic molecules, *Proc. Natl. Acad. Sci. USA*, **108**, 7302-7306 (2011).

[88] M.R. Chowdhury and L.C. Tribedi, Electron impact ionization of O_2 and the interference effect from forward-backward asymmetry, *J. Phys. B: At. Mol. Opt. Phys.*, **50**, 155201 (2017).

[89] M.R. Chowdhury, C.R. Stia, C.A. Tachino, O.A. Fojón, R.D. Rivarola and L.C. Tribedi, Coherent electron emission from O_2 in collisions with fast electrons, *Eur. Phys. J. D*, **71**, 218 (2017).

[90] T. Schlathölter, R. Hoekstra, R. Morgenstern, Charge driven fragmentation of biologically relevant molecules, *Int. J. Mass Spec.* , **233**, 173-179 (2004).

[91] S. Denifl, B. Sonnweber, G. Hanel, P. Scheier and T.D. Märk, Threshold electron impact ionization studies of uracil, *Int. J. Mass Spec.* , **238**, 47-53 (2004).

[92] NIST Chemistry web book, http://webook.nist.gov.

[93] S. M. Hecht, A.S. Gupta and N.J. Leonard , Position of uridine thiation: the identification of minor nucleosides from transfer RNA by mass spectrometry, *Biochim. Biophys. Acta*, **182**, 444-448 (1969).

[94] J.M. Rice, G.O. Dudek and M. Barber, Mass spectra of nucleic acid derivatives: pyrimidines, *J. Am. Chem. Soc.*, **87** 4569-4576 (1965).

[95] A.N. Agnihotri and L.C. Tribedi, Ph.D dissertation (unpublished): The theoretical calculations used for 51 MeV C^{6+} and 72 MeV O^{8+} are obtained as private communications from R.D. Rivarola and details are to be published.
[96] C. Champion, M.E. Galassi, O. Fojón, H. Lekadir, J. Hanssen, R.D. Rivarola, P.F. Weck, A.N. Agnihotri, S. Nandi and L.C. Tribedi, Ionization of RNA-uracil by highly charged carbon ions, *J. Phys.:Conf. Ser.*, **373**, 012004 (2012).
[97] E.C. Montenegro, G.M. Sigaud and R.D. DuBois, Projectile and target scaling of the total ionization cross sections of atoms and molecules, *Phys Rev A*, **87** 012706 (2013).
[98] C. Champion, M.E. Galassi, P.F. Weck, S. Incerti, R.D. Rivarola, O. Fojón, J. Hanssen, Y. Iriki, A. Itoh, Proton-induced ionization of isolated uracil molecules: a theory/experiment confrontation, *Nucl Instr Meth Phys Res B*, **314**, 66-70 (2013).
[99] A.H. Kelkar, L. Gulyás, and L.C. Tribedi , Angle-differential observation of plasmon electrons in the double-differential cross-section spectra of fast-ion-induced electron ejection from C_{60}, *Phys. Rev. A*, **92**, 052708 (2015).
[100] L.C. Tribedi *et al.* (to be published).
[101] N. Bohr, K. Dan. Vidensk. Selsk, Mat. Fys. Medd. **18**, 1-144 (1948).
[102] H. A. Bethe, Ann. d. Physik **5**, 325-400 (1930).
[103] R.E. Olson, Electron capture and impact-ionization cross sections for multiply charged ions colliding with helium, *Phys. Rev A*, **18**, 2464-2469 (1978).
[104] R.E. Olson, K.H. Berkner, W.G. Graham, R.V. Pyle, A.S. Schlachter and J.W. Stearns, Charge-state dependence of electron loss from H by collisions with heavy, highly stripped ions, *Phys. Rev. Lett.*, **41**, 163-166 (1978).
[105] A.S. Schlachter, K.H. Berkner, W.G. Graham, R.V. Pyle, P.J. Schneider, K.R. Stalder, J.W. Stearns, J.A. Tanis and R.E. Olson, Ionization of rare-gas targets by collisions of fast highly charged projectiles, *Phys. Rev. A*, **23**, 2331-2339 (1981).
[106] C. Illescas, L.F. Errea, L. Méndez, B. Pons, I. Rabadán and A. Riera, Classical treatment of ion-H_2O collisions with a three-center model potential, *Phys. Rev. A*, **83**, 052704 (2011).
[107] G.H. Gillespie, A scaling cross section for the ionisation of atomic hydrogen by fast, highly-stripped ions, *J. Phys. B*, **15**, L729-L732 (1982).
[108] W. Wu, E.F. Deveney, S. Datz, D.D. Desai, H.F. Krause, J.M. Sanders, C.R. Vane, C.L. Cocke and J.P. Giese, Scaling rule for target ionization by highly charged ions at low-to-intermediate velocities, *Phys. Rev. A*, **53**, 2367-2370 (1996).
[109] R.D. DuBois, A.C.F. Santos and S.T. Manson, Empirical formulas for direct double ionization by bare ions: $Z = -1$ to 92, *Phys. Rev. A*, **90**, 052721 (2014).

Chapter 4

Young-type interference effects in heavy-ion collisions with diatomic molecules

R.D. Rivarola[1*], C.A. Tachino[1], O.A. Fojón[1] and M.F. Ciappina[2]

[1] *Laboratorio de Colisiones Atómicas, Facultad de Ciencias Exactas, Ingeniería y Agrimensura, and Instituto de Física Rosario (CONICET-UNR), 2000 Rosario, Argentina*

[2] *Institute of Physics of the ASCR, ELI-Beamlines Project, Na Slovance 2, 182 21 Prague, Czech Republic*

We review the theoretical and experimental progress in the field of coherent electron emission from simple molecules colliding with energetic charged particles. In particular, we outline the different theoretical approaches to tackle single ionization of diatomic molecules by impact of heavy ions. Furthermore, comparisons with the experimental measurements are made. Throughout, we emphasize the phenomenology behind these molecular processes and discuss the advantages and drawbacks of the different theoretical methods.

Contents

1. Introduction

It is well-known that the dispute existing in the physics community has persisted for centuries about the nature of light, i.e. whether light is composed

*Corresponding author. Email: rivarola@ifir-conicet.gov.ar

of corpuscles or if it possesses intrinsically a wave-like nature. It was Newton who postulated that light rays were built up by very small corpuscles emitted from shining bodies. This hypothesis was in disagreement with some known phenomena (at Newton's time) that we nowadays attribute to interferences. Interestingly, at the same time, Huygens and others concluded that light consists of traveling waves. Nevertheless, the overpowering influence of Newton prevailed until the double-slit experiment performed by Young at the beginning of the 19th century conclusively demonstrated the wave character of light [1]. Young determined that light, upon passing through two closely separated lines traced on an opaque glass creates, when projected to a wall, an interference pattern, i.e. an alternation of bright and dark segments. However, it was Planck in 1901 who postulated the quantized nature of electromagnetic radiation to explain the black-body radiation spectra. Soon after, in 1905, Einstein reinforced Planck's conjecture, with his description of the photoelectric effect, but simultaneously giving support to the corpuscular concept of light. An additional contribution to the controversy appears in 1923, when Compton performed a series of scattering experiments with high-energy radiation backing the idea that light is made up of particles that, after Lewis [2], we nowadays call photons. The settlement between all the above cited contradictory findings came after the wave-particle duality principle, first introduced by de Broglie in 1924. As is known, this assumption is valid not only for photons, but for every quantum object. This duality constitutes the cornerstone of quantum mechanics, being one of the most notable conceptual departures from classical physics. In fact, the famous experiment of Davisson and Germer in 1927 confirmed the de Broglie hypothesis.

A large set of electron diffraction or double-slit type experiments was performed since the 1960s with electrons [3–8] and even larger quantum objects such as buckyballs (C_{60} fullerenes) [9]. To paraphrase Feynman, "... electron diffraction contains the only mystery; ... telling the way it works would entail conveying the intricacies of the basic peculiarities of all quantum mechanics". The Heisenberg uncertainty principle, indeed, lies behind all these experiments. As is well-known, it enforces constraints to the precision of simultaneous measurements of the position and momentum of quantum objects, as opposed to classical physics where this simultaneity is perfectly possible. In order to obtain an interference pattern, the momentum of the quantum object must be so precisely measured that its position is delocalized by more than the slit width. If this delocalization is dropped out, decoherence occurs and the interference pattern vanishes [10].

Whether it is possible to determine through which slit an object passes without losing interference patterns is a never ending proposition that continues to be the matter of research and maybe controversy in the present day too. In particular, a thought experiment proposed by Feynman in 1965 has been already presented [11].

A different and fascinating mechanism, related to double-slit experiments, is the coherent superposition of quantum objects emitted from spatially separated positions. This is often named as 'molecular double-slit scenario' [10]. Here, a homonuclear diatomic molecule, such as molecular hydrogen [12] or nitrogen [13], is ionized by irradiation with light or by impact of charged particles such as electrons or heavy ions. These processes have received particular attention for decades due to their intrinsic importance in areas such as astrophysics and biology (see e.g. [14] for more details). The appearance of interference effects in these collisions processes was first recognized by Cohen and Fano in 1966 in theoretical studies of photoionization of N_2 and O_2 molecules [15]. They showed that an oscillating pattern was present in the partial photoionization cross section (PCS) when a two-center electron wave function was employed to describe the coherent electron emission from both atomic sites. Electrons can be emitted coherently from both of the atoms in these molecules in such a way that the electron waves could be either in phase or out of phase. As a consequence, these systems should exhibit an interference behavior equivalent to that seen in macroscopic double-slit experiments. The first clear experimental demonstration for these interference phenomena, however, was given more than three decades later, in 2001, in single ionization of H_2 molecules by impact of heavy-ions, Kr^{34+} [12]. Starting from this seminal work, where the coherent electron emission from H_2 molecules by impact of highly charged ions was established, there has been a persistent activity in the field.

Thanks to the consistent advance of experimental capabilities in atomic physics, measurements on few-body fragmentation processes, such as single and double ionization by charged particle impact or mutual ionization of both collision partners, have become much more complex. They are nowadays routinely performed in many laboratories worldwide. The workhorse is the cold target recoil ion momentum spectroscopy (COLTRIMS) technique, also known as a reaction microscope, which has led to a true new generation of atomic collision experiments (for a review see e.g. [16]). The so-called kinematically complete experiments are now feasible for a broad range of projectiles and kinematical conditions. Additionally, they allow studies about the coherent emission of electrons from atomic centers. These investigations are of great interest in diverse areas and could indeed trigger new theoretical and experimental developments [17].

One additional area that deserves to be considered, in principle theoretically, is the possibility to perform single ionization experiments with aligned (fixed in space) molecules. In our view, the pump-probe techniques, that are used extensively in strong laser driven experiments, have come of age and they could be adapted to align molecules, before heavy-ions or electron projectile beams initiate the electron dynamics (see e.g. [18]).

This chapter is organized as follows. In Section 2, we present the theoretical description of single ionization of simple molecules by heavy-ion impact. Special emphasis is put on the distorted wave models and comparisons with the most relevant experiments are carried out. Within this section, we include both processes involving dielectronic and multi-electronic molecules. Additionally, we consider single and multiple charged ions as projectiles. Conclusions and an outline of possible future developments are given in Section 3. Atomic units are used throughout unless otherwise stated.

2. Theoretical approach

Here we review the theory employed to describe single ionization of diatomic molecules by impact of charged heavy particles, with the focus on the study of Young-type interference effects.

Consider a laboratory frame of reference whose origin is fixed at the center of mass of the molecule, with the z-axis along the incidence direction. A bare ion of nuclear charge Z_P impinges with velocity v such that the impact energy belongs to the intermediate-to-high (non-relativistic) energy regime. In this context, the collision time is smaller than the roto-vibrational times of the molecule. This justifies the assumption that the target nuclei remain frozen in their initial positions during the ionization process. For example, for small molecules and the projectile energies under consideration, the collision time τ is of the order of the sub-fs to -as (1 fs=10^{-15} s, 1 as=10^{-18} s). On the other hand, the typical vibration and rotational frequencies are in the range of 10^{13} ~10^{14} Hz and 10^{10} ~10^{12} Hz, respectively (see e.g. Ref. [19]). The adequacy of these assumptions has been tested for numerous aggregates, including atoms and molecules [20, 21].

The presence of several target electrons may represent a stumbling block difficult to overcome. Regarding this aspect of the problem, it is possible, under certain conditions, to reduce the study of a multi-electronic system to an equivalent one electron problem. In the first place, it should be noted that according to the present impact energy regime, the collision time is smaller than that of the electronic relaxation. Then, the passive electrons (those which are not ionized) may be supposed to remain frozen in their initial orbitals during the collision. Simultaneously, the active electron (the one to be ionized) evolves independently in an effective mean field of the residual target. The validity of this approximation was demonstrated for the case of single ionization of atomic targets by Fainstein *et al.* [22], who followed the main ideas introduced by Rivarola *et al.* regarding the influence of the static potential in electron capture collisions [23]. This approximation was later extended and applied to treat electron capture [24] and single ionization [25] processes involving molecular targets. The geometry of the

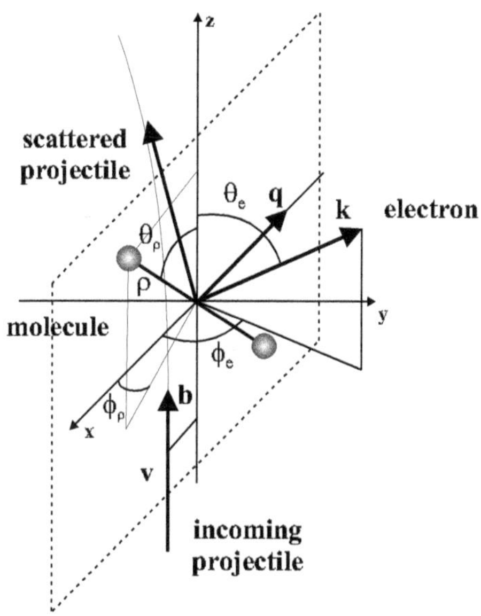

Fig. 1. Coordinate system, angles and quantities used for the case of heavy ions. The z-axis is parallel to the direction of the incoming projectile and the $x - z$ plane defines the so-called scattering plane. The internuclear axis ρ of the diatomic molecule subtends a polar angle θ_ρ (angle with respect to the z-axis) and an azimuthal angle ϕ_ρ (angle with respect to the x-axis). Reproduced from Ref. [18] with the permission of the original Publisher, IOP.

collision system is shown in Fig. 1. The projectile, that initially impinges with the momentum $\boldsymbol{K}_i = K_i\,\hat{\boldsymbol{z}}$, is scattered after the collision process on the

x-z plane (i.e. the scattering plane) with the final momentum $\boldsymbol{K_f}$. Then the momentum transfer is given by the difference $\boldsymbol{q} = \boldsymbol{K}_i - \boldsymbol{K}_f$. The momentum $\boldsymbol{k}$ of the ionized electron is represented in usual spherical coordinates by the magnitude $k = |\boldsymbol{k}|$ as well as angles θ_e and ϕ_e. Finally, θ_ρ defines the polar orientation angle of the molecular internuclear vector $\boldsymbol{\rho}$, whereas ϕ_ρ represents the azimuthal angle (i.e. the angle with respect to the x axis).

To investigate the existence of interference effects due to coherent emission of electrons from the proximity of the target nuclei, three-fold and two-fold differential cross sections are calculated. The triply differential cross section (TDCS), which depends on the energy E_k and the subtended solid angle $\Omega_k = (\theta_e, \phi_e)$ of the emitted electron as well as of the molecular orientation $\Omega_\rho = (\theta_\rho, \phi_\rho)$, can be obtained by integrating the post- or the prior-versions of the transition amplitude $A_{i,f}(\boldsymbol{b}, \Omega_\rho)$ over the impact parameter vector $\boldsymbol{b}$ and over the molecular orientation:

$$\sigma^{(3)}(E_k, \Omega_k, \Omega_\rho) = \frac{\mathrm{d}\sigma}{\mathrm{d}E_k\, \mathrm{d}\Omega_k\, \mathrm{d}\Omega_\rho} = N_e\, k \int \mathrm{d}\boldsymbol{b} \left|A^{\pm}_{i,f}(\boldsymbol{b}, \Omega_\rho)\right|^2. \tag{1}$$

Here, the sign + and − refer to the post- and prior-versions of the transition amplitude, respectively and N_e is the number of electrons. The TDCS may also be calculated by employing the corresponding T-matrix element which is related to the transition amplitude $A^{\pm}_{i,f}(\boldsymbol{b}, \Omega_\rho)$ through the two-dimensional Fourier transform:

$$T^{\pm}_{i,f}(\boldsymbol{\eta}, \Omega_\rho) = i\, v \int \mathrm{d}\boldsymbol{b} \exp(i\boldsymbol{\eta}\cdot\boldsymbol{b})\, A^{\pm}_{i,f}(\boldsymbol{b}, \Omega_\rho). \tag{2}$$

Symbol v denotes the projectile velocity and $\boldsymbol{\eta}$ is the vectorial component of the momentum transferred by the projectile to the target perpendicular to the direction of incidence. In this manner, and by using the Parseval identity, it can be demonstrated that:

$$\sigma^{(3)}(E_k, \Omega_k, \Omega_\rho) = N_e\, k\, \frac{(2\pi)^4}{v^2} \int \mathrm{d}\boldsymbol{\eta} \left|T^{\pm}_{i,f}(\boldsymbol{\eta}, \Omega_\rho)\right|^2. \tag{3}$$

In the following, and in order to describe the dynamics of the active electron, the main results obtained within the continuum distorted wave – eikonal initial state (CDW-EIS) approximation will be revisited. The CDW-EIS approach was first introduced by Crothers and McCann [26], who applied it to the study of monoelectronic atoms. Subsequently, this method has been extended to the case of multi-electronic targets [22, 27]. Using the straight line version of the impact parameter formalism within the distorted wave model, the first-order approximation of the transition amplitude for the active electron in the post- and prior-versions can be written as:

$$A^{+}_{i,f}(\boldsymbol{b},\Omega_\rho) = \int_{-\infty}^{+\infty} \mathrm{d}t \left\langle \chi_f^- \middle| W_f^\dagger \middle| \chi_i^+ \right\rangle, \tag{4}$$

$$A^{-}_{i,f}(\boldsymbol{b},\Omega_\rho) = \int_{-\infty}^{+\infty} \mathrm{d}t \left\langle \chi_f^- \middle| W_i \middle| \chi_i^+ \right\rangle, \tag{5}$$

respectively. Here, χ_i^+ and χ_f^- are the distorted wave functions that satisfy the correct ongoing and incoming asymptotic boundary conditions, respectively. This is a fundamental aspect of atomic collisions to take into account in the perturbative and non-perturbative formulations alike. In fact, in the exact eikonal transition amplitude for rearranging collisions (charge exchange, ionization, ...), it is critical to satisfy the correct boundary conditions (CBC) in both the entrance and exit channels, as analyzed in e.g. Ref. [28–35]. With the neglect of the CBC, the so-called disconnected diagrams in perturbative treatments inevitably occur with the ensuing divergence of the transition amplitudes [30, 31]. A specific example of divergence can also be traced back to the intermediate elastic channel within the perturbation series expansions of the exact transition amplitude using the Coulomb Green function as the propagator [31]. In Eqs. (4) and (5), quantities W_f and W_i are the perturbation operators corresponding to the final and initial channel, respectively.

A short note about the validity of the straight line version of the impact parameter approach or the semi-classical approximation (SCA) to treat coherence phenomena should be made. The ionization is dictated by the Coulomb interaction between the incoming projectile and the active electron. Meanwhile, the coherent electron emission is a consequence of the quantum delocalization of the electron wave function among the molecular nuclei. In this way, the quantum description of the electron is key for describing interference effects. On the contrary, one can use either a classical or a quantum characterization of the projectile without affecting the interference phenomena. It is this proper SCA that will be adopted in our models.

In the CDW-EIS approximation, χ_i^+ is chosen as:

$$\chi_i^+(\boldsymbol{x},t) = \frac{\exp\left(i\,\boldsymbol{K}_i\cdot\boldsymbol{R}_i\right)}{(2\pi)^{3/2}}\,\phi_i(\boldsymbol{x})\,\exp(-i\,\varepsilon_i\,t)\,\mathcal{L}_i^+(\boldsymbol{s}), \tag{6}$$

where the plane wave in the curly brackets describes the incident particle and ϕ_i represents the bound state of the active electron in the entrance channel, ε_i is the initial orbital energy. Quantities $\boldsymbol{s}$ and $\boldsymbol{x}$ are the position vectors of the active electron with respect to the projectile nucleus and the molecular center of mass, respectively, and $\mathcal{L}_i^+$ is the eikonal distortion function given by:

$$\mathcal{L}_i^+(\boldsymbol{s}) = \exp\left[-i\,\nu\,\ln(v\,s + \boldsymbol{v}\cdot\boldsymbol{s})\right], \tag{7}$$

where $\nu = Z_P/v$. As can be seen from Eqs. (6) and (7), the electron is described simultaneously in the combined fields of the target and the projectile. In this sense, the distorted wave function presents intrinsically a two-center character. Also, the use of two-center wave functions avoids the possible divergences in the corresponding perturbative series associated with the presence of disconnected diagrams [36].

In order to describe the bound state of the active electron, we use a linear combination of atomic orbitals (AO) centered on each target nuclei. However, some questions arise from the fact that a molecular target may have several molecular orbitals (MO), and electron emission can occur from any of them. Also, it should be recalled that the s-type AOs have spherical symmetry, so that they remain unchanged under a spatial rotation. On the other hand, in the p-type AOs, the presence of angular factors in the corresponding wave functions implies a rupture of this symmetry. Thus, in a randomly oriented molecule, the p-type AOs associated with each target center will depend on the molecular spatial coordinates and the corresponding wave functions must be modified. The AOs should then be described in a molecular reference system $x'y'z'$, which is chosen in such a way that its z'-axis lies along the internuclear molecular axis and whose origin coincides with those of the laboratory reference system. Therefore, the initial state ϕ_i of the active electron bound to a given MO should be written as:

$$\phi_{i,\mathrm{MO}}(\boldsymbol{x}') = \sum_{j,h} \omega_{j,h}\,\psi_{j,h}(\boldsymbol{x}'_j), \tag{8}$$

where $\omega_{j,h}$ is the normalization factor. Here, the index j identifies the nucleus of the molecule in which the atomic wave functions are centered, the subscript h represents a set of quantum numbers nlm, the vectors $\boldsymbol{x}'$ and $\boldsymbol{x}'_j$ are the electron coordinates with respect to the target center of mass and to the j-th molecular center, respectively. The change of coordinates from the laboratory system to the molecular one is given by the relation $\boldsymbol{x}' = \mathcal{T}\,\boldsymbol{x}$, where $\mathcal{T}$ is the rotation matrix. The transformation is made using the Euler angles [37].

The CDW-EIS wave function in the final channel of the ionization process is chosen as:

$$\chi_f^-(\boldsymbol{x},t) = \frac{\exp\left(i\,\boldsymbol{K}_f\cdot\boldsymbol{R}_i\right)}{(2\pi)^{3/2}}\,\phi_f(\boldsymbol{x})\,\exp\!\left(-i\,k^2\,t/2\right)\mathcal{L}_f^-(\boldsymbol{s}), \tag{9}$$

where ϕ_f is the continuum state of the active electron in the exit channel, and $\mathcal{L}_f^-$ the continuum distortion function given by:

$$\mathcal{L}_f^-(\boldsymbol{s}) = N(\zeta)\,{}_1F_1\left[-i\,\zeta;1;-i\,(p\,s+\boldsymbol{p}\cdot\boldsymbol{s})\right], \tag{10}$$

where $N(a) = \exp(\pi\,a/2)\,\Gamma(1+i\,a)$ with $\Gamma(z)$ being the Gamma function, ${}_1F_1(a;b;c)$ is the Kummer confluent hypergeometric function, $\boldsymbol{p} = \boldsymbol{k} - \boldsymbol{v}$ is the momentum of the electron in the final channel which is taken with respect to the projectile nucleus, and $\zeta = Z_P/p$.

After introducing the CDW-EIS approximation, it is worth mentioning certain points related to the continuum distorted wave (CDW) method. The original impact parameter version of the CDW method for charge exchange [28] has been extended to its quantum-mechanical counterpart for ionization in Ref. [29]. In particular, the CDW-EIS method employs, in the exit channel, the total scattering wave function from the CDW method through Eqs. (9) and (10). Moreover, the CDW-EIS approximation uses, in the entrance channel, the Coulomb logarithmic phase given in Eq. (7), as a large distance asymptotic approximation of the corresponding full Coulomb wave function distortion from the CDW method $\mathcal{L}_i^+(\boldsymbol{s}) = N^+(\nu)\,{}_1F_1(i\,\nu;1;iv\,s+i\boldsymbol{v}\cdot\boldsymbol{s})$, where $N^+(\nu) = \exp(\pi\nu/2)\Gamma(1-i\nu)$. Thus, overall, for ionizing collisions, the CDW-EIS model appears to be a further approximation to the CDW method.

In order to describe the final continuum wave function ϕ_f, we employ the two-effective center (TEC) approximation, introduced by Wang *et al.* [38] for electron capture by impact of bare ions on homonuclear and heteronuclear molecules. This approximation has successfully been used to describe single ionization of molecules by photon, electron and ion impact. Then, within the TEC approach, ϕ_f is taken as:

$$\begin{aligned}\phi_f(\boldsymbol{x}) = {} & \frac{1}{(2\pi)^{3/2}}\exp\left[i\,\boldsymbol{k}\cdot(\boldsymbol{x}_j+\boldsymbol{\rho}_j)\right] \\ & \times N(\xi)\,{}_1F_1\left[-i\,\xi;1;-i\,(k\,x_j+\boldsymbol{k}\cdot\boldsymbol{x}_j)\right].\end{aligned} \tag{11}$$

This is used when considering the component $\psi_{j,h}$ of the initial orbital wave function, as per Eq. (8)). In Eq. (11), vector $\boldsymbol{\rho}_j$ ($\boldsymbol{x}_j$) is the position of the j-nucleus with respect to the center of mass of the molecule (the j-th target nucleus) relative to the laboratory reference frame, and $\xi = Z_T^{\text{eff}}/k$ where $Z_T^{\text{eff}} = \sqrt{-2\,n^2\varepsilon_i}$ is an effective target charge with n being the corresponding principal quantum number. Then, within this framework, it can be shown that the scattering matrix element $T^{\pm}_{i,f}$ for a given MO is given by:

$$\begin{aligned} T^{\pm}_{\text{MO}}(\boldsymbol{\eta},\Omega_\rho) &= \sum_j \exp\left[-i\,(\boldsymbol{k}-\boldsymbol{q})\cdot\boldsymbol{\rho}_j\right] \sum_h \omega_{j,h}\, T^{\pm(\text{eff})}_{j,h_{(\text{MO})}}(\boldsymbol{\eta},\Omega_\rho) \\ &= \sum_j \exp\left[-i\,(\boldsymbol{k}-\boldsymbol{q})\cdot\boldsymbol{\rho}_j\right] T^{\pm(\text{eff})}_{j_{(\text{MO})}}(\boldsymbol{\eta},\Omega_\rho), \end{aligned} \tag{12}$$

where $T^{\pm(\text{eff})}_{j,h_{(\text{MO})}}$ is an effective scattering matrix element associated with the atomic orbital $\psi_{j,h}$ and the MO. In Eq. (12), quantity $T^{\pm(\text{eff})}_{j_{(\text{MO})}}$ is an effective one-center scattering matrix element associated with the basis set of AOs centered on the j-th target nucleus. Then, we can define the TDCS for single ionization from a given MO as:

$$\sigma^{(3)}_{\text{MO}}(E_k,\Omega_k,\Omega_\rho) = N_e\, k\, \frac{(2\pi)^4}{v^2} \int \mathrm{d}\boldsymbol{\eta}\, \left|T^{\pm}_{\text{MO}}(\boldsymbol{\eta},\Omega_\rho)\right|^2. \tag{13}$$

Adding all the partial contributions given by (13), the TDCS for the complete molecule can be obtained via:

$$\sigma^{(3)}(E_k,\Omega_k,\Omega_\rho) = \sum_{\text{MO}} \sigma^{(3)}_{\text{MO}}(E_k,\Omega_k,\Omega_\rho). \tag{14}$$

The DDCS are calculated by averaging the preceding expressions over the solid angle Ω_ρ:

$$\sigma^{(2)}(E_k,\Omega_k) = \sum_{\text{MO}} \sigma^{(2)}_{\text{MO}}(E_k,\Omega_k) = \sum_{\text{MO}} \frac{1}{4\pi} \int \mathrm{d}\Omega_\rho\, \sigma^{(3)}_{\text{MO}}(E_k,\Omega_k,\Omega_\rho). \tag{15}$$

2.1. *Dielectronic molecules: H_2 and HeH^+*

Let us first focus our attention on single ionization of dielectronic diatomic molecules. As has earlier been mentioned, the first experimental evidence on the existence of interference effects due to coherent electron emission was given at the beginning of the current century, when double differential spectra of electrons emitted from the H_2 target by 60 MeV/u Kr^{34+} ion impact were measured as a function of the ejected electron energy for fixed observation angles [12]. In the same study, a model calculation based on a theoretical formalism analogous to that developed for electron scattering [40] was employed with the aim of providing a theoretical background and to give an explanation of the experimentally observed effects.

According to the theory presented in [12], interference effects were expected to be revealed through small oscillations appearing in the electronic spectra. However, the measured cross sections showed a strong decreasing behavior (by several orders of magnitude) with the ionized electron velocity. As a result, no evidence of oscillations in cross section data could be detected. Thus, in an attempt to enhance the visibility of possible interference structures, the measured cross sections were divided by a doubled theoretical cross section from the CDW-EIS approach for ionization from an effective hydrogenic atom with the nuclear charge $Z_T^{\rm eff}$ =1.05 (which follows from the orbital binding energy). Oscillations associated with interference effects were thus observed as clear evidence of coherent electron emission.

The pioneering work of Stolterfoht *et al.* [12] motivated further experiments [41–43] and this gave rise to the development of different theoretical models for describing the physical mechanisms leading to the appearance of interference patterns in the electronic spectra. Among the theoretical models, the TEC approximation has been applied by Galassi *et al.* [25, 39] to study single ionization of the H_2 targets by 60 MeV/u Kr^{34+} ion impact. It should be pointed out that H_2 is a homonuclear molecule with only one MO in its fundamental state $^1\Sigma_g^+$. Its two electrons with opposite spins are bound to the target in a $\sigma_g 1s$ MO. In this case, the initial bound state of the active electron was chosen according to Eq. (8) as:

$$\phi_i(\boldsymbol{x}) = \omega_1(\rho)\,\xi(\boldsymbol{x}_1) + \omega_2(\rho)\,\xi(\boldsymbol{x}_2). \tag{16}$$

Here, $\xi(\boldsymbol{x}_j)\,(j=1,2)$ are the 1s-type AOs defined by the variational single-zeta functions:

$$\xi(\boldsymbol{x}_j) = \frac{(Z_T^{\rm eff})^{(3/2)}}{\pi^{1/2}} \exp\left(-Z\,x_j\right), \tag{17}$$

where $Z_T^{\rm eff} = 1.193$ is an atomic effective charge for hydrogen. The normalization factors $\omega_j(\rho) = 0.5459\,(j = 1,2)$ correspond to an equilibrium internuclear distance $\rho = 1.4$ a.u. The orbital energy was taken from the experimental value obtained for single ionization of H_2, i.e. $\varepsilon_i = -0.566$ a.u. The electrons bound to the projectile were considered to be entirely placed onto the corresponding nucleus, so that a net charge of $Z_P = 34$ was employed in the calculation. As was mentioned before, the s-type AOs have spherical symmetry and hence they do not depend on the orientation of the molecule. For this reason, vectors $\boldsymbol{x}$ and $\boldsymbol{x}_j\ (j = 1,2)$ are given with respect to the laboratory reference frame instead of the molecular one.

Thus, considering Eq. (16), it can be shown that the squared modulus of the transition matrix element reads as:

$$\left|T_{\rm H_2}(\boldsymbol{\eta},\Omega_\rho)\right|^2 = 2\left[1+\cos\left((\boldsymbol{k}-\boldsymbol{q})\cdot\boldsymbol{\rho}\right)\right]\left|T_{\rm H}^{\rm eff}(\boldsymbol{\eta})\right|^2, \tag{18}$$

where $T_{\rm H}^{\rm eff}(\boldsymbol{\eta})$ is an effective one-center scattering matrix element corresponding to ionization from an effective hydrogen atom. Hereafter, the superscript $\pm$ will be omitted for the sake of clarity and simplicity in the notation. The number 2 in Eq. (18) indicates the presence of two identical hydrogen nuclei. The oscillatory factor appearing in Eq. (18) represents the signature of interference patterns coming from the coherent electron emission from both effective hydrogen atoms. For a randomly oriented molecule, Eq. (18) must be integrated over $\boldsymbol{\eta}$ and averaged over all possible molecular orientations, so that:

$$\begin{aligned}\sigma^{(2)}(E_k,\Omega_k) &= 2\,N_{\mathrm{H}_2}\,k\,\frac{(2\pi)^4}{v^2}\left\{\int \mathrm{d}\boldsymbol{\eta}\;|T_{\mathrm{H}}^{\mathrm{eff}}(\boldsymbol{\eta})|^2\right.\\ &\qquad\left.+\int \mathrm{d}\boldsymbol{\eta}\;\frac{\sin\left(|\boldsymbol{k}-\boldsymbol{q}|\,\rho\right)}{|\boldsymbol{k}-\boldsymbol{q}|\,\rho}\;|T_{\mathrm{H}}^{\mathrm{eff}}(\boldsymbol{\eta})|^2\right\}\\ &= 2\,S_{\mathrm{H}^+}^{(\mathrm{dir})}(E_k,\Omega_k)+2\,S_{\mathrm{H}_2}^{(\mathrm{int})}(E_k,\Omega_k),\end{aligned} \tag{19}$$

where $N_{\mathrm{H}_2} = 2$ is the number of electrons in the H_2 molecule. Here, $S_{\mathrm{H}^+}^{(\mathrm{dir})}(E_k,\Omega_k)$ represents the contribution to the DDCS given by one effective H center and $S_{\mathrm{H}_2}^{(\mathrm{int})}(E_k,\Omega_k)$ describes the interference contribution obtained by integrating only the oscillatory part of expression (18). In Fig. 2, the ratios between the theoretical molecular DDCS and the doubled theoretical atomic DDCS are plotted as a function of the electron velocity for emission angles θ_e =20° and 30°. The ratios obtained by dividing the experimental DDCS for the H_2 target and the doubled theoretical DDCS for atomic hydrogen are also shown. Calculations using a semiclassical version of a first Born peaking approximation [44] are also included for the purpose of comparison. Regarding the validity of the theoretical model from Ref. [44], it can be easily seen that, within the TEC approach, the frequency of the oscillations depends on the ejection angles:

$$|\boldsymbol{k}-\boldsymbol{q}|^2 = k^2 + k\left[\eta\,\sin(\theta_e)\,\cos(\phi_\eta-\phi_e)+q_{\mathrm{min}}\cos(\theta_e)\right]+q^2. \tag{20}$$

Therefore, the validity of utilizing the simple analytical forms to describe the interference structures, like the models employed by Stolterfoht *et al.* [41] and Nagy *et al.* [44] is questionable, as has been discussed in Ref. [39]. In Eq. (20), quantity $q_{\mathrm{min}} = \Delta\varepsilon/v$ is the minimum momentum transfer, which corresponds to $\eta = 0$. Both experimental as well as theoretical ratios exhibit characteristic oscillations associated with the coherent emission of electrons. If no effects due to the structure of the molecule are expected, it could be assumed that, at the high impact energy considered here, the ratio will give a value close to unity. This value may be slightly different from unity due to the different binding energies of H_2 and H, the corresponding effective charges, and the normalization of the respective bound-state

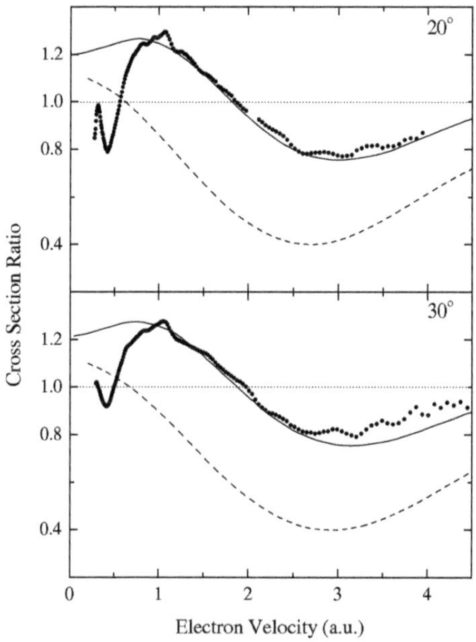

Fig. 2. Ratio between the DDCS for single ionization of H_2 and two times the DDCS for ionization of H as a function of electron velocity at 20° and 30° emission angles, for 60 MeV amu^{-1} Kr^{34+} impact. Solid circles: experiment from Ref. [12] Theory: solid line, CDW-EIS calculation [39], dashed line: the results from Ref. [44]. Reproduced from Refs. [12, 39, 44] with the permission of the original Publishers, APS and IOP.

wave functions. In Fig. 2, an abrupt fall in the experimental data, as the electron energy decreases, is seen although it does not correspond to interference effects; rather this is associated with electron correlation and/or screening effects, as has been shown for photoionization of the H_2 targets [45]. These effects are very sensitive to the representation used for the bound and continuum wave functions. However, at higher velocities, the theoretical results show, as do the experimental data, a distinctive interference pattern. This oscillation can be studied in more detail by looking at the contributions to the DDCS from the terms $S^{(\mathrm{dir})}_{\mathrm{H}^+}(E_k, \Omega_k)$ and $S^{(\mathrm{int})}_{\mathrm{H}_2}(E_k, \Omega_k)$, that are plotted in Fig. 3 together with the experimental and theoretical DDCS ratios. The former quantity being independent of the internuclear vector $\boldsymbol{\rho}$ (and hence of the molecular orientation), shows a monotonous increase. On the contrary, $S^{(\mathrm{int})}_{\mathrm{H}_2}(E_k, \Omega_k)$ contains an oscillatory factor, depending on both $\boldsymbol{q}$ and $\boldsymbol{k}$, as well as on the internuclear distance ρ. It is seen that $S_{H_2}^{(int)}(E_k, \Omega_k)$ shows a damped oscillatory behavior with the ejected electron velocity, demonstrating that interference patterns arise from coherent emission of electrons from the vicinities of both molecular centers.

The TEC approximation introduced by Galassi *et al.* [39] for single ionization of H_2 was subsequently compared with the experimental cross section ratios in the cases of the 1.5 MeV/u F^{9+} and 1 MeV/u C^{6+} ions impacting on atomic and molecular hydrogen targets [46]. Once again, the presence of interference patterns due to coherent electron emission from the molecular centers was found, confirming the previous results.

The influence of the molecular orientation on interference patterns has also been investigated. To this end, the TDCS, obtained by integrating Eq. (18) over η, was evaluated for impact of the 13.7 MeV/u C^{6+} projectiles on H_2 [50]. A full coplanar geometry where the molecule, the emitted electron and the projectile are all in the same plane, was considered. The resulting

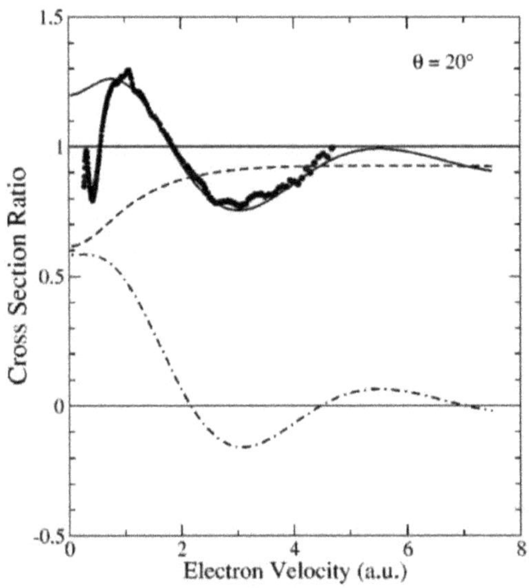

Fig. 3. Experimental-to-theoretical cross section ratios for single ionization of the H_2 molecules by the 60 MeV amu^{-1} Kr^{34+} projectiles. Solid circles: experiment from Ref. [12]. Theory: solid line, the CDW-EIS model; dashed-line, the contribution from $S_{dir}(E_k, \Omega_k)$; dot-dashed line, the contribution from $S_{int}(E_k, \Omega_k)$ [39]. Reproduced from Refs. [12, 39] with the permission of the original Publisher, APS.

angular distributions are shown in Fig. 4 for a 100 eV fixed ejection electron energy, as a function of the emission angle and for three different molecular orientations.

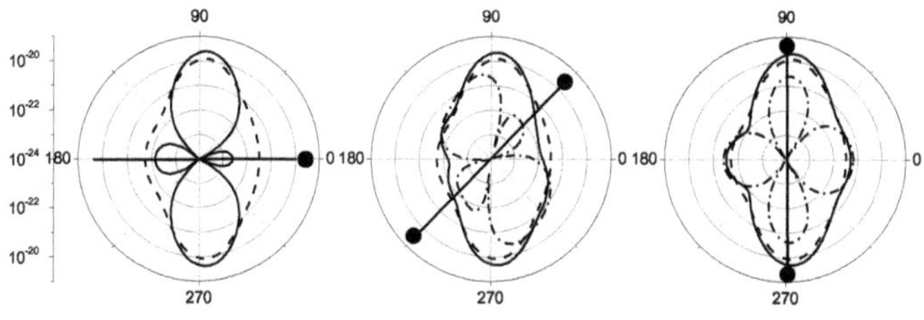

Fig. 4. Angular distribution of emitted electrons at 100 eV for different orientations of the H_2 molecule: θ_ρ = 0°, 45° and 90°. All the curves present the theoretical CDW-EIS calculations. Full curve, full molecular calculation; chained curve, the molecular calculation with the neglect of the contribution from η to the interference factor (see the main text for details); dashed curve, the results for the two-effective H atoms. Reproduced from Ref. [50] with the permission of the original Publisher, IOP.

For a molecular orientation parallel to the impact velocity, the transverse momentum transfer $\boldsymbol{\eta}$ is perpendicular to the internuclear vector $\boldsymbol{\rho}$, so the argument of the oscillatory factor in Eq. (18) is reduced to the expression $(\boldsymbol{k}\cdot\boldsymbol{\rho} - q_{\min}\,\rho)$, containing the minimum momentum transfer, $q_{\min}$. As can be seen, the dependence on $\boldsymbol{\eta}$ is contained only in the term $|T_{\mathrm{H}}^{\mathrm{eff}}(\boldsymbol{\eta})|^2$, i.e. in the atomic effective scattering matrix element. Thus, the corresponding oscillatory behavior observed in Fig. 4 is dominated by the cosine factor in Eq. (18), showing four lobes with a predominance of the electron emission in the direction perpendicular to the initial projectile velocity. As it was pointed out in Ref. [50], for $k\,\rho$ <1, the electron wavelength is much larger than the internuclear distance and, therefore, the two nuclei appear as one. In contrast, for $k\,\rho$ >1, the de Broglie wavelength is smaller than the internuclear separation, allowing for the interference between the outgoing waves from both centers. In this case, for the considered electron velocity, we have $k\,\rho$ ≈3.8, which satisfies the condition $k\,\rho$ >1. As the orientation of the molecule is modified from the position at $k\,\rho$ ≈3.8, different values of η begin to contribute to the interference factor. This gives a contribution to the calculated TDCS, with a tendency to wash out the presence of interferences. To give a further support to this analysis, the cross sections with a neglect of the contribution from η in the cosine term were also computed and the corresponding results plotted in Fig. 4 for the three selected molecular orientations. In the three cases, the characteristics of the four interference lobes appear showing the preference of electron emission in the direction perpendicular to the molecular axis. The TDCS results obtained by completely neglecting the oscillatory factor in Eq. (18) are also shown in Fig. 4. Obviously, no fingerprints of interferences are observed in the latter results.

Regarding the case of heteronuclear dielectronic molecules, the possible existence of interference effects has been theoretically analyzed by Tachino *et al.* [51–53], who considered single ionization of HeH^+ by ion impact. It should be pointed out that, in the ground state, the two electrons of the HeH^+ molecule occupy the 1σ MO, one with spin up and the other with spin down. This MO is described as a linear combination of the Slater-type orbitals (STO). Herein, one $1s$-STO is centered on the He^{2+} nucleus and another $1s$-STO is centered on the H^+ nucleus:

$$\phi_{1\sigma}(\boldsymbol{x}) = \omega_1 \, \psi_{1_{1s}}^{\mathrm{STO}}(\boldsymbol{x}_1) + \omega_2 \, \psi_{2_{1s}}^{\mathrm{STO}}(\boldsymbol{x}_2) \tag{21}$$

where the subscripts 1 and 2 denote the He^{2+} and H^+ nuclei, respectively. This is equivalent to expanding the MO in a minimal basis set of the STO functions. The quantum chemistry program Gaussian 98 [54] was employed to calculate the different parameters that characterize the ground state of the target. Optimized values of both the ω_j $(j = 1, 2)$ coefficients and the STO exponents in Eq. (21) were obtained by using an STO-6G basis set within the HF approximation, where each STO is represented by a linear combination of six Gaussian type orbitals (GTO) [55, 56]. Although this minimal representation of the 1σ MO of the HeH^+ is convenient to evaluate the collision transition amplitudes, accurate values of the orbital energy ε_i and the equilibrium internuclear distance ρ can only be obtained by going beyond the minimal basis set approximation. Thus, both ρ and ε_i have been determined by using a much larger 6-311 G* basis set. The obtained results are $\rho = 0.771$ Å and $\varepsilon_i = \pm 1.63317$ au.

The HeH^+ is a highly asymmetric system where the electrons have a large preference to be placed in the proximity of the alpha particle in the bound MO. The TDCS calculations have been carried out within a coplanar geometry (the molecule, the emitted electron and the projectile are all in the same plane) for impact of 1 MeV protons and for molecular orientations parallel and perpendicular to the incident velocity vector. Fixed electron energies E_k =100 eV and E_k =400 eV have been considered. As in the case of H_2, to emphasize the presence of interferences associated with the coherent electron emission from both target centers, a ratio between the molecular TDCS and the atomic TDCS should be calculated. Taking into account that the electronic cloud is located mainly around the He^{2+} nucleus, we chose to define this ratio in the following way:

$$R_{\mathrm{HeH}^+}(E_k,\Omega_k,\Omega_\rho) = \frac{\sigma^{(3)}_{\mathrm{HeH}^+}(E_k,\Omega_k,\Omega_\rho)}{\sigma^{(3)}_{\mathrm{He}^{2+}}(E_k,\Omega_k)}. \tag{22}$$

Here, $\sigma^{(3)}_{\mathrm{He}^{2+}}$ is an effective atomic TDCS corresponding to the first addend in the MO wave function. As can be seen from Fig. 5, in spite of the asymmetry of the electronic cloud in the initial state, the signatures of coherent electron emission from the neighborhood of the two atomic centers are clearly identified in the calculated angular distributions. For a parallel orientation (i.e. for $\boldsymbol{\eta}\cdot\boldsymbol{\rho}$ =0), the interference factor does not depend on $\boldsymbol{\eta}$ and, thus, it is not affected by the integration over $\boldsymbol{\eta}$. The number of lobes in the cross section ratios associated with interference patterns increases with the emitted electron velocity. For the perpendicular orientation, on the contrary, interferences are less visible than for the parallel case, which is due to the fact that integration over the transverse momentum transfer vector washes out these structures. Coherence due to emission from the vicinity of the nuclei composing the target is clearly visible even if the ionization cross section is averaged over all the molecular orientations. In this case, the DDCS reads:

$$\sigma^{(2)}_{\mathrm{HeH}^+}(E_k,\Omega_k) = S^{(\mathrm{dir})}_{\mathrm{He}^{2+}}(E_k,\Omega_k) + S^{(\mathrm{dir})}_{\mathrm{H}^+}(E_k,\Omega_k) + S^{(\mathrm{int}_1)}_{\mathrm{HeH}^+}(E_k,\Omega_k). \tag{23}$$

Here, the first and second terms, i.e. $S^{(\mathrm{dir})}_{\mathrm{He}^{2+}}$ and $S^{(\mathrm{dir})}_{\mathrm{H}^+}$, correspond to the direct contributions from the electronic distribution around molecular centers He^{2+} and H^+, respectively. The third term, $S^{(\mathrm{int}_1)}_{\mathrm{HeH}^+}$, contains the information about the interference effect arising from coherent electron emission. In order to investigate how the partial localization of the electron around the alpha particle influences the oscillatory structures related to interferences, the cross sections ratio obtained by dividing the molecular cross section $\sigma^{(2)}_{\mathrm{HeH}^+}$ by the sum of the direct terms appearing in Eq. (23) is analyzed.

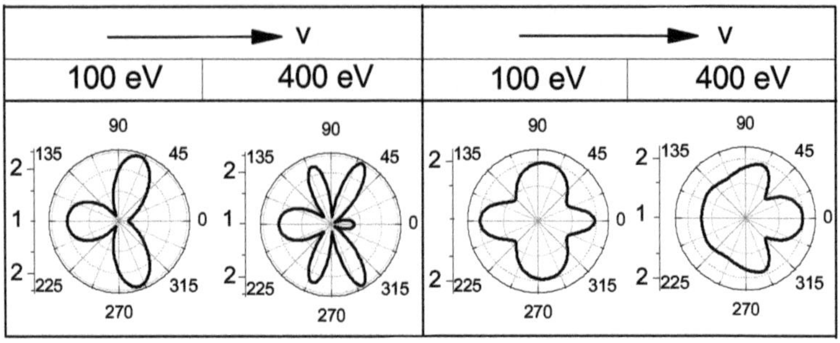

Fig. 5. The TDCS ratios for single ionization of HeH^+ by impact of 1 MeV protons and for two different orientations of the molecule with respect to the ion beam. Right panel: parallel orientation ($\theta_\rho = 0°$); the He^{2+} nucleus is located at the position corresponding to 0° and the H^+ one is located at the position corresponding to 180°. Left panel: perpendicular orientation ($\theta_\rho = 90°$); the He^{2+} nucleus is located at the position corresponding to 90° and the H^+ one is located at the position corresponding to 270° [51]. Reproduced from Ref. [51] with the permission of the original Publisher, IOP.

After some algebra, we have:

$$R_{\mathrm{HeH}^+}(E_k,\Omega_k) = 1 + \frac{S^{(\mathrm{int}_1)}_{\mathrm{HeH}^+}}{S^{(\mathrm{dir})}_{\mathrm{He}^{2+}}(E_k,\Omega_k) + S^{(\mathrm{dir})}_{\mathrm{H}^+}(E_k,\Omega_k)}. \tag{24}$$

If we set $T_1^{\mathrm{eff}}(\eta) = T_2^{\mathrm{eff}}(\eta)$, the DDCS ratio corresponding to the hydrogen molecule is recovered:

$$R_{\mathrm{H}_2}(E_k,\Omega_k) = 1 + \frac{S^{(\mathrm{int}_1)}_{\mathrm{H}_2}}{S^{(\mathrm{dir})}_{\mathrm{H}^+}(E_k,\Omega_k)}. \tag{25}$$

In order to better visualize the presence of interference patterns, the quantity $R_{i_{\mathrm{HeH}^+}} = R_{\mathrm{HeH}^+} - 1$ is plotted. The results are shown in Fig. 6, as a function of the final electron energy at fixed emission angles. The case of proton as a projectile with an impact energy of 100 MeV is considered. As it can be appreciated, interference patterns appear under the form of smooth oscillations with the feature that for a fixed collision energy the ratio frequency is larger for backward scattering than for forward scattering,

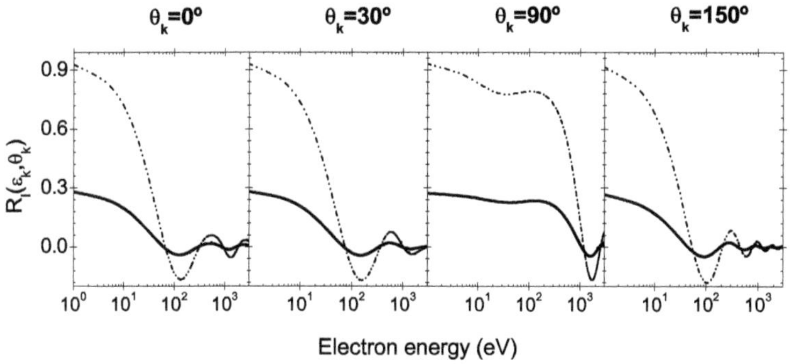

Fig. 6. Comparison between the cross sections ratios for single ionization of HeH^+ by proton impact, for impact energy of 13.7 MeV/u and different values of θ_e. Solid lines: R_{i,HeH^+}; dot-dot-dashed lines: R_{i,H_2} [52]. Reproduced from Ref. [52] with the permission of the original Publisher, IOP.

as was previously observed for the H_2 targets. For the sake of comparison, the ratios $R_{i_{\mathrm{H}_2}} = R_{\mathrm{H}_2} - 1$ for the hydrogen molecule were also included in Fig. 6. It can be seen that the shape of the curves for H_2 is very similar to those corresponding to the HeH^+ molecule, but with larger amplitudes. The difference between the amplitudes may be attributed to the partial localization of the emitted electron around the alpha particle. A recent investigation using a first-order version of the Born approximation has confirmed these results [57]. It is interesting to point out that although we are comparing two targets with different electron distribution symmetries, the frequency of the oscillations in the R_i ratios is similar for the homonuclear molecule and for the heteronuclear molecular ion. In particular, the positions of the respective maxima and minima are almost coincident.

2.2. *Multi-electronic molecules: N_2*

Let us now focus on diatomic homonuclear molecules with more than two bound electrons in the initial state. In particular, the nitrogen molecule will be considered. Its electronic configuration in the ground state $^1\Sigma_g^+$ is $(\sigma_g 1s)^2(\sigma_u^* 1s)^2(\sigma_g 2s)^2(\sigma_u^* 2s)^2(\pi_u 2p)^4(\sigma_g^* 2p)^2$. As can be seen, the presence of several MOs renders the study of single ionization more complicated since electron emission can take place from any of these orbitals. By reference to Eq. (8), the expression for the initial bound wave function corresponding to a given MO is:

$$\phi_{\mathrm{MO}}(\boldsymbol{x}') = \sum_{j=1}^{2} \omega_{j,h}\, \psi_{j,h}(\boldsymbol{x}'_j). \tag{26}$$

As was done before for the HeH^+ case, a basis set of the STOs centered on each molecular nucleus is employed. Both the STO characteristic exponents and the coefficients $\omega_{j,h}$ are optimized through the employment of the quantum chemistry program Gaussian 98 [54]. Specifically, a minimal STO-6G basis set within the HF approximation was considered. The equilibrium internuclear distance ρ and the orbital energies ε_i for each of the target MOs were obtained by employing a larger 6-311 G* basis set. In this case, the σ-type MO wave functions are described as a linear combination of $1s$, $2s$ and $2p_z$ STOs. The wave functions corresponding to the π-type MO are expressed as a linear combination of the $2p_x$ and $2p_y$ STOs. As stated, the s-type wave functions have spherical symmetry and the angular position of the molecule in space does not affect their expression. This is not the case for the p_x, p_y and p_z atomic wave functions since they depend on the orientation of the molecule in space. This implies that the corresponding effective scattering matrix elements $T^{\mathrm{eff}}_{j,2p_{x,y,z}}$ will depend also on the molecular orientation Ω_ρ.

With the aim of investigating the presence of interference effects due to coherent electron emission in molecules with more than one MOs, the TDCS and DDCS for the single ionization of N_2 molecules in the ground state by impact of charged ions were evaluated within the TEC approach [58]. First, the contributions from each MO are evaluated by computing the corresponding TDCS for impact of 1 MeV protons and an emission energy E_k =100 eV as per Eq. (13). The corresponding results are presented in Fig. 7 for two different target orientations, when a molecule is aligned in the parallel or perpendicular orientation relative to the direction of the incident beam. The lobes associated with interference effects are present in the plots corresponding to the parallel orientation. These structures appear mainly in the $\sigma_g 1s$, $\sigma_u^* 1s$ and $\pi_u 2p$ distributions. However, for the complete molecule, these patterns are less visible in the perpendicular orientation spectra [58]. Moreover, for the inner shells $\sigma_g 1s$ and $\sigma_u^* 1s$ it seems that in the corresponding angular spectra the positions of the maxima and the minima are reversed. As a consequence, it has been verified by Tachino *et al.* [58] that the TDCS for the complete molecule do not present oscillations associated with interference effects. This explains the suppression of

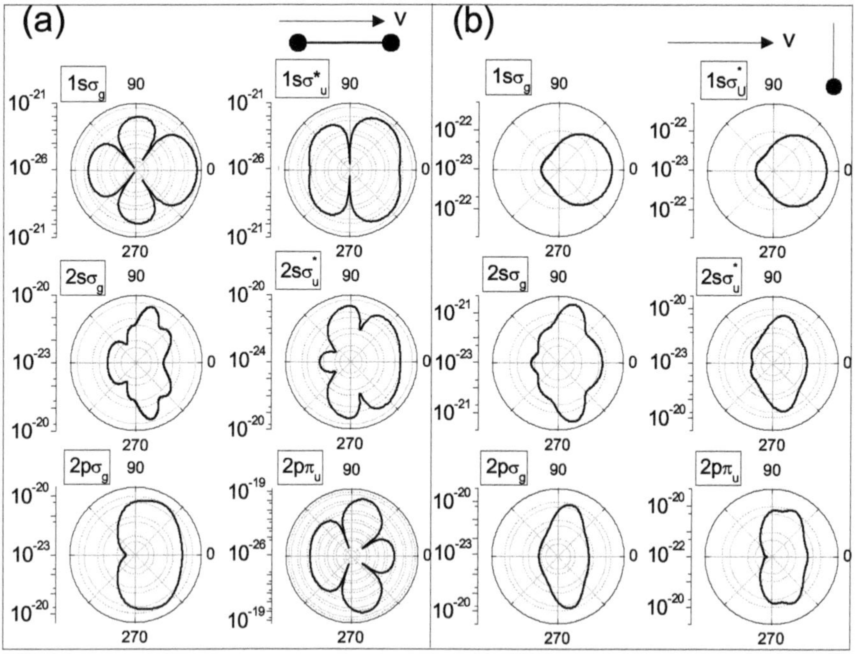

Fig. 7. The TDCS for each MO of N_2 as a function of the emission angle θ_e. The final electron energy is E_k =100 eV. The molecule is aligned in the parallel (a) or perpendicular (b) orientation with respect to the direction of the incoming projectile. The energy of the incident proton is 1 MeV. The TDCS are given in units of $eV^{-1}Sr^{-2}cm^2$ [58]. Reproduced from Ref. [58] with the permission of the original Publisher, IOP.

primary electron interferences in the experiments carried out by Baran *et al.* [59, 60] who measured the DDCS for single ionization of N_2 molecules by proton impact. They suggest that this suppression of interference effects could come from the delocalization of several MOs. A similar behavior has also been observed in Refs. [61] and O_2 [62] for N_2 and O_2 molecules, respectively. Furthermore, Winkworth *et al.* [63, 64] have obtained a comparable pattern for impact of the H^+ and $O^{5+,8+}$ projectiles on the O_2 targets.

In order to reveal the influence of electron interference effects in molecular spectra, the DDCS ratios calculated by dividing the theoretical DDCS for the complete molecule and the doubled theoretical counterparts corresponding to ionization of an effective nitrogen atom are plotted in Fig. 8.

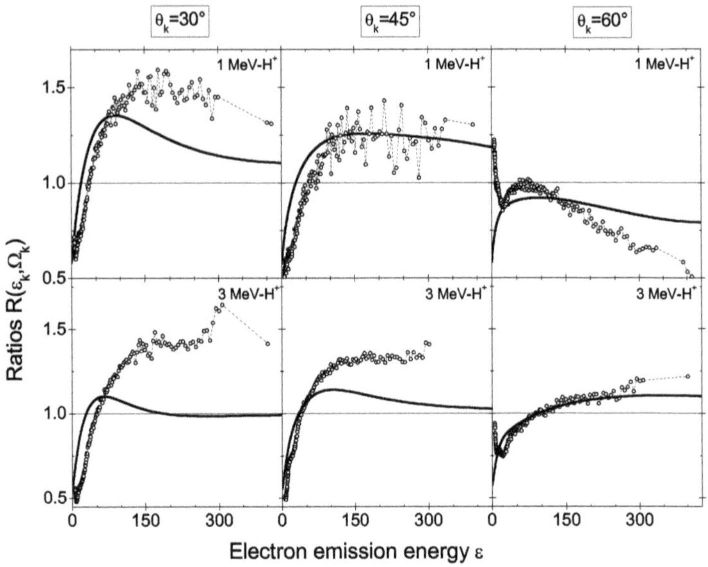

Fig. 8. Theoretical (solid lines) and experimental (open circles) DDCS ratios for impact of protons at 1 and 3 MeV on N_2 molecules. Experimental Auger structures have been eliminated for a better comparison with the theory [58]. Reproduced from Ref. [58] with the permission of the original Publisher, IOP.

The displayed results correspond to proton impact at 1 and 3 MeV, and electron emission angles θ_e =30°, 45° and 60°. For comparison, the ratios of the experimental DDCS for N_2 and the doubled theoretical DDCS for N are also shown. No signatures of regular oscillating structures are seen in the considered energy domain either in the theoretical or in the experimental ratios. This can be explained by the fact that although the contributions of each MO to this ratio present oscillatory patterns, they do not have a marked regularity and the superposition of these quantities gives no regular structures that could be associated with interference effects [58].

However, if the DDCS ratios are redefined, some interesting facts can be observed, particularly for the two inner MOs. We redefine the DDCS ratios by dividing the MO-DDCS $\sigma^{(2)}_{\rm MO}(E_k, \Omega_k)$ by the same quantity but excluding the interference term [58]. This ratio will be denominated by $C_{\rm MO}(E_k, \Omega_k)$. In Fig. 9, the $C_{\rm MO}(E_k, \Omega_k)$ ratios at 1 MeV proton impact and electron emission angle θ_e =30° for the innermost MOs are presented. It can be seen that the corresponding curves show regular oscillating structures, with

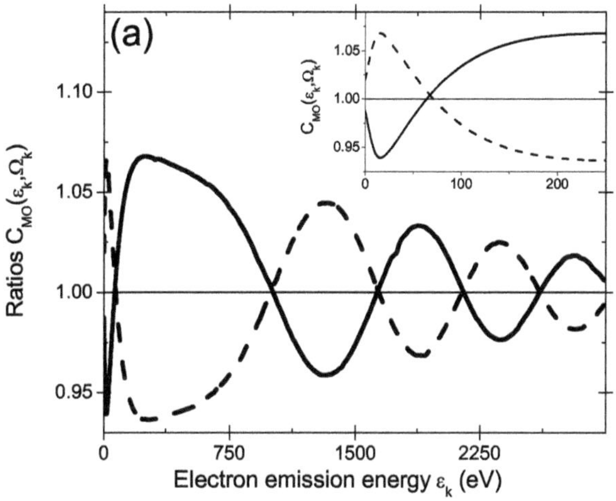

Fig. 9. The $C_{\sigma_g 1s}$ (solid line) and the $C_{\sigma_u^* 1s}$ (dashed line) ratios for the innermost orbitals of the nitrogen molecule, for impact of protons at 1 MeV and for the electron emission angle $\theta_e = 30°$. Reproduced from Ref. [58] with the permission of the original Publisher, IOP.

frequencies that increase with the increasing values of the emission energies. Also, the oscillation amplitudes become smaller as the values of E_k are larger. The two curves are out of phase, since the position of the maxima (minima) corresponding to the $\sigma_g 1s$ orbital is similar to the position of the minima (maxima) appearing in the $\sigma_u^* 1s$ MO curve. This behavior can be explained by considering that the N_2 initial bound wave functions may be approximated as:

$$\phi_{\sigma_g 1s}(\boldsymbol{x}) \approx \omega\, \psi_{1_{1s}}^{\rm STO}(\boldsymbol{x}_1) + \omega\, \psi_{2_{1s}}^{\rm STO}(\boldsymbol{x}_2), \tag{27}$$

$$\phi_{\sigma_u^* 1s}(\boldsymbol{x}) \approx \omega'\, \psi_{1_{1s}}^{\rm STO}(\boldsymbol{x}_1) - \omega'\, \psi_{2_{1s}}^{\rm STO}(\boldsymbol{x}_2), \tag{28}$$

where ω =0.70484 and ω' =0.70464. The coefficients of the remaining terms are at least two orders of magnitude smaller than ω and ω', so that in principle these terms which correspond to the $2s$ and $2p$ AOs can be neglected. With regard to the orbital energy ε_i, the values for both inner orbitals are very similar. Therefore, it can be assumed that $\sigma_g 1s$ and $\sigma_u^* 1s$ are almost-degenerate orbitals. The corresponding phases are determined by the sign of the second term in (27) and (28) according to the gerade or ungerade character of the MO.

3. Conclusions and outlook

In this chapter, we have studied the coherent electron emission from simple molecules by impact of energetic heavy charged particles. We have analyzed particularly how an interference pattern emerges in different observable quantities. We have made a summary of the theoretical approach and explained the most important details and features. Furthermore, regarding the interference effects, we have emphasized both their similarities and differences for protons and highly-charged ions as the projectiles. Generally speaking, the theoretical models are able to account to some degree the available experimental data and to give a clear interpretation of the interference phenomena. The latter have been shown to be universal, in the sense that interference patterns seem to be present independently of the nature of the projectiles, once the adequate range of impact energies and final electron as well as projectile variables are chosen. Interestingly, we can identify the electron emission from the different molecular centers as the atomic counterpart of the double-slit experiment.

We can foresee that experiments using highly aligned molecules in space, as targets will provide a demanding test for the theories and will open new theoretical and experimental pathways. For instance, the use of the now widespread COLTRIMS techniques may open the way to molecular spectroscopy in which Young-type interferences could be observed in collisions of atoms with fixed-in-space molecules. In analogy with optics, the variation of the fringe separation may be related to a change in the internuclear distance (equivalent to the slit separation) possibly allowing determination of the electronic internal degrees of freedom of the molecule. The feasibility for these potential applications was shown for collisions of molecular hydrogen ions and helium atoms [65, 66]. However, the complexities of such collisions may still need more theoretical and experimental work to be implemented in the near future.

Moreover, the COLTRIMS technique was revealed to be very useful to fully analyze two-slit interferences in experiments of photodissociation of hydrogen molecules [67]. Interestingly, interference patterns may serve to extract conclusions about entanglement and coherence phenomena. In particular, these experiments show that a system comprised merely of two electrons could suffice to detect transitions from a quantum interference to a classical particle-like pattern. However, being encoded in the entangled two-electron system, quantum coherence will not be destroyed.

In addition, recent experiments with X-rays and a double slit made on molecular oxygen paved the road to a special kind of measurements known in quantum mechanics as *Gedanken* experiments in quantum mechanics [68, 69]. Herein, the oxygen molecule is excited (by soft X-rays coming from a synchrotron) to a repulsive state leading to dissociation. While dissociating, the molecule may decay into a variety of O_2^+ electronic states by emission of an outer electron and subsequent filling of a hole by an inner one. By employing current electron-ion coincidence techniques, a full momentum determination of all charged particles may be measured. Since the outer electron is ejected in the presence of the two molecular nuclei, interference effects may be observed. The momentum delivered to the nuclei would be different for the electron emitted soon after excitation or with some notable delay. Consequently, the interference effect would be either related or unrelated to a double slit pattern. In this way, Bohr's complementarity principle, resulting from the Einstein-Rosen-Podolsky paradox, may be tested at the quantum level with this molecular-like double slit experiments. Similar tests concerning the debates leading to the foundation of quantum mechanics may also be carried out through kinematically complete experiments of collisions of HD^+ molecules with He atoms [70].

Acknowledgments

Supported by the project Advanced research using high intensity laser produced photons and particles (CZ.02.1.01/0.0/0.0/16_019/0000789) from European Regional Development Fund (ADONIS). RDR, OAF and CAT acknowledge financial support from PICT-2015-3392 (Agencia Nacional de Promoción Científica y Tecnológica and Ministerio de Ciencia, Tecnología e Innovación Productiva, Argentina) and PIP 0784 CONICET (Consejo Nacional de Investigaciones Científicas y Técnicas, Argentina). CAT acknowledges support from PICT-2015-3341 (Agencia Nacional de Promoción Científica y Tecnológica and Ministerio de Ciencia, Tecnología e Innovación Productiva, Argentina).

References

[1] T. Young, *A course of Lectures on Natural Philosophy and the Mechanical Arts* (London: J. Johnson, 1807).

[2] G.N. Lewis, The conservation of photons, *Nature* **118**, 874-875 (1926).

[3] C. Jönsson, Elektroneninterferenzen an mehreren künstlich hergestellten Feinspalten, *Z. Phys.* **161**, 454-474 (1961).

[4] A. Tonomura, J. Endo, T. Matsuda, T. Kawasaki and H. Ezawa, Demonstration of single-electron buildup of an interference pattern, *Am. J. Phys.* **57**, 117-120 (1989).
[5] B. Barwick, G. Gronniger, Y. Lu, S.Y. Liou and H. Batelaan, A measurement of electron-wall interactions using transmission diffraction from nanofabricated gratings, *J. Appl. Phys.* **100**, 074322 (2006).
[6] S. Frabboni, G.C. Gazzai and G. Pozzi, Young's double-slit interference experiment with electrons, *Am. J. Phys.* **75**, 1053-1055 (2007).
[7] S. Frabboni, G.C. Gazzai and G. Pozzi, Nanofabrication and the realization of Feynman's two-slit experiment, *Appl. Phys. Lett.* **93**, 073108 (2008).
[8] S. Frabboni, A. Gabrielli, G.C. Gazzadi, F. Giorgi, G. Matteucci, G. Pozzi, N.S. Cesari, M. Villa and A. Zoccoli, The Young-Feynman two-slits experiment with single electrons: Build-up of the interference pattern and arrival-time distribution using a fast-readout pixel detector, *Ultramicroscopy* **116**, 73-76 (2012).
[9] M. Arndt, O. Nairz, J. Vos-Andreae, C. Keller, G. van der Zouw and A. Zeilinger, Wave-particle duality of C_{60} molecules, *Nature* **401**, 680-682 (1999).
[10] W. Becker, Molecular physics: Matter-wave interference made clear, *Nature* **474**, 586-587 (2011).
[11] R. Bach, D. Pope, S. Liou and H. Batelaan, Controlled double-slit electron diffraction, *New J. Phys.* **15**, 033018 (2013).
[12] N. Stolterfoht, B. Sulik, V. Hoffmann, B. Skogvall, J.Y. Chesnel, J. Rangama, F. Frémont, D. Hennecart, A. Cassimi, X. Husson, A.L. Landers, J.A. Tanis, M.E. Galassi, and R.D. Rivarola, Evidence for interference effects in electron emission from H_2 colliding with 60 MeV/u Kr^{34+} oons, *Phys. Rev. Lett.* **87**, 023201 (2001).
[13] D. Rolles, M. Braune, S. Cvejanović, O. Geßner, R. Hentges, S. Korica, B. Langer, T. Lischke, G. Prümper, A. Reinköster, J. Viefhaus, B. Zimmermann, V. McKoy and U. Becker, Isotope-induced partial localization of core electrons in the homonuclear molecule N_2, *Nature* **437**, 711-715 (2005).
[14] G.W.F. Drake (Ed.), *Springer Handbook of Atomic, Molecular and Optical Physics* (Berlin, Springer, 2005).
[15] H.D. Cohen and U. Fano, Interference in the photo-ionization of molecules, *Phys. Rev.* **150**, 30-33 (1966).
[16] J. Ullrich, R. Moshammer, A. Dorn, R. Dörner, L.Ph. H. Schmidt and H. Schmidt-Böcking, Recoil-ion and electron momentum spectroscopy: reaction microscopes, *Rep. Prog. Phys.* **66**, 1463 (2003).
[17] M. Schulz and D.H. Madison, Studies of the few-body problem in atomic break-up processes, *Int. J. Mod. Phys. A* **21**, 3649-3672 (2006).
[18] M.F. Ciappina, O.A. Fojón and R.D. Rivarola, Coherent electron emission from simple molecules by impact of energetic charged particles, *J. Phys. B: At. Mol. Opt. Phys.* **47**, 042001 (2014).
[19] B.H. Bransden and C.J. Joachaim, *Physics of Atoms and Molecules* (UK: Longman Group, 1983).

[20] P.D. Fainstein, V.H. Ponce and R.D. Rivarola, Two centre-effects in ionization by ion impact, *J. Phys. B: At. Mol. Phys.* **24**, 3091-3119 (1991).
[21] N. Stolterfoht, R.D. DuBois and R.D. Rivarola, *Electron Emission in Heavy Ion-Atom Collisions* (Berlin: Springer, 1997).
[22] P.D. Fainstein, V.H. Ponce, and R.D. Rivarola, A theoretical model for ionization in ion-atom collisions: Application for the impact of multicharged projectiles on helium, *J. Phys. B: At. Mol. Opt. Phys.* **21**, 287-299 (1988).
[23] R.D. Rivarola, R.D. Piacentini, A. Salin and Dž. Belkić, The influence of the static potential in high-energy K-Shell electron capture collisions, *J. Phys. B: At. Mol. Opt. Phys.* **13**, 2601-2609 (1980).
[24] S.E. Corchs, R.D. Rivarola and J.H. McGuire, Impact-parameter formulation for electron capture from molecular targets, *Phys. Rev.* A **47**,3937-3944 (1993).
[25] M.E. Galassi and R.D. Rivarola, Multicenter character in single-electron emission from H_2 molecules by ion-impact, *Phys. Rev.* A **70**, 032701 (2004).
[26] D.S.F. Crothers and J.F. McCann, Ionization of atoms by ion impact, *J. Phys. B: At. Mol. Opt. Phys.* **16**, 3229-3242 (1983).
[27] L. Gulyás, P.D. Fainstein and A. Salin, CDW-EIS theory of ionization by ion impact with Hartree-Fock description of the target, *J. Phys. B: At. Mol. Opt. Phys.* **28**, 245-257 (1995).
[28] I.M. Cheshire, Continuum distorted wave approximation; resonant charge transfer by fast protons in atomic hydrogen, *Proc. Phys. Soc.* **84**, 89-98 (1964).
[29] Dž. Belkić, Quantum theory of ionization in fast collisions between ions and atomic systems, *J. Phys. B: At. Mol. Opt. Phys.* **11**, 3529-3552 (1978).
[30] Dž. Belkić, R. Gayet and A. Salin, Electron capture in high-energy ion-atom collisions, *Phys. Rep.* **56**, 279-369 (1979).
[31] D.P. Dewangan and J. Eichler, A first-order Born approximation for charge exchange with Coulomb boundary conditions, *J. Phys. B: At. Mol. Opt. Phys.* **19**, 2939-2944 (1986).
[32] Dž. Belkić, R. Gayet, J. Hanssen and A. Salin, The first Born approximation for charge transfer collision, *J. Phys. B: At. Mol. Opt. Phys.* **19**, 2945-2953 (1986).
[33] Dž. Belkić, I. Mančev and J. Hanssen, Four-body methods for high-energy ion-atom collisions, *Rev.Mod. Phys.* **80**, 249-314 (2008).
[34] Dž. Belkić, *Quantum Theory of High-Energy Ion-Atom Collisions* (London: Taylor & Francis, 2008).
[35] Dž. Belkić, Review of theories on ionization in fast ion-atom collisions with prospects for applications to hadron therapy, *J. Math. Chem.* **47**, 1366-1419 (2010).
[36] R. Gayet, Charge exchange scattering amplitude to first order of a three body expansion, *J. Phys. B: At. Mol. Opt. Phys.* **8**, 483-491 (1972).
[37] H. Goldstein, C.P. Poole and J.L. Safko, *Classical Mechanics*, 3rd edn. (Addison-Wesley, 2001).

[38] Y.D. Wang, J.H. McGuire and R.D. Rivarola, Impact parameter treatment of high-velocity electron capture from diatomic molecules at fixed orientation, *Phys. Rev.* A **40**, 3673-3680 (1989).

[39] M.E. Galassi, R.D. Rivarola, P.D. Fainstein and N. Stolterfoht, Young-type interference patterns in electron emission spectra produced by impact of swift ions on H_2 molecules, *Phys. Rev.* A **66**, 052705 (2002).

[40] A. Messiah, *Quantum Mechanics*, Vol. II, pp. 848-852. (North–Holland, 1970).

[41] N. Stolterfoht *et al.*, Interference effects in electron emission from H_2 by 68-MeV/u Kr^{33+} impact: Dependence on the emission angle, *Phys. Rev.* A **67**, 030702(R) (2003).

[42] S. Hossain, A.S. Alnaser, A.L. Landers, D.J. Pole, H. Knutson, A. Robinson, B. Stramper, N. Stolterfoht and J.A. Tanis, Interference effects in electron emission from H_2 by 3 and 5 MeV H^+ impact, *Nucl. Instrum. Methods Phys. Res.* B **205**, 484-487 (2003).

[43] S. Hossain, A.L. Landers, N. Stolterfoht and J.A. Tanis, Interference phenomena associated with electron-emission from H_2 by (1-5)–MeV H^+ impact, *Phys. Rev.* A **72**, 010701(R) (2005).

[44] L. Nagy, L. Kocbach, K. Porá and J. P. Hansen, Interference effects in the ionization of H_2 by fast charged projectiles, *J. Phys. B: At. Mol. Opt. Phys.* **35**, L453-L459 (2002).

[45] O.A. Fojón, J. Fernández, A. Palacios, R.D. Rivarola and F. Martín, Interference effects in H_2 photoionization at high energies, *J. Phys. B: At. Mol. Opt. Phys.* **37**, 3035-3042 (2004).

[46] D. Misra, U. Khadane, Y.P. Singh, L.C. Tribedi, P.D. Fainstein and P. Richard, Interference effects in electron emission in heavy ion collisions with H_2 detected by comparison with the measured electron spectrum from atomic hydrogen, *Phys. Rev. Lett.* **92**, 153201 (2004).

[47] D. Misra, A. Kelkar, U. Kadhane, A. Kumar, L.C. Tribedi and P.D. Fainstein, Influence of Young-type interferences on the forward-backward asymmetry in electron emission from H_2 in collisions with 80–MeV bare C ions, *Phys. Rev.* A **74**, 060701(R) (2006).

[48] D. Misra, A. Kelkar, U. Kadhane, A. Kumar, Y.P. Singh, L.C. Tribedi and P.D. Fainstein, Angular distribution of low-energy electron emission in collisions of 6-MeV/u bare carbon ions with molecular hydrogen: Two-centre mechanism and interference effect, *Phys. Rev.* A **75**, 052712 (2007).

[49] N. Stolterfoht *et al.*, Evidence for two-centre effects in the electron emission from 25 MeV/u Mo^{40+}+He collisions: Theory and experiment, *Europhys. Lett.* **4**, 899-904 (1987).

[50] G. Laurent, P.D. Fainstein, M.E. Galassi, R.D. Rivarola, L. Adoui and A. Cassimi, Orientation and interference effects in single ionization of H_2 by fast ions, *J. Phys. B: At. Mol. Opt. Phys.* **35**, L495-L501 (2002).

[51] C.A. Tachino, M.E. Galassi, F. Martín and R.D. Rivarola, Coherence in collisionally induced electron emission from diatomic heteronuclear molecules, *J. Phys. B: At. Mol. Opt. Phys.* **42**, 075203 (2009).

[52] C.A. Tachino, M.E. Galassi, F. Martín and R.D. Rivarola, Partial localization in coherent electron emission from asymmetric diatomic molecules, *J. Phys. B: At. Mol. Opt. Phys.* **43**, 135203 (2010).

[53] C.A. Tachino, M.E. Galassi, F. Martí and R.D. Rivarola, Coherence and partial localization in electron emission from asymmetric diatomic molecules, *J. Phys.: Conf. Series* **288**, 012026 (2011).

[54] M.J. Frisch *et al.*, *Gaussian 03, Revision C.02.* Gaussian, Inc., Wallingford, CT, (2004).

[55] K. O-ohata, H. Taketaand S. Huzinaga, Gaussian expansions of atomic orbitals, *J. Phys. Soc. Japan* **21**, 2306-2313 (1966).

[56] W.J. Hehre, R.F. Stewart and J.A. Pople, Self-consistent molecular-orbital methods. I. Use of Gaussian expansions of Slater-type atomic orbitals, *J. Chem. Phys.* **51**, 2657-2664 (1969).

[57] K. Nagy-Póra, L. Czipa and L. Nagy, Two-.centre effects in the ionization of heteronuclear molecules, *Stud. UBB Phys.* **57**, 49-58 (20012).

[58] C.A. Tachino, F. Martín and R. D. Rivarola, Theoretical study of interference effects in single electron ionization of N_2 molecules by proton impact, *J. Phys. B: At. Mol. Opt. Phys.* **45**, 025201 (2012).

[59] J. L. Baran, S. Das, F. Járay-Szábo, L. Nagy and J. Tanis, Interferences in electron emission spectra from 1, 3 and 5 MeV H^{+}+N_2 collisions, *J. Phys.: Conf. Series* **58**, 215-218 (2007).

[60] J. L. Baran, S. Das, F. Járay-Szábo, K. Póra, L. Nagy and J. Tanis, Suppression of primary electron interferences in the ionization of N_2 by 1–5 MeV/u protons, *Phys. Rev.* A **78**, 012710 (2008).

[61] S. Nandi, S. Biswas, C.A. Tachino, R.D. Rivarola and L.C. Tribedi, Double differential electron emission from N_2 under impact of fast C^{6+} ions and Young-type interference, *Eur. Phys. J. D* **69**, 192 (2015).

[62] S. Nandi, A. N. Agnihotri, C.A. Tachino, R.D. Rivarola, F. Martín and L.C. Tribedi, Electron double differential cross sections for ionization of O_2 under fast C^{6+} ion impact and interference oscillation, *Phys. Scr.* **T156**, 014038 (2013).

[63] M. Winkworth, P.D. Fainstein, M.E. Galassi, J. Baran, B. S. Dassanayake, S. Das, A. Kayani and J.A. Tanis, Interferences in electron emission from O_2 by 30 MeV $O^{5,8+}$ impact, *Nucl. Instrum. Methods Phys. Res. B* **267**, 373-376 (2009).

[64] M. Winkworth, P.D. Fainstein, M.E. Galassi, J. Baran, S. Das, B.S. Dassanayake, A. Kayani and J. A. Tanis, Interference effects in electron emission spectra for 3 MeV/u H^{+}+O_2 collisions, *J. Phys.: Conf. Series* **163**, 012044 (2009).

[65] L.Ph.H. Schmidt, S. Schössler, F. Afaneh, M. Schöffler, K.E. Stiebing, H. Schmidt-Böcking and R. Dörner, Young-type interference in collisions between hydrogen molecular ions and helium, *Phys. Rev. Lett.* **101**, 173202 (2008).

[66] S.F. Zhang, D. Fischer, M. Schulz, A.B. Voitkiv, A. Senftleben, A. Dorn, J. Ullrich, X. Ma and R. Moshammer, Two center interferences in dielectronic transitions in H_2^{+}+He collisions, *Phys. Rev. Lett.* **112**, 023201 (2014).

[67] D. Akoury, K. Kreidi, T. Jahnke, Th. Weber, A. Staudte, M. Schöffler, N. Neumann, J. Titze, L.Ph.H. Schmidt, A. Czasch, O. Jagutzki, R.A. Costa Fraga, R.E. Grisenti, R. Díez Muiño, N.A. Cherepkov, S.K. Semenov, P. Ranitović, C.L. Cocke, T. Osipov, H. Adaniya, J.C. Thompson, M.H. Prior, A. Belkacem, A.L. Landers, H. Schmidt-Böcking and R. Dörner, The simplest double slit: Interference and entanglement in double photoionization of H_2, *Science* **318**, 949-952 (2007).

[68] F. Martín, Quantum mechanics: Thought experiments made real, *Nat. Phot.* **9**, 76-77 (2015).

[69] X.-J.Liu, Q. Miao, F. Gel'mukhanov, M. Patanen, O. Travnikova, C. Nicolas, H. Ågren, K. Ueda and C. Miron, Einstein-Bohr recoiling double-slit gedanken experiment performed at the molecular level, *Nat. Phot.* **9**, 120-125 (2015).

[70] L.Ph.H. Schmidt, J. Lower, T. Jahnke, S. Schössler, M.S. Schöffler, A. Menssen, C. Lévêque, N. Sisourat, R. Taïeb, H. Schmidt-Böcking and R. Dörner, Momentum transfer to a free floating double slit: Realization of a thought experiment from the Einstein-Bohr debates, *Phys. Rev. Lett.* **101**, 173202 (2008).

Chapter 5

Wave packet scattering in intermediate-energy p-He collisions

L. Nagy[1*], F. Járai-Szabó[1], S. Borbély[1], T. Arthanayaka[2], B.R. Lamichhane[2], A. Hasan[2] and M. Schulz[2]

[1]*Babeş-Bolyai University, Faculty of Physics str. Kogălniceanu 1, 400084 Cluj, Romania*

[2]*Department of Physics and LAMOR, Missouri University of Science & Technology, Rolla, MO 65409, USA*

In this joint theoretical and experimental work, the fully differential cross sections for the ionization of helium by 75 keV proton projectiles have been determined. We have calculated and measured the cross sections for different coherence widths of the wave packet associated to the projectile. In spite of the fact that the agreement between theory and experiment is not always good, they are consistent in the strong dependence of the cross sections on the coherence widths of the projectile. We have demonstrated that if cross sections differential in the scattering angle or momentum transfer are concerned, the coherence properties of the projectile have to be taken into account carefully.

Contents

*Corresponding author. Email: lnagy@phys.ubbcluj.ro

1. Introduction

Since the foundation of quantum mechanics, it is well known that particles have also a wavelike character [1–3], and that a wave packet may be associated with a particle [4]. This is also the starting assumption in the case of the projectile (e.g. an ion) when quantum scattering theory is developed [5, 6]. However, in practical calculations, it is assumed that the momentum of the projectile is well defined, resulting (according to Heisenberg's uncertainty principle) in a delocalization of the projectile. Generally, this means that a projectile is described by a plane wave, which has infinite coherence length. This model is used in most of the time-independent quantum scattering calculations [6].

Another approach for inelastic scattering is the semi-classical approximation (SCA) [7], where the projectile is assumed to move on a classical trajectory, and quantum mechanics is applied only for the electron system. In the case of fast ion projectiles, this description was considered to be justified, because the de Broglie wavelength of the projectile is much smaller than the atomic dimensions [8]. This method was widely used for theoretical descriptions of ionization and excitation of atoms by fast ions [9–17]. Because the wave properties of the projectile are not taken into account, there is no coherence among different trajectories. It was demonstrated [8] that under the conditions of the validity of the SCA, the two approaches lead to the same result. However, the demonstration was done for total cross sections, or at least to those that are not differential in the projectile scattering angle.

While it was well established that in a scattering experiment, the projectile has to be described as a wave packet, no attention was given to its coherence length (longitudinal or transverse). This neglect seemed to be justified, because both theories (quantum or semi-classical) described correctly the ionization or excitation processes. However, until the year 2000, no experiment was performed to measure fully differential cross sections (FDCS) for ionization by ion impact. This kind of kinematically complete experiments became possible only with the development of the cold target recoil ion momentum spectroscopy (COLTRIMS) [18, 19].

One of the first papers published on a kinematically complete experiment for ionization by fast ions [20] came with a surprise: the FDCS for the ionization of helium by 100 MeV/u C^{6+} ions showed some features that were not consistent with the theory. In the scattering plane, the agreement between theory and experiment was good, as expected. However, in the plane perpendicular to the momentum transfer, huge disagreement occurred. Therein, the theory predicted very small and almost constant cross sections as a function of the angle relative to the projectile velocity. By contrast, the experimental data showed two pronounced maxima in the direction perpendicular to the projectile velocity. This was not anticipated, because it was common knowledge, that for this high energy of the projectile, even the simplest first Born approximation should produce reliable results. Yet this was not the case in a state-of-the-art theoretical model for ion-atom collisions, the continuum distorted wave approach with the Hartree-Fock wave functions for the active electron (CDW-HF), used in Ref. [20] alongside a plane wave for the initial state of the projectile.

Several attempts to explain this disagreement were published. The authors of the original paper [20] suggested that higher-order processes involving the projectile-target nucleus interaction may be responsible for the structures observed in the perpendicular plane. The importance of taking into account the internuclear interactions was emphasized also in other papers [21]. The first Born approximation convoluted with elastic scattering [22] reproduced the structures observed in the perpendicular plane. There were some suggestions [23] that the sophisticated CDW-HF calculations are correct, and the discrepancy is caused only by the low experimental resolution. However, after a careful analysis [24], this possibility was excluded by the experimentalists: they have shown that experimental resolution is only partly responsible for the discrepancy.

Meanwhile, two of the authors of the present chapter (F.J-Sz. and L.N.) have published a calculation for the same ionization process based on an SCA method [25, 26], which reproduced the observed structures in the perpendicular plane. At that time, it was not clear, why the pronounced differences between SCA and full quantum calculations occurred, but the finding was confirmed also by other authors [27].

The explanation of the discrepancy between the experimental data versus quantum and SCA calculations arose with the publication of a study [28], where the coherence width of the projectile could be experimentally manipulated. The authors of Ref. [28] have shown that the cross section differential in the projectile scattering angle for ionization of H_2 by proton impact depends on the coherence width of the wave packet associated with the projectile. The coherence width of the wave packet can be influenced by modifying the distance between the collimating slit of the projectile and the target. It can be expressed as $\Delta d_x = \lambda L/(2a)$, where λ is the de Broglie wavelength of the projectile, L the target-slit distance, while a stands for the width of the slit.

With the idea of the dependence of FDCS on the projectile's coherence width in mind, Wang *et al.* [29] have repeated the experiment for ionization of helium, not with 100 MeV/u C^{6+} projectiles, but with 3 MeV protons. This measurement, performed at the Heidelberg storage ring, was a multi-pass experiment, in contrast to the single-pass experiment for C^{6+} projectiles. Taking into account that the perturbation parameter $\eta = Z_p/v_p \approx 0.09$ is the same (small) for both projectiles, one would expect to obtain similar cross sections (Z_p and $v_p \equiv |\boldsymbol{v}_p|$ are the nuclear charge and velocity of the projectile, respectively). However, the results were different, mainly in the direction perpendicular to the projectile velocity vector $\boldsymbol{v}_p$ and momentum transfer $\boldsymbol{q}$. The full quantum calculations by the continuum distorted wave – eikonal initial state (CDW-EIS) reproduced well the H^+ data. The origin of the differences between the two sets of the experimental data can be explained by the different coherence properties of the projectiles. While the coherence width of the C^{6+} was estimated to be 10^{-3} a_0, for the proton projectiles, much larger coherence width was achieved, around 4 a_0. This large coherence is due to the fact that in the storage ring, the beam was very cold and the collision occurs far from the source. To simplify, the projectiles in the C^{6+} [20] and H^+ [29] experiments can be called incoherent and coherent, respectively. This is consistent with the observation that the SCA calculations [25, 26], where the projectiles are treated incoherently, describe well the C^{6+} experimental data [20]. Likewise, the CDW-EIS calculations, with a coherent plane wave for the incoming projectile, are in good agreement with the H^+ data [29].

This explanation for the discrepancy between full quantum theory and experiment for the C^{6+} was not satisfactory for every member of the scientific community. Thus, Kouzakov *et al.* [30] have sustained the idea that the experimental uncertainties, including those due to a velocity spread of the He gas atoms in a supersonic jet, can be responsible for the observed disagreement. Schulz *et al.* have explained again in a comment [31], that the experimental resolution is not the main contributor to the discrepancies between the measurement and quantum theory. Their explanation relies upon the very small coherence width of the C^{6+} projectile. However, in a reply [32], as well as in a very recent article, Kouzakov *et al.* [33], starting from the wave packet description of the scattering process, have argued that, while the uncertainty in the transverse momentum of the C^{6+} projectile is large, the uncertainty in the transverse component of the velocity is small (≈ 0.04 a.u.), i.e. negligible relative to the incident velocity (58.6 a.u.).

Kouzakov *et al.* [33] claimed that, due to a small velocity spread, the velocity function can be replaced by the exact value, in which case the wave function of the projectile reduces to a plane wave. However, comparison of the spread in the transverse component of the velocity to its longitudinal component is not relevant. This is the case because the scattering t matrix depends on the scalar product between the momentum transfer and projectile velocity, $\boldsymbol{q} \cdot \boldsymbol{v}_p$, where $\boldsymbol{q}$ is almost perpendicular to the velocity $\boldsymbol{v}_p$.

In the last few years more and more theoretical and experimental studies sustain the importance of projectile coherence properties. In a more detailed study, the experimental group from the Missouri University of Science and Technology have again confirmed a pronounced projectile coherence effect in the scattering angle dependence of doubly differential cross sections for ionization of H_2 by proton impact [34]. On the theoretical side, Járai-Szabó and Nagy [35] have performed incoherent and coherent calculations for the ionization of He starting from the same probability amplitude, and obtained good agreement with the incoherent C^{6+} ion [20] and the coherent proton [29] data. Recently, we have completed this study assigning a variable coherence width to the projectile [36]. A very conclusive paper was published by our group for the ionization of He by 75 keV proton impact [37]. In this latter joint experimental and theoretical study the coherence width of the projectile was manipulated by modifying the slit-target distance [28], and the FDCSs were measured for two different coherence widths. The pronounced structures obtained in the coherent over incoherent cross section ratios showed again the strong influence of the coherence properties on the FDCSs. Describing theoretically the projectile as a Gaussian wave packet, similar structures in the cross section ratios were obtained. Another very convincing experiment regarding coherence effects was performed for dissociative capture for proton-H_2 collisions [38]. Here coherent and incoherent cross sections were measured simultaneously in the same experiment under otherwise identical experimental conditions. Therein, any observed difference in these latter two kinds of cross sections cannot be attributed to experimental artifacts and, therefore, would represent a clear signature of coherence. More on the experimental evidence about the coherence effects can be found in a recent review article [39] as well as in a further theoretical support from Ref. [40, 41].

Nevertheless, the debate about this issue still continues. Thus, Feagin and Hargreaves [42] claimed that varying the slit-target distance in the experiment influences the divergence of the beam, and that this should be the source of the different results, not the coherence properties. However, this claim was refuted by the experimentalists [34, 39]. An experimental study on 1 MeV proton-helium collisions [43] obtained data in very good agreement with the quantum theory associating plane wave to the projectile, and did not observe differences in FDCSs depending on the coherence properties. However, the coherence length was too large to observe significant de-coherence effects [41].

In the present study, we analyze in more detail the effects of coherence width of the projectile on FDCSs for ionization of He by 75 keV proton impact. We improve the theory by describing the projectile as a two-dimensional Gaussian wave packet in coordinate space [36] and compare the results with the experimental data on an absolute scale.

2. Theory

2.1. *Impact parameter formulation*

In the impact parameter formulation or the SCA for electronic transitions induced by ion impact, the projectile is described classically, moving on a well defined trajectory. The electron system is governed by quantum mechanics and, according to the Born-Oppenheimer approximation, it moves in the time-dependent potential created by the nuclei (target and projectile). This approximation is justified, if: 1) the associated de Broglie wavelength of the projectile is much smaller than the atomic dimensions, 2) the energy transfer to the target is much smaller than the projectile's kinetic energy and 3) the scattering angle of the projectile is small [8].

The ionization probability, differential in the ejected electron energy E and angle Ω_e for a given impact parameter $\boldsymbol{b}$, is given by:

$$\frac{d^3\sigma}{dE d\Omega_e d\boldsymbol{b}} = |a(\boldsymbol{b}, E, \Omega_e)|^2, \tag{1}$$

where $a(\boldsymbol{b}, E, \Omega_e)$ is the transition probability amplitude. The impact parameter cannot be measured experimentally, but it may be linked to the measurable projectile scattering angle, assuming classical scattering. In previous SCA studies, usually a screened Coulomb scattering was considered [9, 10, 14]. In our calculation we treat separately the interaction of the projectile with the ejected electron and with the residual ion. The perpendicular (to the projectile's initial momentum) component of the momentum transfer may be obtained from the transverse momentum balance as the sum of the momentum transfer to the target ($\boldsymbol{p}_{T\perp}$) and to the ejected electron ($\boldsymbol{p}_{e\perp}$) as:

$$\boldsymbol{q}_{\perp} = \boldsymbol{p}_{T\perp} + \boldsymbol{p}_{e\perp}. \tag{2}$$

We assume that the impact parameter influences mainly the momentum transfer to the residual ion, and we calculate the projectile scattering angle in the center of mass frame due to the collision with the residual ion from the integral [8]:

$$\theta_{pTc} = \pi - 2b_0 \int_{r_0}^{\infty} \frac{dr}{r^2} \left(1 - \frac{U(r)}{E_c} - \frac{b_0^2}{r^2}\right)^{-\frac{1}{2}}. \tag{3}$$

Here, $U(r)$ is the potential created by the residual ion, E_c the projectile's energy (calculated with the reduced mass), b_0 the impact parameter corresponding to θ_{pTc}, while r_0 stands for the closest approach between the projectile and the nucleus. Passing to the laboratory frame, the relationships between the scattering angles and the transverse momentum transfers are:

$$\theta_{pT} = \frac{q_{T\perp}}{p_0}; \qquad \theta_p = \frac{q_{\perp}}{p_0}, \tag{4}$$

where p_0 is the initial momentum of the projectile. Starting from the formula (1), expressing the element of the solid angle around the direction into which the projectile is scattered by:

$$d\Omega_p = \sin\theta_p d\theta_p d\varphi_p, \tag{5}$$

and taking into account the relation

$$d\boldsymbol{b} = b_0 db d\varphi_p, \tag{6}$$

one obtains the ionization FDCS via

$$\frac{d^3\sigma}{dE d\Omega_e d\Omega_p} = b_0 |a(\boldsymbol{b}_0, E, \Omega_e)|^2 \frac{1}{\sin\theta_p} \left|\frac{db}{d\theta_p}\right|. \tag{7}$$

Alternatively, instead of the projectile's scattering angle, the FDCS can be expressed also as a function of the transverse momentum transfer $d\boldsymbol{q}_{\perp}$

$$\frac{d^3\sigma}{dE d\Omega_e d\boldsymbol{q}_{\perp}} = b_0 |a(\boldsymbol{b}_0, E, \Omega_e)|^2 \left|\frac{db}{d\boldsymbol{q}_{\perp}}\right|. \tag{8}$$

Because in the framework of the SCA for a given recoil ion and electron momentum, the projectile has a well defined classical trajectory, in the presented model, the coherence width of the projectile may be considered as being equal to zero.

2.2. *Plane wave description*

In quantum scattering theory, the cross section may be obtained from the scattering amplitude: $R(\boldsymbol{q}_\perp, E, \Omega_e)$ [6, 44]

$$\frac{d^3\sigma}{dEd\Omega_e d\boldsymbol{q}_\perp} = p_0|R(q_\perp, E, \Omega_e)|^2. \tag{9}$$

In the derivation of this formula [6], it is assumed that the wave packet associated with the projectile in momentum space is sharply peaked around $\boldsymbol{p}_0$, meaning that it is approximated by a plane wave, or at least a wave with very large spatial dimensions. In this model, the coherence width of the projectile is assumed to be infinite. The scattering amplitude, under the conditions of the validity of the SCA from the previous section, may be obtained as an inverse Fourier transform of the probability amplitude [8, 44]:

$$R(\boldsymbol{q}_\perp, E, \Omega_e) = \frac{1}{2\pi}\int d^2\boldsymbol{b} e^{i\boldsymbol{b}\cdot\boldsymbol{q}_\perp} b^{2i\frac{Z_P Z_T}{v_P}} a(\boldsymbol{b}, E, \Omega_e), \tag{10}$$

where $b^{2iZ_P Z_T/v_P}$ is an eikonal factor accounting for the interaction of the projectile and Z_T is the target nuclear charge. Obviously, if one is interested in the FDCS as a function of momentum transfer, this can be calculated in a more direct way from the quantum scattering formula (9) than from the SCA (8). The SCA provides the transition probability as a function of the impact parameter, such that the relationship between the impact parameter and scattering angle (or momentum transfer) has to be determined.

It should be noted that Eq. (10) represents a fully quantum-mechanical approach. The use of an impact parameter in calculations is not per se inconsistent with quantum-mechanics. Only the notion that a well-defined impact parameter can be determined from the scattering angle (i.e. the notion of well-defined trajectories) is a classical concept. However, Eq. (10) integrates over a broad spectrum of impact parameters contributing to a given scattering angle and it thus fully acknowledges the quantum-mechanical uncertainty inherent in the impact parameter.

It was demonstrated [8, 44] for cross sections which are not differential in the momentum transfer (or the projectile scattering angle), that under the conditions of validity of the SCA (listed in the previous subsection), the two methods lead to the same result. However, it is clear that the two methods (SCA and plane wave) operate with totally different coherence properties of the projectile. A more general treatment of the collision process, allowing to cover all scenarios between the extremes of zero (SCA, Eq. (8)) and infinite coherence length (plane wave, Eqs. (9) and (10)), is provided by a wave packet description.

2.3. *Wave packet description*

In the quantum theory of scattering, one associates with a projectile a wave packet which is usually described in the momentum space $\phi(\boldsymbol{p}_p)$. The momentum space representation is useful in the time-independent treatment of scattering, because the integration over the impact parameter may be performed analytically [6]. This representation was used also in a recent article treating the topic of projectile coherence [33]. However, at the end of these calculations, $\phi(\boldsymbol{p}_p)$ was approximated by a Dirac-delta function and the projectile wave packet was replaced by a plane wave. In such an approach, an incoherent projectile beam was ruled out right from the onset.

Instead of these traditional treatments, we describe the wave packet associated with the projectile in the coordinate space. We assume that for a certain perpendicular momentum transfer $\boldsymbol{q}_\perp$, corresponding (by classical scattering) to the impact parameter $\boldsymbol{b}_0$, the probability amplitude $a(\boldsymbol{b}, E, \Omega_e)$ in the transformation (10) has to be weighted by the spacial extension of the wave packet $\Psi(\boldsymbol{b}-\boldsymbol{b}_0)$. In a previous work [37], we considered the wave packet with a Gaussian profile, depending only on the value of the impact parameter. In these conditions the transformation (10) becomes:

$$R(\boldsymbol{q}_\perp, E, \Omega_e) = \frac{N_1}{2\pi}\int d^2\boldsymbol{b}e^{i\boldsymbol{b}\cdot\boldsymbol{q}_\perp}b^{2i\frac{Z_P Z_T}{v_P}}a(\boldsymbol{b}, E, \Omega_e)e^{-\frac{(b-b_0)^2}{2\sigma_b^2}}. \quad (11)$$

Here, N_1 is a normalization constant and σ_b is the standard deviation of the Gaussian. The coherence width of the projectile is considered to be the full width at half maximum (FWHM) of the Gaussian $\Delta b = 2.355\ \sigma_b$.

This description had the shortcoming that took into account the distribution only in the radial direction, resulting in a ring-shaped wave packet. In order to improve such a description, we presently consider a Gaussian distribution in two dimensions [36]:

$$R(\boldsymbol{q}_\perp, E, \Omega_e) = \frac{N_2}{2\pi}\int d^2\boldsymbol{b}e^{i\boldsymbol{b}\cdot\boldsymbol{q}_\perp}b^{2i\frac{Z_P Z_T}{v_P}}a(\boldsymbol{b}, E, \Omega_e)e^{-\frac{(b_x-b_{0x})^2}{2\sigma_x^2}-\frac{(b_y-b_{0y})^2}{2\sigma_y^2}}, \quad (12)$$

where N_2 is a normalization constant. Here, $\{b_x, b_y\}$ and $\{b_{0x}, b_{0y}\}$ are the components of $\boldsymbol{b}$ or $\boldsymbol{b}_0$, respectively, while σ_x and σ_y stand for the standard deviations in the corresponding directions. In this model, we may vary separately the coherence width of the wave packet in the x (parallel to $\boldsymbol{q}_\perp$) as well as y (perpendicular to $\boldsymbol{q}_\perp$ and $\boldsymbol{p}_0$) directions.

2.4. *The calculation of the transition amplitude*

All the above methods are based on the calculation of the impact-parameter-dependent transition amplitude $a(\boldsymbol{b}, E, \Omega_e)$. In some of our previous works [25, 35], we have applied the independent electron approximation and the first-order perturbation theory. In the present case, because of the relatively low energy of the proton projectile (75 keV), this treatment does not provide the results of the desired accuracy. Instead of these approximations, we have performed *ab initio* calculations based on the numerical solution of the time-dependent Schrödinger-equation (TDSE) for the two-electron system, which fully accounts for the electron-electron correlations and for the projectile-electron post-collision interactions. However, since the set of the basis functions does not contain any projectile states, capture is not taken into accounted.

The time-dependent Schrödinger equation for the two target electrons, which are moving in the combined Coulomb field of the target core and the projectile, is solved numerically using the time-dependent close-coupling (TDCC) method [45]. In our implementation [17], the fully correlated two-electron wave function is represented in the basis of the symmetrized coupled spherical harmonics:

$$\Psi(\boldsymbol{r}_1, \boldsymbol{r}_2, t) = \sum_{l_1 l_2 LM} \frac{R_{l_1 l_2}^{LM}(r_1, r_2, t)}{r_1 r_2} \Upsilon_{l_1 l_2}^{LM}(\Omega_1, \Omega_2), \tag{13}$$

where we have exploited the planar reflection symmetry of the studied system relative to the collision plane.

The radial wave functions $R_{l_1 l_2}^{LM}(r_1, r_2, t)$ are represented by the finite element discrete variable representation (FEDVR) method [46, 47], where each radial coordinate is divided into segments (finite elements) with variable length. Inside each segment, the wave function is represented by a local polynomial basis built on top of a Gauss-Lobatto quadrature ensuring the continuity of the wave function at the finite element boundaries. The wave function is propagated in time using the short iterative Lanczos method [48, 49], where in each time step, the time evolution operator is calculated in the Krylov subspace generated by the repeated action of the Hamiltonian on the wave function $\Psi(t)$.

The accuracy of the propagated wave function $\Psi(\boldsymbol{r}_1, \boldsymbol{r}_2, t)$ is ensured by performing rigorous numerical convergence tests. The convergence as a function of the radial grid size and density, and as a function of the angular basis size is explicitly verified. The target relative error for each Lanczos propagation step is kept below 10^{-20} with the help of the automatic time-step-size control, while the size of the Krylov subspace was limited to 15 in order to ensure the orthogonality of the generated basis vectors.

At each impact parameter the $a(\boldsymbol{b}, E, \Omega_e)$, ionization probability amplitude is calculated directly from the time-dependent wave function

$\Psi(\boldsymbol{b}, \boldsymbol{r}_1, \boldsymbol{r}_2)$ by its projecting onto single continuum eigenstates. Since the fully correlated three-body continuum eigenstates are not known, we use the following uncorrelated ansatz:

$$\Psi^{SC}(\boldsymbol{r}_1, \boldsymbol{r}_2, E) = \frac{1}{\sqrt{2}}\left[\Psi^G(\boldsymbol{r}_1)\Psi^C(\boldsymbol{r}_2, E) + \Psi^G(\boldsymbol{r}_2)\Psi^C(\boldsymbol{r}_1, E)\right], \quad (14)$$

where $\Psi^G(\boldsymbol{r})$ and $\Psi^C(\boldsymbol{r}, E)$ are the respective ground and the continuum eigenstate of He^+.

Consequently, the errors contained in the $a(\boldsymbol{b}, E, \Omega_e)$ transition amplitudes are accumulated during the projection of the accurate time-dependent wave function $\Psi(\boldsymbol{r}_1, \boldsymbol{r}_2, t)$ onto the approximate continuum states. Firstly, these continuum states do not account for the electron-electron and electron-projectile interactions. This introduces a projection error. Such an error can be significantly reduced by propagating our collision system until the interparticle distances become large, where the interparticle interactions are weak. Secondly, the employed single continuum states are not orthogonal to the exit-channel states for describing electron capture by the projectile. This leads to contamination of the ionization transition amplitudes by the contribution from electron capture by the projectile. Unfortunately, this contamination is not reduced by the increase of the interparticle distances (i.e. with the increase of the time-propagation length). Nevertheless, the magnitude of this contamination can be estimated as follows.

In the differential ionization cross section (differential in electron momentum), the magnitude of the said contamination can be estimated from the experimental total electron capture cross sections [50]. Within this contamination, the electron momentum distribution closely resembles the Compton profile of the hydrogenic $1s$ electron centered on the projectile velocity. Based upon this approach, at any electron momentum value, the contamination from electron capture can be calculated and compared to the ionization cross section. Under the conditions of the present paper (75 keV projectile energy and 0.63 a.u. continuum electron velocity), the error introduced by the contamination from electron capture varies between 1 % (for electron emission in the direction opposite to the outgoing projectile) and 5 % (for electron emission in the direction of the outgoing projectile). This error is significantly lower than the observed discrepancies between the theory and experiments.

3. Experimental set-up

The experiment was performed at the ion accelerator of Missouri University of Science and Technology. A proton beam with an energy spread of much less than 1 eV was produced using a hot cathode source and accelerated to an energy of 75 keV. The beam was collimated with a pair of slits, each with a width of 150 μm, placed at varying distance from the target.

Data were obtained for slit distances of $L_1 = 50$ cm, corresponding to a transverse coherence length $\Delta b_x \approx 3.5$ a.u. and $L_2 = 6.5$ cm. In the case of a smaller slit, the distance Δb_x is not determined by the slit itself, but rather by the exit aperture at the accelerator terminal. Here, the transverse coherence length is $\Delta b_x \approx 1.0$ a.u.

The collimated proton beam was crossed with a very cold helium beam (T ≈1-2 K) from a supersonic gas jet. The recoil ions produced in the collision were extracted by a weak electric field of about 8 V/cm. These ions traversed a field-free region about twice as long as the extraction region, and were guided onto a two-dimensional position sensitive detector. The scattered projectiles were charge-analyzed by a switching magnet and decelerated to an energy of 5 keV.

Finally, the protons were energy-analyzed using an electrostatic parallel-plate analyzer [51] which was set for a pass energy corresponding to a projectile energy loss of $\Delta E = 30$ eV. Protons with this energy loss were detected by another two-dimensional position-sensitive detector.

The analyzer slits were very narrow (75 μm) in the vertical direction (y-direction), but long in the horizontal direction (x-direction) so that only projectiles scattered in the x-direction were recorded. Both detectors were set to operate in coincidence.

From the time-of-flight of the recoil ions from the collision region to the detector, which is contained in the coincidence time, the x-component of the recoil-ion momentum was determined. The y- and z-components (defined by the projectile beam direction) were calculated from the position information.

The x-component of the momentum transfer $\boldsymbol{q}$ from the projectile to the target atom was obtained from the projectile position information, while that due to the analyzer slit geometry q_y was kept fixed at zero. The z-component of $\boldsymbol{q}$ was determined by ΔE as $q_z = \Delta E / v_p$. The ejected electron momentum was obtained using momentum conservation law as $\boldsymbol{p}_e = \boldsymbol{q} - \boldsymbol{p}_T$.

4. Results and discussion

In order to study the projectile coherence effects in ion-atom collisions, we have measured and calculated the FDCSs for ionization of helium by 75 keV proton impact for two different coherence widths in the x direction of the wave packet associated with the projectile. In the y direction, the coherence width was constant in the experiment, $\Delta b_y = 3.5\ a_0$. For the representation of the data, we keep the x component of the recoil ion momentum (p_{Tx}) at a constant value. The energy transfer to the ejected electron is fixed at 30 eV. We note here, that because in our calculations we link the impact parameter $\boldsymbol{b}_0$ to $\boldsymbol{p}_{T\perp}$, for a given p_{Tx}, the angle and the value of the impact parameter varies with the variation of the electron ejection angle.

Figures 1-3 represent the FDCSs for ionization of helium for three different values of p_{Tx} for various angles θ_e of the ejected electron (relative to the projectile initial trajectory in the z axis) as a function of the azimuthal ejection angles φ_e. The experimental data are shown for a smaller ($\Delta b_x = a_0$, called incoherent) and a larger ($\Delta b_x = 3.5\ a_0$, called coherent) coherence width of the projectile wave packet in the x direction. The theoretical curves are obtained using the expression (12) for these two coherence widths (keeping $\Delta b_y = 3.5\ a_0$) and carrying a fully coherent calculation based on (10). Both the experimental and the theoretical results are given on an absolute scale.

Figure 1 shows the results in the case of the recoil ion momentum $p_{Tx} = 0.2$ a.u. for three different polar angles of the ejected electron, 25°, 45° and 65°. Data for the φ_e angles close to 180° are missing. This occurs because for the small value $p_{Tx} = 0.2$ a.u. of the recoil ion momentum $p_{e\perp} > p_{Tx}$, and according to Eq. (2) projected onto the x direction, $q_\perp$ would be negative for the corresponding angle. Since, by our definition, the x axis is pointing in the direction of $\boldsymbol{q}_\perp$, instead of changing the coordinate system for the φ_e angles close to 180°, we have omitted these data. For the two smaller polar angles, there is not much difference between the coherent and incoherent experimental data. This is reproduced by the theory, in spite of the fact that with the absolute scale for 45°, the calculated value underestimates the experiment by a factor of 2. At $\theta_e = 65°$, the coherent experiment shows a minimum for $\varphi_e = 0$, which is not reproduced by the theory. Herein, however, the cross section decreases with the increasing coherence width, in some accordance with the experimental observation around $\varphi_e = 0$.

For a larger recoil ion momenta, $p_{Tx} = 0.7$ a.u. (Fig. 2) and $p_{Tx} = 1.25$ a.u. (Fig. 3), the calculated cross sections are more sensitive to the

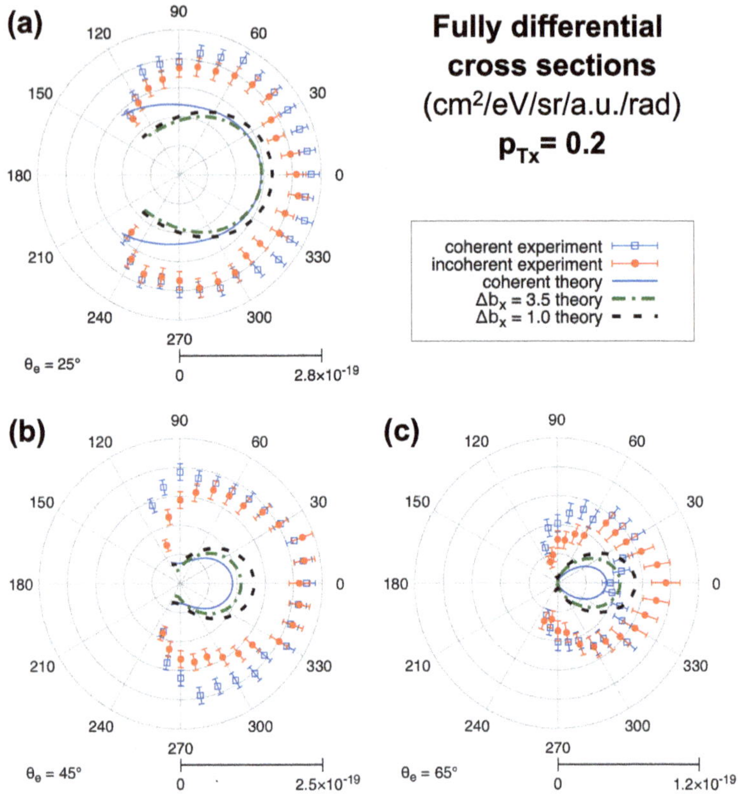

Fig. 1. Fully differential cross sections for the ionization of helium by 75 keV proton impact as a function of the azimuthal angle of the ejected electron. The projectile energy loss is fixed at 30 eV. The x component or recoil ion momentum (p_{Tx}) is fixed at 0.2 a.u., whereas the polar angles of the ejected electron are fixed at 25° (a), 45° (b) and 65° (c). The theoretical results for different coherence widths of the projectile are compared to the experimental data.

coherence properties of the projectile than are the experimental data. In the case of $\theta_e = 25°$, as one may observe on Figure 2a, the theory for the $\Delta b_x = 1\ a_0$ and the fully coherent calculation overestimates the measured data, while the theory for $\Delta b_x = 3.5\ a_0$ is in good agreement with the experiment. On the other hand, for $\theta_e = 45°$, a strange behavior is observed: while for $\varphi_e = 180°$, the theory and experiment are in good agreement, for $\varphi_e = 0$ (in the direction of the binary peak), the theoretical results are much

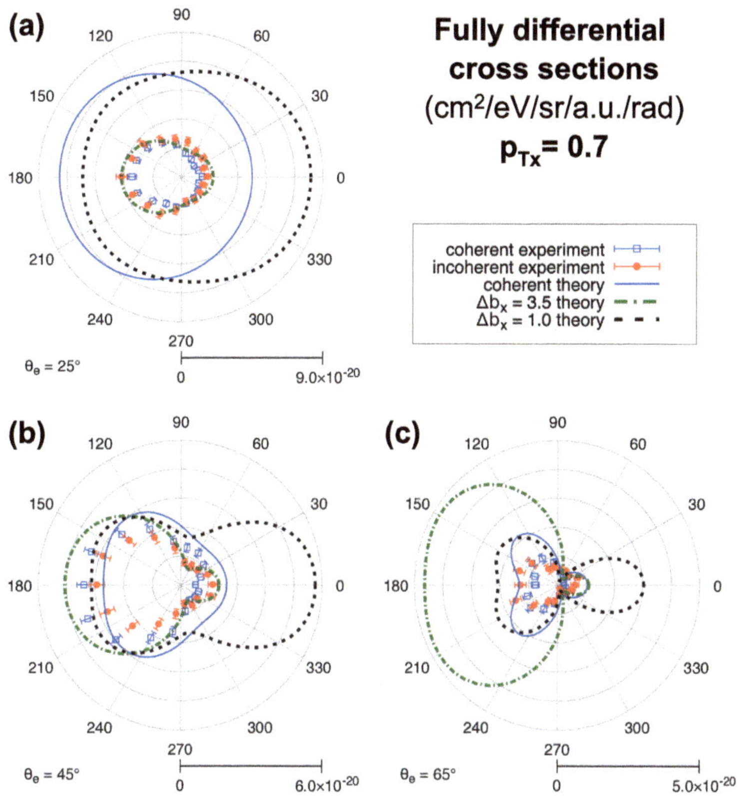

Fig. 2. Same as Figure 1, but for $p_{Tx} = 0.7$ a.u.

larger than the experimental data. Qualitatively, however, the theoretically obtained dependence on the coherence (larger cross sections for smaller coherence width) is correct.

Figure 2b compares the experimental results for $\Delta b_x = 3.5$ with the fully coherent theoretical calculations, and close agreement is seen. One would then think that a wave packet of width equal to 3.5 a.u. scattered on helium could be considered as coherent. However, the coherence width of the wave packet should not be compared to the atomic dimensions, but rather to the impact parameter interval which gives a significant contribution.

On Fig. 3, we have presented the FDCS as a function of the azimuthal angle of the ejected electron for a large recoil ion momentum ($p_{Tx} = 1.25$). For this momentum too we observe pronounced differences in the results

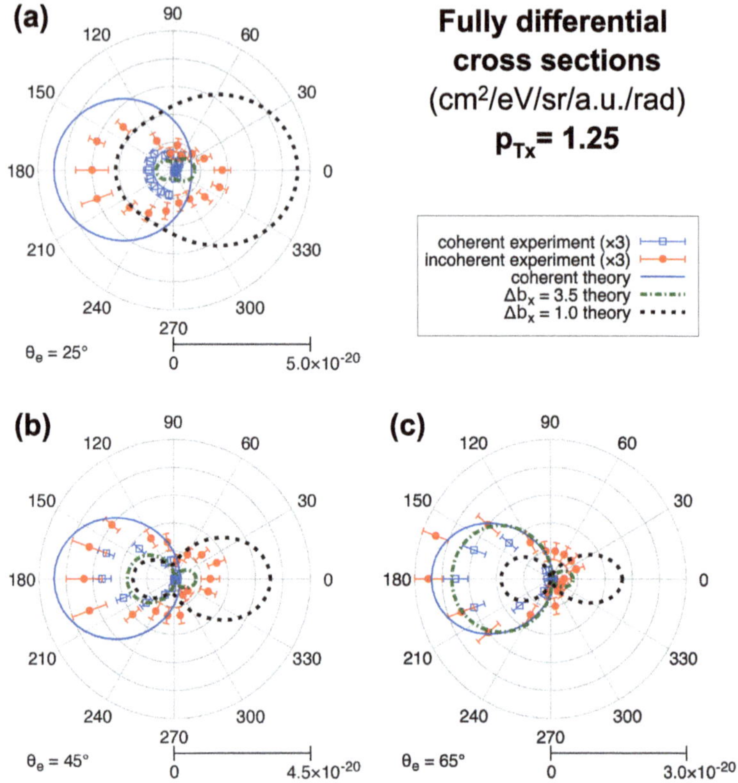

Fig. 3. Same as Figure 1, but for p_{Tx} = 1.25 a.u.

(both experimental and theoretical) obtained with different coherence width of the projectile wave packet. Typically, the theoretical cross sections are larger than the experimental ones, but they show the same tendency in the dependence on the coherence. Again, the theoretical results show that the coherence widths of the order $\Delta b_x = 3.5\ a_0$ are far from a fully coherent (plane wave) description.

Figure 4 shows how the cross section varies if we increase the coherence width (in both the x and y directions) from 3.5 to 20 a.u. One may observe, that the variation of the cross section is dramatic, and full coherence is not reached even at width as wide as 20 a.u. This observation shows that, in fact, the experimental coherence width of 3.5 a.u. cannot be considered as

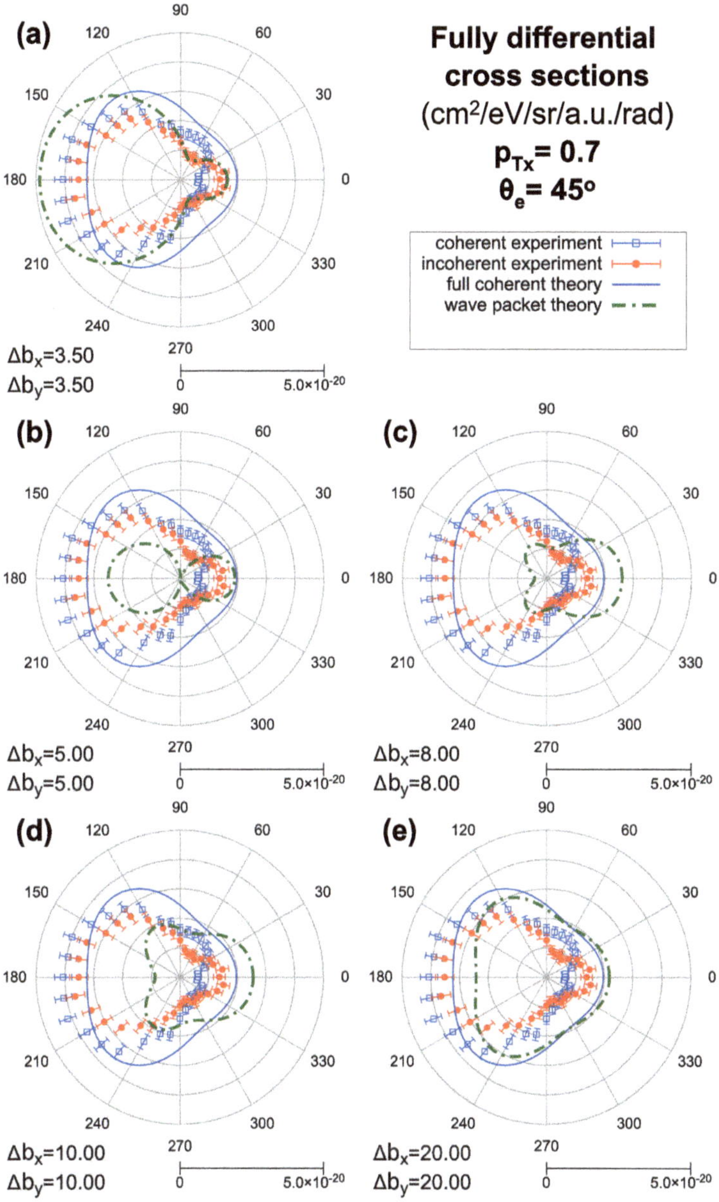

Fig. 4. Same as Figure 1 for $p_{Tx} = 0.7$ a.u. and $\theta_e = 45°$, showing how theoretical results change by increasing the coherence width of the projectile from 3.5 a_0 to 20 a_0.

fully coherent. Figure 2c (for ejection angle $\theta = 65°$) also illustrates a strong dependence of the FDCS on the coherence width of the projectile. The results for $\Delta b_x = 1\ a_0$, $\Delta b_x = 3.5\ a_0$ and with the fully coherent calculations show very different behavior. In this case too the theoretical calculations overestimate the experimental data. Overall, Fig. 4 for $p_{Tx} = 0.7$ a.u. and $\theta_e = 45°$ illustrate how the theoretical results change by increasing the coherence width of the projectile from 3.5 a_0 to 20 a_0.

In order to emphasize the effect of the coherence width on the cross sections, we display in Fig. 5 the cross section ratios for the larger projectile coherence width (3.5 a.u.) over its smaller counterpart (1 a.u.). Often these are labeled as coherent over incoherent cross section ratios, but this is not correct due to the huge differences between the fully coherent results and those with $\Delta b_x = \Delta b_y = 3.5$. Generally, we can state that agreement

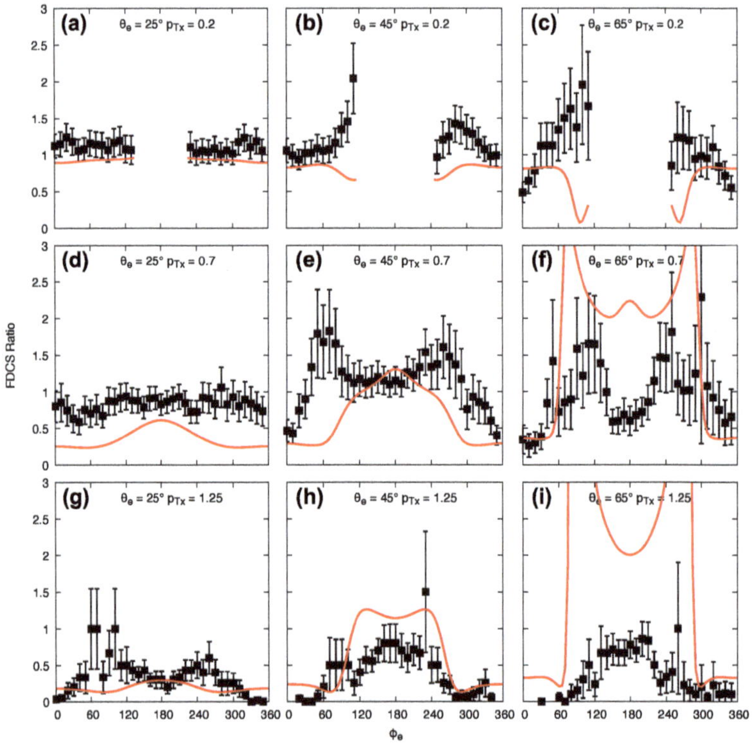

Fig. 5. Coherent ($\Delta b_x = 3.5$) over incoherent ($\Delta b_x = 1$) experimental and theoretical cross section ratios for different p_{Tx} and θ_e values.

of theory with experiment for the ratios is better than for the corresponding absolute cross sections. While good agreement is observed for several combinations of the p_{Tx} components of recoil ion momenta and for the θ_e ejection angles, in some cases the experimental and theoretical ratios have different behavior. For example, for p_{Tx} = 0.2 and θ_e = 65° (Fig. 5c), the theoretical ratio exhibits minima which are absent from the experimental data. Further, the experimental ratios show two maxima for p_{Tx} = 0.7 and θ_e = 45° (Fig. 5e) but these structures are not reproduced by the theory. Moreover, for θ_e = 65° with p_{Tx} = 0.7 (Fig. 5f) or p_{Tx} = 1.25 (Fig. 5i), the theoretical maxima are too large relative to their experimental counterparts.

A possible explanation for these discrepancies is that in the theoretical calculations we did not take into account the experimental resolution. This effect could be important mainly for the ejection angles, which are measured indirectly and, consequently, may lead to larger errors. Looking at Figs. 5e and 5f, the discrepancy can be explained, if the experimental resolution for the θ_e angle is larger than 20°. Thus, the large maxima at θ_e = 65° are spread out by way of a low resolution. These maxima are also experimentally observed at θ_e = 45°, while they are missing in the theory. However, the experimental uncertainties do not wash out the differences between the coherent and incoherent calculations, because the observed structures are much larger than the resolution. In a recent article [36], for 100 MeV/amu C^{6+} projectiles, we have convoluted our theoretical FDCSs with the experimental uncertainties. We have shown that this procedure modified the results. However, for realistic resolutions, the difference between the coherent and incoherent calculations still persists. Another possible source of error is the uncertainty of the experimental coherence lengths. Since the theoretical FDCSs are very sensitive to the coherence length, even a small inaccuracy in the experimentally determined value could have a large effect on the FDCS.

As a general observation, for the low recoil ion momentum p_{Tx} = 0.2, the coherence width of the projectile has less effect on the cross sections (the ratio is around 1). On the other hand, for the larger recoil ion momenta, the effect is more pronounced. A minimum is observed always for ϕ_e = 0, so in this direction the coherent cross section is much smaller than the incoherent one.

5. Conclusions

We have presented three methods for calculating the FDCS for ionization of atoms by fast ion impact. In the impact parameter formulation (or SCA), the projectile is treated classically and, therefore, its associated coherence width is zero. Applying an inverse Fourier transform (10) to the impact parameter-dependent probability amplitude, one obtains the scattering amplitude, in agreement with the full quantum description of the scattering process. In this description, a plane wave with infinite coherence width is associated with the projectile. In order to obtain a more realistic wave packet description of the projectile, in the Fourier transform, the transition probability amplitude has been convoluted by a two-dimensional Gaussian with varying width (12). In this way, we could consider wave packets associated with the projectile having variable transverse coherence lengths.

The FDCSs for ionization of helium by 75 keV proton impact have been experimentally determined with two different slit-target distances, resulting in two different coherence widths of the projectile. Pronounced differences between the two sets of measurements have been observed, mainly for higher values of the recoil ion momenta. The theoretical calculations with variable coherence widths confirm this observation. Specifically, the calculated FDCSs are very sensitive to the coherence width of the projectile. For some combinations of the recoil ion momenta and ejection polar angles of the electron, agreement between theory and experiment is good, even on an absolute scale. However, there are other combinations for which significant differences exist between experiment and theory. Agreement between experiment and theory is much better for the ratio between cross sections obtained with the larger and the smaller coherence widths of the projectile. There may be three reasons for the discrepancy: 1) the assignment of a scattering angle to a certain impact parameter by classical scattering for inelastic collisions is not unique; 2) we did not take into account in our calculations the experimental resolution for the ejection angles of the electrons, which may modify the obtained results; 3) the experimental coherence lengths have some uncertainty and the FDCSs are very sensitive to the coherence length.

In the present chapter, we have demonstrated, both theoretically and experimentally, that the FDCSs for ionization of atoms by ion impact are sensitive to the coherence width of the projectile. The projectile paths for different impact parameters add up coherently, and the interference patterns depend on the dimensions of the wave packet. The infinite coherence width description of the projectile is no longer valid, whenever we are concerned with FDCSs for fast ion projectiles.

Acknowledgments

This work was supported by a Grant of the Romanian CNCS UEFISCDI, project No. PN-II-ID-PCE-2011-3-0192. The *ab initio* transition amplitudes were calculated using Babeş-Bolyai University's high performance computing resources. M.S. acknowledges support from the National Science Foundation under grant nos PHY-1401586 and PHY-1703109.

References

[1] L. De Broglie, Recherches sur la théorie des quanta, *Ann. Phys. (Paris).* **10**, 22–128 (1925).

[2] E. Schrödinger, Quantisierung als eigenwertproblem, *Ann. Phys. (Berlin).* **384**, 361–376 (1926).

[3] E. Schrödinger, An undulatory theory of the mechanics of atoms and molecules, *Phys. Rev.* **28**, 1049–1070 (1926).

[4] E. Schrödinger, Der stetige übergang von der mikro- zur makromechanik, *Naturwissenschaften.* **14**, 664–666 (1926).

[5] M. Born, Zur quantenmechanik der stoßvorgänge, *Z. Phys.* **37**, 863–867 (1926).

[6] J. Taylor, *Scattering theory.* Dover Publications Inc. (2000).

[7] J. Bang and J. Hansteen, *Mat. Fys. Medd. Dan. Vid. Selsk.* **31**, 13 (1959).

[8] M. McDowell and J. Coleman, *Introduction to the theory of ion-atom collisions.* North-Holland Publishing Company, Amsterdam-London (1970).

[9] P. A. Amundsen and L. Kocbach, Electronic relativistic effects in the K-shell Coulomb ionization of heavy atoms by massive charged particles, *J. Phys. B.* **8**, L122–L126 (1975).

[10] P. A. Amundsen, Electronic relativistic effects and Coulomb deflection of the projectile in L-shell ionization of atoms by light ions, *J. Phys. B.* **10**, 1097–1112 (1977).

[11] J. F. Chemin, J. Roturier, B. Saboya, Q. T. Dien, and J. P. Thibaud, Dependence of the K-shell ionization probability on small impact parameters with 1- and 2-MeV protons, *Phys. Rev. A.* **11**, 549–555 (1975).

[12] J. F. Reading and A. L. Ford, The forced impulse method applied to the double ionisation of helium by collision with high-energy protons, antiprotons and alpha particles, *J. Phys. B.* **20**, 3747–3769 (1987).

[13] J. F. Reading and A. L. Ford, Double ionization of helium by protons and antiprotons in the energy range 0.30 to 40 MeV, *Phys. Rev. Lett.* **58**, 543–546 (1987).

[14] J. F. Reading, A. L. Ford, and X. Fang, Projectile angular dependence of the ratio of double to single ionization produced by fast protons in collision with helium, *Phys. Rev. Lett.* **62**, 245–248 (1989).

[15] L. Nagy, J. Wang, J. C. Straton, and J. McGuire, Ionization-excitation of helium by fast charged particles, *Phys. Rev. A.* **52**, R902–R904 (1995).
[16] L. Nagy and Á. Benedek, An improved calculation for the ionization-excitation of helium, *J. Phys. B.* **35**, 491–499 (2002).
[17] S. Borbély, J. Feist, K. Tőkési, S. Nagele, L. Nagy, and J. Burgdörfer, Ionization of helium by slow antiproton impact: Total and differential cross sections, *Phys. Rev. A.* **90**, 052706 (2014).
[18] R. Moshammer, J. Ullrich, M. Unverzagt, W. Schmidt, P. Jardin, R. E. Olson, R. Mann, R. Dörner, V. Mergel, U. Buck, and H. Schmidt-Böcking, Low-energy electrons and their dynamical correlation with recoil ions for single ionization of helium by fast, heavy-ion impact, *Phys. Rev. Lett.* **73**, 3371–3374 (1994).
[19] J. Ullrich, R. Moshammer, A. Dorn, R. Dörner, L. P. H. Schmidt, and H. Schmidt-Böcking, Recoil-ion and electron momentum spectroscopy: reaction-microscopes, *Rep. Prog. Phys.* **66**, 1463–1545 (2003).
[20] M. Schulz, R. Moshammer, D. Fischer, H. Kollmus, D. H. Madison, S. Jones, and J. Ullrich, Three-dimensional imaging of atomic four-body processes, *Nature (London).* **422**, 48–50 (2003).
[21] M. F. Ciappina and W. R. Cravero, Fully differential cross sections for C^{6+} single ionization of helium: the role of nucleus-nucleus interaction, *J. Phys. B.* **39**, 2183–2194 (2006).
[22] M. Schulz, M. Dürr, B. Najjari, R. Moshammer, and J. Ullrich, Reconciliation of measured fully differential single ionization data with the first Born approximation convoluted with elastic scattering, *Phys. Rev. A.* **76**, 032712 (2007).
[23] J. Fiol, S. Otranto, and R. E. Olson, Critical comparison between theory and experiment for C^{6+}+He fully differential ionization cross sections, *J. Phys. B.* **39**, L285–L290 (2006).
[24] M. Dürr, B. Najjari, M. Schulz, A. Dorn, R. Moshammer, A. B. Voitkiv, and J. Ullrich, Analysis of experimental data for ion-impact single ionization of helium with monte carlo event generators based on quantum theory, *Phys. Rev. A.* **75**, 062708 (2007).
[25] F. Járai-Szabó and L. Nagy, Semiclassical description of kinematically complete experiments, *J. Phys. B.* **40**, 4259–4267 (2007).
[26] F. Járai-Szabó and L. Nagy, Impact parameter method calculations for fully differential ionization cross sections, *Nucl. Instrum. Meth. Phys. Res. B.* **267**, 292–294 (2009).
[27] M. S. Pindzola, F. Robicheaux, and J. Colgan, Single and double ionization in C^{6+} + He collisions, *Phys. Rev. A.* **82**, 042719 (2010).
[28] K. N. Egodapitiya, S. Sharma, A. Hasan, A. C. Laforge, D. H. Madison, R. Moshammer, and M. Schulz, Manipulating atomic fragmentation processes by controlling the projectile coherence, *Phys. Rev. Lett.* **106**, 153202 (2011).
[29] X. Wang, K. Schneider, A. LaForge, A. Kelkar, M. Grieser, R. Moshammer, J. Ullrich, M. Schulz, and D. Fischer, Projectile coherence effects in single ionization of helium, *J. Phys. B.* **45**, 211001 (2012).

[30] K. A. Kouzakov, S. A. Zaytsev, Y. V. Popov, and M. Takahashi, Singly ionizing 100-MeV/amu C^{6+}+He collisions with small momentum transfer, *Phys. Rev. A.* **86**, 032710 (2012).
[31] M. Schulz, R. Moshammer, D. Fischer, and J. Ullrich, Comment on "Singly ionizing 100-MeV/amu C^{6+} + He collisions with small momentum transfer", *Phys. Rev. A.* **87**, 046701 (2013).
[32] K. A. Kouzakov, S. A. Zaytsev, Y. V. Popov, and M. Takahashi, Reply to "Comment on 'Singly ionizing 100-MeV/amu C^{6+}+He collisions with small momentum transfer' ", *Phys. Rev. A.* **87**, 046702 (2013).
[33] K. A. Kouzakov, Theoretical analysis of the projectile and target coherence in coltrims experiments on atomic ionization by fast ions, *Eur. Phys. J. D.* **71**, 63 (2017).
[34] S. Sharma, T. P. Arthanayaka, A. Hasan, B. R. Lamichhane, J. Remolina, A. Smith, and M. Schulz, Complete momentum balance in ionization of H_2 by 75-keV-proton impact for varying projectile coherence, *Phys. Rev. A.* **89**, 052703 (2014).
[35] F. Járai-Szabó and L. Nagy, Theoretical investigations on the projectile coherence effects in fully differential ionization cross sections, *Eur. Phys. J. D.* **69**, 4 (2015).
[36] L. Nagy, F. Járai-Szabó, and S. Borbély, The effect of projectile wave packet width on the fully differential ionization cross-sections, *J. Phys. B.* **51**, 144005 (2018).
[37] T. Arthanayaka, B. R. Lamichhane, A. Hasan, S. Gurung, J. Remolina, S. Borbély, F. Járai-Szabó, L. Nagy, and M. Schulz, Fully differential study of wave packet scattering in ionization of helium by proton impact, *J. Phys. B.* **49**, 13LT02 (2016).
[38] B. R. Lamichhane, T. Arthanayaka, J. Remolina, A. Hasan, M. F. Ciappina, F. Navarrete, R. O. Barrachina, R. A. Lomsadze, and M. Schulz, Fully differential study of capture with vibrational dissociation in $p+H_2$ collisions, *Phys. Rev. Lett.* **119**, 083402 (2017).
[39] M. Schulz, Chapter eight - the role of projectile coherence in the few-body dynamics of simple atomic systems, *Adv. At. Mol. Phys.* **66**, 507–543 (2017).
[40] L. Sarkadi, I. Fabre, F. Navarrete, and R. O. Barrachina, Loss of wave-packet coherence in ion-atom collisions, *Phys. Rev. A.* **93**, 032702 (2016).
[41] F. Navarrete, M. Ciappina, L. Sarkadi, and R. Barrachina, The role of the wave packet coherence on the ionization cross section of He by p^+ and C^{6+} projectiles, *Nucl. Instrum. Meth. Phys. Res. B.* **408**, 165–168 (2017).
[42] J. M. Feagin and L. Hargreaves, Loss of wave-packet coherence in stationary scattering experiments, *Phys. Rev. A.* **88**, 032705 (2013).
[43] H. Gassert, O. Chuluunbaatar, M. Waitz, F. Trinter, H.-K. Kim, T. Bauer, A. Laucke, C. Müller, J. Voigtsberger, M. Weller, J. Rist, M. Pitzer, S. Zeller, T. Jahnke, L. P. H. Schmidt, J. B. Williams, S. A. Zaytsev, A. A. Bulychev, K. A. Kouzakov, H. Schmidt-Böcking, R. Dörner, Y. V. Popov, and M. S. Schöffler, Agreement of experiment and theory on the single ionization of helium by fast proton impact, *Phys. Rev. Lett.* **116**, 073201 (2016).

[44] R. Rivarola and P. Fainstein, Electron emission in collisions of highly charged ions with atoms and diatomic molecules, *Nucl. Instrum. Meth. Phys. Res. B.* **205**, 448–463 (2003).
[45] M. Foster, J. Colgan, and M. S. Pindzola, Fully correlated electronic dynamics for antiproton impact ionization of helium, *Phys. Rev. Lett.* **100**, 033201 (2008).
[46] B. I. Schneider and L. A. Collins, The discrete variable method for the solution of the time-dependent Schrödinger equation, *J. Non-Cryst. Solids.* **351**, 1551–1558 (2005).
[47] T. N. Rescigno and C. W. McCurdy, Numerical grid methods for quantum-mechanical scattering problems, *Phys. Rev. A.* **62**, 032706 (2000).
[48] T. J. Park and J. C. Light, Unitary quantum time evolution by iterative Lanczos reduction, *J. Chem. Phys.* **85**, 5870–5876 (1986).
[49] B. I. Schneider, J. Feist, S. Nagele, R. Pazourek, S. Hu, L. A. Collins, and J. Burgdörfer, *Recent Advances in Computational Methods for the Solution of the Time-Dependent Schrödinger Equation for the Interaction of Short, Intense Radiation with One and Two-Electron Systems: Application to He and* H_2^+ *(Quantum Dynamic Imaging, CRM Series in Mathematical Physics).* Springer (2011).
[50] M. E. Rudd, R. D. DuBois, L. H. Toburen, C. A. Ratcliffe, and T. V. Goffe, Cross sections for ionization of gases by 5-4000-kev protons and for electron capture by 5-150-kev protons, *Phys. Rev. A.* **28**, 3244–3257 (1983).
[51] A. D. Gaus, W. T. Htwe, J. A. Brand, T. J. Gay, and M. Schulz, Energy spread and ion current measurements of several ion sources, *Rev. Sci. Instrum.* **65**, 3739–3745 (1994).

Chapter 6

Double ionization of helium by fast electrons

M.J. Ambrosio[1]* and L.U. Ancarani[2]

[1]*Department of Physics, Kansas State University, Manhattan, KS 66506, USA*

[2]*Théorie, Modélisation, Simulation, SRSMC, UMR CNRS 7565, Université de Lorraine, 57078 Metz, France*

The double ionization of helium by fast electrons (2000 and 5599 eV) is investigated for a variety of kinematical and coplanar geometrical configurations that correspond to small momentum transfers. In a first order treatment, the resulting three-body problem in the final state is transformed into a driven Schrödinger equation which is solved with the generalized Sturmian functions *ab initio* method. From the asymptotic behavior of the constructed double continuum scattering solution, we extract fully differential cross sections that we compare with experimental and other theoretical results.

Contents

*Corresponding author. Email: mj_ambrosio@phys.ksu.edu

1. Introduction

Scattering problems with few-body dynamics are still one of the great challenges in physics and continue to attract considerable interest. The investigation of double ionization of atoms and molecules by charged particles relies on the understanding of the role of electron-electron correlations in such two-electron process.

From an experimental perspective, cross sections are much smaller for double than for single ionization, and thus measurements are more difficult for the former collision. This is particularly true if one focuses on fully differential cross sections (FDCS), which provide the most detailed information on the process. Theoretically, the proper inclusion of electron-electron correlations and mastering few-body boundary conditions when solving the Schrödinger equation are very challenging, both from the formal and computational point of view. Helium is the most studied target, both experimentally and theoretically. A charged particle colliding with such an atom is a four-body problem. For double ionization, in particular, the final state is a pure four-body problem (the helium nucleus, the scattered projectile and two ejected electrons). When all particles are detected in coincidence (e.g. in an (e,3e) or (p,p2e) experiment), the FDCS can be measured. The cross section is differential in the direction and energy of the projectile and of the two electrons in the exit channel. This provides a very stringent test to any theoretical description.

For double ionization by photon impact, on the other hand, the exit channel is a pure three-body problem (the helium nucleus and two ejected electrons). Double photoionization of helium has been thoroughly investigated, and a high level of detail has been reached. Moreover, a very satisfactory experiment-theory agreement has been reported on practically the whole energy range, for fully differential as well as total cross sections (see, e.g. Ref. [1]).

In the case of double ionization of helium by sufficiently fast incident and scattered charged particle, the four-body problem reduces (in a first-order treatment of the projectile-target interaction) to a Coulomb three-body problem. These processes have been investigated in their fully differential both experimentally and theoretically, using electrons [2–20] as well as a number of positively charged projectiles [21–26]. The experiment-theory agreement is by far not as good as for photon impact, and a number of issues still need to be understood. A variety of kinematical and geometrical configurations have been explored. Even in the coplanar geometry (when the projectile and the two ejected electrons are in the same scattering plane), the available theoretical descriptions seem to provide an incomplete picture of the double ionization process.

In this chapter, we focus on the FDCS for the double ionization of the helium ground state by fast incident and scattered electrons. We consider incident energies that have been selected in (e,3e) experiments (2000 and 5599 eV) and excess energies of less than 50 eV (shared symmetrically or not between the ejected electrons). In such kinematical configurations, a first-order treatment of the projectile-target interaction should be valid and we, thus, need to solve a pure three-body problem in the exit channel. The latter is solved here with the *ab initio* approach based on generalized Sturmian functions (GSF). With the GSF, a double continuum of two electrons escaping from the helium nucleus is constructed first. Subsequently, the FDCS are extracted from the hyperspherical asymptotic behavior of these continua. The major part of the present study is focused on coplanar geometric situations where the momentum transfer q is small (less than 0.5 a.u.), with only one configuration within the impulsive regime (of 2.0 a.u.). Within this geometry and in the low q range (i.e. within the optical limit), we stand relatively close to the dipolar regime (also named quasi-photon regime) and, as such, should reproduce the double photoionization picture with its selection rules. As the momentum transfer increases, however, one loses the dipole character of the transition. The purpose of this chapter is to collect, for a variety of high-incident energy and small momentum transfer coplanar situations, a number of results (old and new) obtained with the GSF method. Comparisons are made with data from (e,3e) measurements as well as from a selection of other *ab initio* calculations, namely the convergent close-coupling (CCC) and the J-matrix.

As described in Section 2, the three-body problem takes the form of a driven Schrödinger equation, which we solve with the GSF method. From a double continuum solution, we extract the double ionization transition amplitude, which then yields the FDCS. A selection of the results is presented in Section 3.

Atomic units ($\hbar = e = m_e = 1$) will be used throughout, unless stated otherwise.

2. Theory

2.1. *Fast projectile formulation*

We denote by $\boldsymbol{r}_1$ the position vector of the projectile of mass m_P (presently $m_P = m_e$=1). Similarly, let $\boldsymbol{r}_i$ be the position vector of the ith helium electron ($i = 2, 3$) with respect to its parent nucleus of mass m_T and charge $Z = 2$. Further, let $r_{ij} = |\boldsymbol{r_i} - \boldsymbol{r_j}|$ be the distance between particles i and j. The reduced masses are denoted by $\mu_{TP} = m_P m_T / (m_P + m_T)$ and $\mu_T = m_T / (m_T + 1)$. For electronic projectiles and infinitely heavy nucleus, we set $\mu_{TP} = \mu_T = 1$.

Let H denote the full four-body Hamiltonian, and $\bar{W}$ the projectile-target interaction:

$$\bar{W} = \frac{Z}{r_1} - \frac{1}{r_{12}} - \frac{1}{r_{13}}. \tag{1}$$

Setting H_0 as the unperturbed Hamiltonian (see below), we have $H = H_0 + \bar{W}$. For a total energy E, the four-body Schrödinger equation with the outgoing (+) type-behavior: $[H - E] \Psi^+ (\boldsymbol{r}_1, \boldsymbol{r}_2, \boldsymbol{r}_3) = 0$ can be transformed into a set of coupled driven equations containing successive orders in $\bar{W}$, as presented in Ref. [16, 25, 27]. The solution is proposed as:

$$\Psi^+ (\boldsymbol{r}_1, \boldsymbol{r}_2, \boldsymbol{r}_3) = \sum_n \Psi^{(n)+} (\boldsymbol{r}_1, \boldsymbol{r}_2, \boldsymbol{r}_3), \tag{2}$$

where each order retains n interactions $\bar{W}$.

With no projectile-target interaction, the zeroth order solution $\Psi^{(0)+} (\boldsymbol{r}_1, \boldsymbol{r}_2, \boldsymbol{r}_3)$ is the product of a free fast incident projectile (a plane wave of momentum $\boldsymbol{k}_i$) and the wave function of the helium ground state $\Phi_i (\boldsymbol{r}_2, \boldsymbol{r}_3)$. This separable solution solves the Schrödinger equation:

$$[H_0 - E] \Psi^{(0)+} (\boldsymbol{r}_1, \boldsymbol{r}_2, \boldsymbol{r}_3) = 0, \tag{3}$$

where H_0 is the unperturbed Hamiltonian

$$H_0 = h_p + h_{He}. \tag{4}$$

Here, h_{He} is the three-body helium Hamiltonian acting on the (2,3) subsystem:

$$h_{He} = \left(-\frac{1}{2\mu_T} \nabla_2^2 - \frac{1}{2\mu_T} \nabla_3^2 - \frac{Z}{r_2} - \frac{Z}{r_3} + \frac{1}{r_{23}} \right), \tag{5}$$

whereas h_p is the free-particle kinetic energy operator associated with the projectile 1 $h_p = -1/(2\mu_{TP}) \nabla_1^2$.

Allowing for one interaction only, one needs to solve the first order non-homogeneous equation:

$$[H_0 - E]\,\Psi^{(1)+}(\boldsymbol{r}_1,\boldsymbol{r}_2,\boldsymbol{r}_3) = -\bar{W}\Psi^{(0)+}(\boldsymbol{r}_1,\boldsymbol{r}_2,\boldsymbol{r}_3)\,, \tag{6}$$

where the perturbation $\bar{W}$ couples the two Hamiltonians h_{He} and h_p. Supposing that the fast projectile is scattered with momentum $\boldsymbol{k}_f$, the first-order solution reads as:

$$\Psi^{(1)+}(\boldsymbol{r}_1,\boldsymbol{r}_2,\boldsymbol{r}_3) = e^{i\boldsymbol{k}_f\cdot\boldsymbol{r}_1}\;\Phi^+_{sc}(\boldsymbol{q},\boldsymbol{r}_2,\boldsymbol{r}_3)\,, \tag{7}$$

where the three-body scattering wave function Φ^+_{sc} characterizes the physics of the two ejected electrons. Here, the dependence of Φ^+_{sc} on the momentum transfer $\boldsymbol{q} = \boldsymbol{k}_i - \boldsymbol{k}_f$ is made explicit.

Upon substitution of Eq. (7) in (6), one arrives at the following three-body driven equation for $\Phi^+_{sc} : (\boldsymbol{q},\boldsymbol{r}_2,\boldsymbol{r}_3)$ [27]:

$$\begin{aligned}&[h_{He} - E_a]\,\Phi^+_{sc}(\boldsymbol{q},\boldsymbol{r}_2,\boldsymbol{r}_3) = \\ &-\frac{4\pi}{q^2}\frac{1}{(2\pi)^3}\left(Z - e^{i\boldsymbol{q}\cdot\boldsymbol{r}_2} - e^{i\boldsymbol{q}\cdot\boldsymbol{r}_3}\right)\Phi_i(\boldsymbol{r}_2,\boldsymbol{r}_3)\,,\end{aligned} \tag{8}$$

where $E_a = E_2 + E_3 = E - k_f^2/(2\mu_{TP})$ is the energy shared by two ejected electrons in their interactions with the nucleus in the final state. For small momentum transfers (dipolar regime), one can expand the exponentials in the driven term and retain the first order. This indeed recovers the dipolar terms $\boldsymbol{r}_2 + \boldsymbol{r}_3$ that mathematically correspond to those encountered in double photoionization of the target [2, 28].

2.2. *Extraction of double ionization FDCS*

When solving the driven equation (8), different ionization channels are simultaneously present and coupled (single ionization with and without excitation, and double ionization). For large hyperradii $\rho = \sqrt{r_2^2 + r_3^2}$, in the so-called Ω_0 region where both r_2 and r_3 are large, the asymptotic behavior of $\Phi^+_{sc}(\boldsymbol{q},\boldsymbol{r}_2,\boldsymbol{r}_3)$ is directly related to the double ionization transition amplitude $T_{\tilde{\boldsymbol{k}}_2,\tilde{\boldsymbol{k}}_3}$ through [29]:

$$|\Phi^+_{sc}(\boldsymbol{q},\boldsymbol{r}_2,\boldsymbol{r}_3)|^2 \underset{\rho\to\infty}{\longrightarrow} (2\pi)\kappa^3|T_{\tilde{\boldsymbol{k}}_2,\tilde{\boldsymbol{k}}_3}|^2\frac{1}{\rho^5}, \tag{9}$$

where $\kappa = \sqrt{2E_a}$ is the hypermomentum. Therefore, in the driven equation approach presented above, the transition amplitude can be extracted from the scattering solution in the double continuum channel.

For two electrons escaping with energies E_2 and E_3 in the solid angles $d\Omega_2$ and $d\Omega_3$, the first-order Born treatment defines the FDCS by:

$$\frac{d^5\sigma}{d\Omega_2 d\Omega_3 d\Omega_f dE_2 dE_3} = (2\pi)^4 \frac{k_f k_2 k_3}{k_i} |T_{\tilde{\boldsymbol{k}}_2,\tilde{\boldsymbol{k}}_3}|^2, \tag{10}$$

where the projectile is scattered in the solid angle $d\Omega_f$.

2.3. *Generalized Sturmian functions approach*

To solve the first-order three-body driven equation (8) for a given $\boldsymbol{q}$, we use a Sturmian approach. Details of the method are described in Ref. [15, 16, 25]. The scattering wave function $\Phi^+_{sc}(\boldsymbol{q}, \boldsymbol{r}_2, \boldsymbol{r}_3)$ is decomposed in total-angular-momentum partial waves. This is subsequently expanded in terms of products of radial GSF, $S^{(+)}_{nl}(r)$, each with outgoing Coulomb behavior and corresponding to an energy E_s chosen to match the excess energy E_a. Explicitly we use:

$$\Phi^+_{sc}(\boldsymbol{q}, \boldsymbol{r}_2, \boldsymbol{r}_3) = \sum_{L,M} \sum_{l_2,l_3} \sum_{n_2,n_3} a^{L,M}_{n_2 l_2, n_3 l_3} \, \mathcal{Y}^{LM}_{l_2 l_3}(\widehat{\boldsymbol{r}}_2, \widehat{\boldsymbol{r}}_3) \frac{S^{(+)}_{n_2 l_2}(r_2)}{r_2} \frac{S^{(+)}_{n_3 l_3}(r_3)}{r_3}. \tag{11}$$

For Φ^+_{sc}, the coupled spherical harmonics $\mathcal{Y}^{LM}_{l_2 l_3}(\widehat{\boldsymbol{r}}_2, \widehat{\boldsymbol{r}}_3)$ generate an overall hyperspherical outgoing asymptotic shape in the large (r_2, r_3) domain (see Fig. 1 for a typical scattering solution).

A discussion on how the method constructs the double continuum scattering solutions is provided in Ref. [14]. The fundamental idea behind this is the fact that the basis set enforces an asymptotic behavior on the wave function at the (r_2, r_3) boundaries, which approximates the proper double-continuum front hyperspherical asymptotic conditions described by Kadyrov *et al.* [29] and Alt and Mukhamedzhanov [30]. Figure 1 shows different regions, labeled from I to IV at the domain boundary. Let $\tan\alpha = r_3/r_2$, $\boldsymbol{e}_2$ and $\boldsymbol{e}_3$ be the unit vectors in the (r_2, r_3) space directed along the axes r_2 and r_3, respectively. Further, let $\kappa \boldsymbol{e}_\rho = \kappa(\boldsymbol{e}_2 \cos\alpha + \boldsymbol{e}_3 \sin\alpha)$ be the hyperspherical front asymptotic wave vector. For $E_s = E_a$, in region I (small α), we have $\kappa \boldsymbol{e}_\rho \approx \kappa \boldsymbol{e}_2$. With this at hand, the basis enforces the proper behavior (and similarly by reversing the roles of r_2 and r_3). Furthermore, the approximation made when moving from regions II and III is equivalent to neglecting the second-order term in $\cos\alpha \approx 1 - (1/2)\alpha^2$. The single-continuum channel information, localized around sector IV, is outside of this chapter's scope (it was discussed in Ref. [12]).

The proposed expansion (11) converts the driven Eq. (8) into a linear system for the coefficients $a^{L,M}_{n_2 l_2, n_3 l_3}$. This is similar to Eq. (21) of Ref. [16]

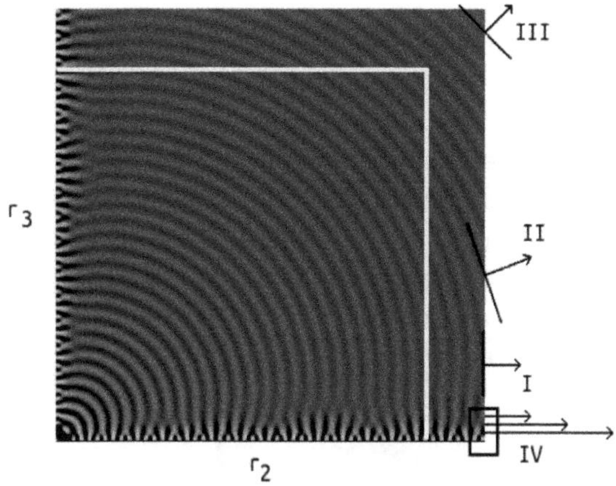

Fig. 1. Real part of the scattering solution as a function of the radial coordinates (r_2, r_3), for a typical coplanar geometry with two electrons sharing equally the excess energy E_a. Through the angular coupling, the GSF spherical basis obtained with an energy $E_s = E_a$ approximates well the proper asymptotic conditions of the double continuum hyperspherical front on its square domain (see the main text for details about the regions labeled I, II, III, IV).

where the said system is solved using the standard methods. In all the kinematical configurations considered below, convergence in the number of partial waves has been verified.

The helium ground state could be described by any sufficiently accurate wave function. For convenience as per Ref. [15, 16], we construct it within the GSF formalism. A GSF basis with negative energies can be used efficiently to obtain two-electron bound states [31–34]. Here, like in Ref. [25], the helium ground state with an energy in the order of -2.9034 a.u. is obtained with the GSF method [33, 34], using 20 Sturmians per coordinate per partial wave, with the individual angular momenta up to 4.

Once $\Phi_{sc}^{+}(\boldsymbol{q}, \boldsymbol{r}_2, \boldsymbol{r}_3)$ is obtained, the quantity $|T_{\tilde{\boldsymbol{k}}_2,\tilde{\boldsymbol{k}}_3}|^2$ is extracted from Eq. (9) at a given, sufficiently large ρ value. Finally, the FDCS is calculated by means of the expression (10). We should point out that this approach is equivalent to the more familiar expression used in the first Born approximation for calculation of the transition matrix element [9].

3. Results

We present our FDCS investigation for equal and unequal energy sharing configurations in two subsections. The results are ordered from lower to higher values of the momentum transfer q. We consider only coplanar geometries where the ejected electrons are emitted within the scattering plane of the projectile. Cross sections will be shown through contour plots, as a function of the ejected angles θ_2 and θ_3. Within a first-order Born approximation, the FDCS should be symmetrical with respect to the direction of the momentum transfer $\boldsymbol{q}$. To facilitate the discussion, we give in Fig. 2 a typical case where several peaks are clearly identified.

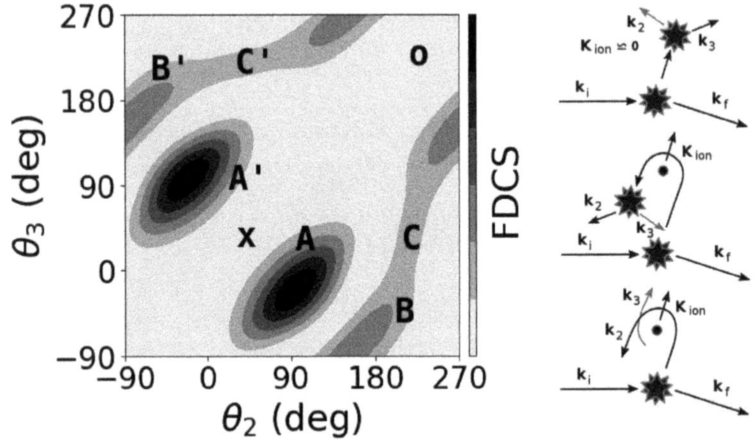

Fig. 2. Left: Binary, recoil, and back-to-back structures are labeled by A, B and C, respectively, with their symmetrical counterparts denoted by primes. The x and $\circ$ symbols denote, respectively, the directions $\pm\boldsymbol{q}$, i.e., $(\theta_2, \theta_3) = (\theta_q, \theta_q)$ and $(\theta_2, \theta_3) = (\theta_q + \pi, \theta_q + \pi)$. Right: Schematic display of the lowest order processes for double ionization. Top: binary peak; middle: recoil peak; bottom: back-to-back peak.

3.1. *Equal energy sharing*

The experiment performed by the Orsay group in 1999, then led by Lahmam-Bennani, provided the first absolute scale FDCS for the double ionization of helium by fast electrons. The kinematic conditions involved a small momentum transfer to the helium target. This is within the dipolar regime, since the perturbation operator behaves approximately linearly with the electron radial coordinates r_2 and r_3. The electronic projectiles were accelerated to energies of 5599 eV, and small deflection events (0.45°) were selected, leading to small momentum transfers, i.e., 0.22 (for $E_a = 4+4$ eV) and 0.24 a.u. (for $E_a = 10+10$ eV). The FDCS structures were, therefore, expected to share similarities with the pure double photoionization case, such as approximate selection rules, and the similarly sized binary and recoil peaks.

Many theoretical attempts have been made to model the measured data presented in Ref. [3]. The first one, employing the CCC method, was published alongside the experimental results in the same article. Other notable approaches include the J-Matrix [10, 18], and the time dependent wavepacket evolution [35]. The shape of the FDCS produced by all the above *ab initio* methods were not in mutual agreement. Moreover, they needed renormalization to match the experimentally observed magnitudes. The exception was the work by Silenou Mengoue *et al.* [18] who, however, admit that convergence with respect to the basis size had not been achieved. The above approaches required different rescaling factors [36]. This is surprising given that all the schemes were solving the same three-body problem for the target subsystem. It was expected, even in the absence of complete agreement with the experiment, that at least the theoretical FDCS are in mutual agreement. However, this was not the case, and such a situation remained unaltered for a few more years. Another J-matrix calculation by Silenou Mengoue [19] showed improved convergence and an agreement with the experimental FDCS in both magnitude and shape. However, this was limited to a small subsample for the 10+10 eV geometry, with no FDCS for the 4+4 eV case. The unexplained situation was further complicated by the results obtained by some other theoretical attempts, made with approximate double continuum descriptions such as the three Coulomb (3C) wave function or its variants (see [36] and references therein). The FDCS were also at odds with the experiments and the *ab initio* results, leading again to different scaling factors. Curiously, when rather simple initial wave functions were used to describe the helium target, an overall magnitude agreement with the measurement was obtained, at least for the 10+10 eV case [37]. It surprising result was later shown to be fortuitous [38]. It triggered some further investigations on the role played by the quality and/or the properties of the initial state description (see, e.g. Ref. [39] and references therein). The global picture remained confusing until 2015 (see below).

For small momentum transfers, the driven term in Eq. (8) will have a similar form as that for the double photoionization. Indeed, a first-order expansion of the complex exponentials in Eq. (8) leads to the linear terms $\boldsymbol{q}\cdot(\boldsymbol{r}_2+\boldsymbol{r}_3)$. Hence, a number of strict zeros in the double photoionization FDCS are carried over as minima to the electron impact cross section, namely for [2]:

$$\boldsymbol{k}_3 = -\boldsymbol{k}_2 \tag{12a}$$

$$\boldsymbol{k}_2 \perp \boldsymbol{q} \wedge \boldsymbol{k}_3 \perp \boldsymbol{q} \tag{12b}$$

$$\boldsymbol{k}_3 = \boldsymbol{k}_2 \tag{12c}$$

$$\boldsymbol{k}_3 + \boldsymbol{k}_2 \perp \boldsymbol{q}. \tag{12d}$$

Here, the first minimum condition corresponds to a zero in the linear term and, therefore, this greatly minimizes back-to-back emission. The second condition also makes the dipole term vanish. The third condition is due to the interelectronic repulsion. The fourth minimum condition stems from a forbidden optical transition (see Ref. [2] and references therein). For momentum transfers $q = 0.24$ or 0.22, one expects to observe such minima in the contour plots.

Figures 3 and 4 show the absolute experimental data for the (10+10) eV configuration. These are compared with the calculated FDCS obtained with the GSF (upscaled by 2.2), the CCC (upscaled by 2.2) and with the J-Matrix (upscaled by 1.2) methods. The GSF and CCC calculations agree remarkably well, not only in cross section shapes, but also in the fact that the very same rescaling factor was needed to bring them up to the experimental amplitude. This agreement, reported in 2015 [15, 16], clarified a part of the controversial picture, with both methods producing very similar FDCS that comply with the minima originating from the quasiphoton regime. In contrast, the J-Matrix calculation produced different results: the rescaling factor is not the same and, moreover, this method predicts some degree of collinear emission in the direction of the momentum transfer, contradicting Eq. (12c) (albeit in an angular range where there are no experimental data to completely verify or reject the validity of the method; $\theta_3 = 319°, 333°$ are the closest angular values). Overall, the shape agreement between theory and experiment is good. It is particularly good for some emission angle configurations ($\theta_3 = 27°, 41°$). However, agreement becomes strikingly worse for some other cases ($\theta_3 = 83°, 97°, 153°, 249°, 263°, 277°$), where the experimental points do not look fully compliant with the minima of Eqs. (12a)–(12d).

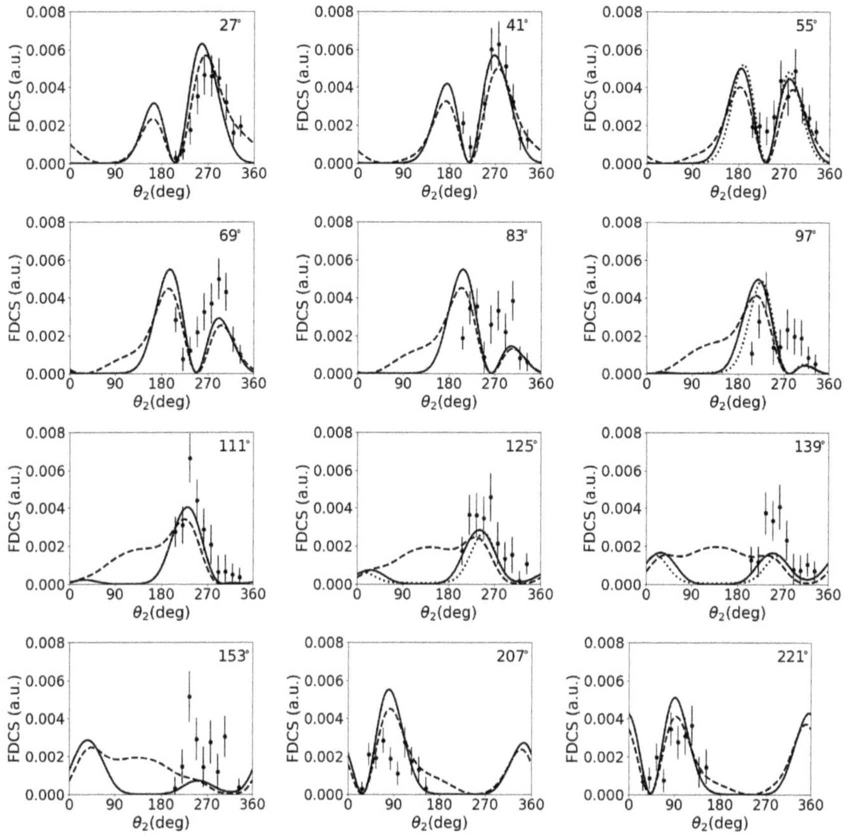

Fig. 3. The FDCS for fixed θ_3 and for the $E_a = 10 + 10$ eV kinematics of Ref. [3]. Continuous line: GSF (rescaled up by a factor 2.2). Dashed line: J-Matrix (rescaled up by a factor 1.2) [10]. Dotted line: CCC (where available, rescaled up by a factor 2.2) . Points: absolute experimental data from Ref. [3]

We now turn to the 4+4 eV excess energy configuration. The shape agreement between theory and experiment, shown in Fig. 5, is slightly worse than that for the 10+10 excess energy. The experimental results, aside from a few configurations ($\theta_3 = 45°, 63°, 135°$), show significantly different peak ratios and nodal structures than those from the theoretical counterparts. On the other hand, the GSF-CCC agreement (see Fig. 6) is now even better than for the 10+10 eV case. This, in turn, reinforces the fact that the first Born treatment of (e,3e) on helium can be considered properly solved, since two independent *ab initio* methods produce cross sections in remarkably good agreement.

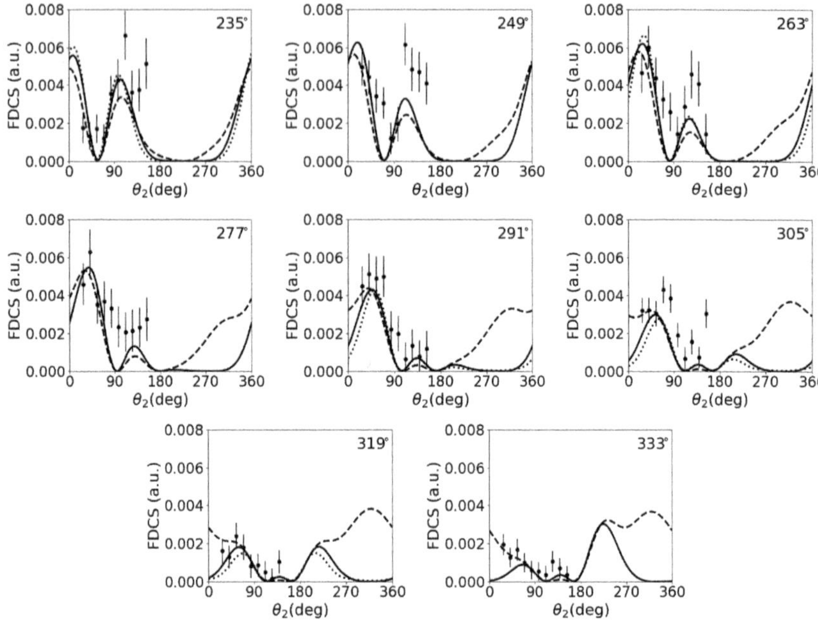

Fig. 4. The FDCS for fixed θ_3 and for the $E_a = 10 + 10$ eV kinematics of Ref. [3]. Continuous line: GSF (rescaled up by a factor 2.2). Dashed line: J-Matrix (rescaled up by a factor 1.2) [10]. Dotted line: CCC (where available, rescaled up by a factor 2.2) . Points: absolute experimental data from Ref. [3]

Overall, it is our belief that the first-order in the Born method can now be considered properly solved. This view is based upon good agreement between the two fundamentally different theories for both experimental excess energies, while complying with the shape restrictions imposed by the small-momentum-transfer regime [2]. There still remains the unexplained lack of agreement (in particular with respect to the magnitude of the FDCS) between the experimental data with the CCC and GSF methods. That discrepancy could be attributed to projectile-target interactions beyond the first-order in the Born method, or to an experimental systematic error (for example, in determining the absolute scale). Questioning an experiment, however, would require stronger evidence from the theoretical side. To this end, it would be advisable to include also the second-order Born term in the given two independent theoretical schemes showing again their good agreement. However, regarding the influence of the second-order Born method, there are two contradicting observations in the literature. On the one hand,

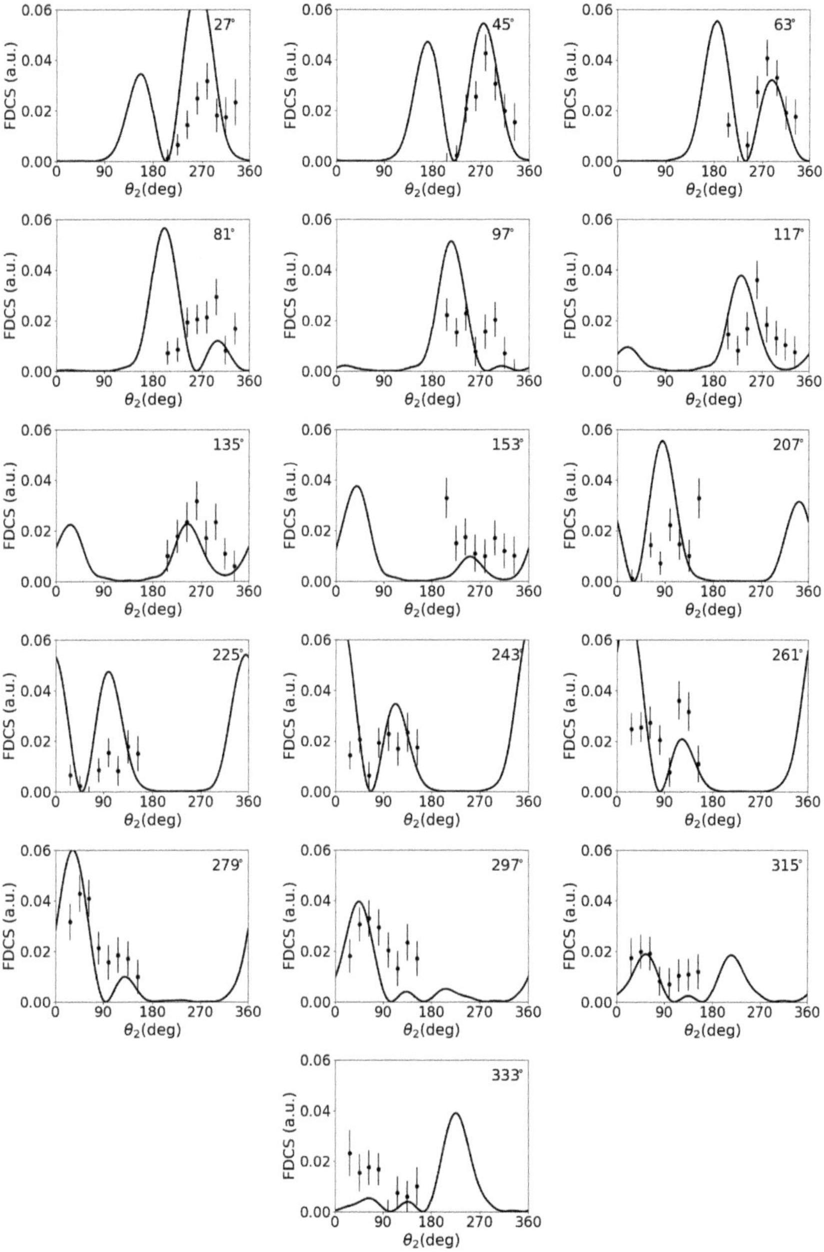

Fig. 5. The FDCS for fixed θ_3 and for the $E_a = 4 + 4$ eV kinematics of Ref. [3]. Continuous line: GSF (rescaled up by a factor 14). Points: absolute experimental data from Ref. [3].

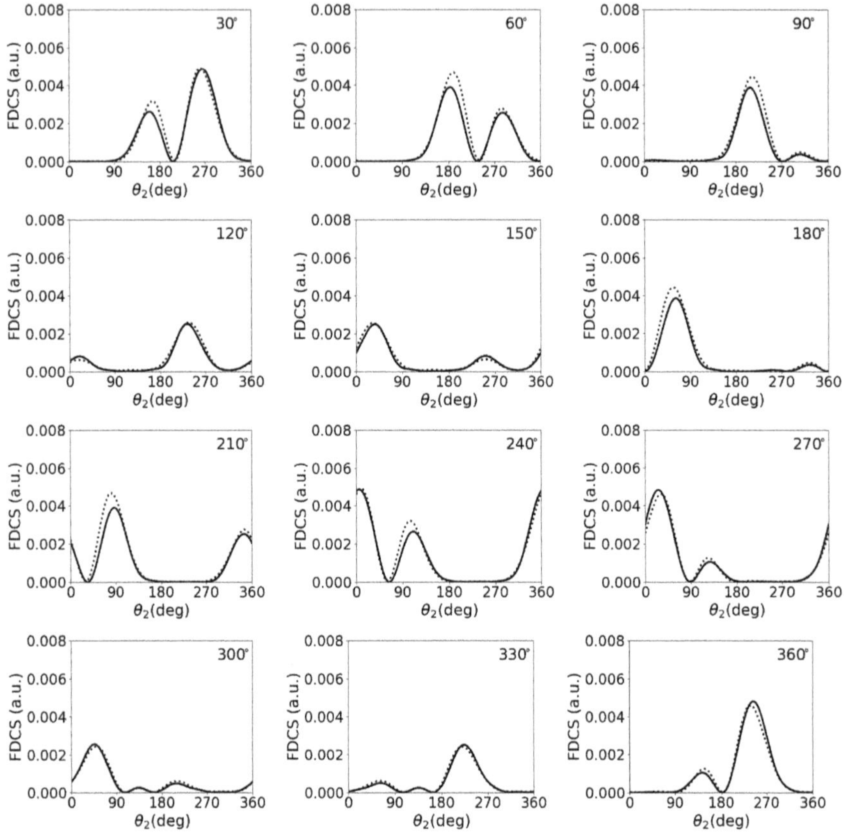

Fig. 6. The FDCS for fixed θ_3 and for the $E_a = 4 + 4$ eV kinematics of Ref. [3]. Continuous line: GSF. Dotted line: CCC [3].

relying upon some other approaches involving asymptotically correct (albeit approximate) descriptions of the initial and final target states [41, 42], it was stated by Kheifets [40] that the second-order corrections only change the FDCS marginally at 5599 eV incident energy. On the other hand, the Green operator expansion formalism by Berakdar [43], suggested more rapidly varying structures for the FDCS, that are not present in the first Born calculations. However, the calculations from Ref. [43] were performed with the approximate solutions for the first-order wave function and, thus, the results should be considered as conditional too.

One other possible source of discrepancy could be related to the description of the initial helium target wave function. We investigated this point within the GSF method. We improved the helium ground state by increasing the number of partial waves from $l_2, l_3 \leq 2$ to $l_2, l_3 \leq 4$. The obtained binding energy improved from -2.90277 a.u. to -2.90334 a.u. However, this did not significantly affect the FDCS [16]. It was shown, in the same publication [16], that if the helium ground state model is too simplistic (using a zero-angular-momentum initial state yielding an energy of -2.8790 a.u.), the unscaled theoretical FDCS fortuitously matched the corresponding experimental data. This curious result is similar to that obtained with approximate double continuum wave functions [37]. The quality of the FDCS with such a simple function, insufficiently correlated ground state wave function is questionable, and it should not be used to validate or invalidate the theoretical treatment of the three-body scattering problem.

Employing a cold target recoil ion momentum spectroscopy (COLTRIMS) apparatus in Heidelberg, Dorn *et al.* [6, 17] measured relative FDCS for electrons of 2 keV incident energy. While no information about the magnitude of the cross sections is available in this measurement, the physical aspects of the involved collision processes, including peak positions and their relative ratios, can be very well understood from comparisons with theoretical FDCS.

Here we present the FDCS for a momentum transfer of $q = 0.5$ a.u. and excess energies E_a of 5+5 eV, 10+10 eV and 25+25 eV. From Fig. 7(a), we can observe that for the 5+5 eV excess energy, the recoil peak is higher than the binary one. A classical interpretation for the recoil peak (the right-middle diagram of Fig. 2) is as follows. The electrons are more likely to undergo a recoil process when a smaller excess energy is available, since this would force the first-impacted electron to stay longer near the nucleus. For the 10+10 and 25+25 eV configurations, there is a strong tendency towards binary emission. The nucleus plays a spectator role in this case. Another trend that emerges from Fig. 7 is that a little back-to-back structure appears with increased excess energy, a type of emission damped by the dipolar-regime emission minima. This can be understood by reference to the exponentials in the driven term of Eq. (8) and the asymptotes of electron coordinates momenta $\boldsymbol{r}_i = (\boldsymbol{k}_i/\kappa)\rho$ for $i = 2, 3$ [29]. When the excess energy is larger, the exponentials can depart from the dipole behavior at smaller ρ even though q can still be regarded as small, leading to the FDCS presenting non-dipole features.

The overall agreement between theory and experiment is very good for the binary structures. The experiments, however, reveal displaced recoil peaks. This puts in evidence that for electron-impact ionization of helium,

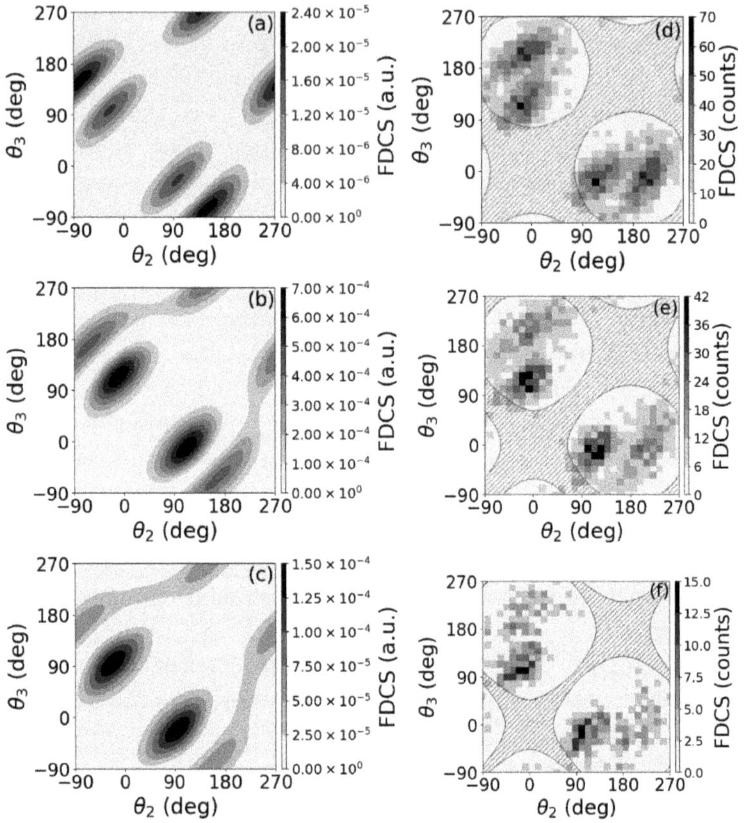

Fig. 7. The FDCS dependence on both ejection angles (θ_2, θ_3) for a momentum transfer $q = 0.5$ a.u. and for excess energies (a) and (d): 5+5 eV, (b) and (e): 10+10 eV, and (c) and (f): 25+25 eV from Ref. [17]. The shaded areas denote the detector dead zones. The contour plots on the left panels show the results from the GSF calculations, whereas those on the right are the experimental data [6, 17].

even at 2 keV incident energies, the second Born effects may be present [5]. The reason is that in the FDCS from the first Born approximation, the maxima of both the binary and recoil peaks are expected to be perfectly aligned. Although this goes beyond the scope of the present chapter, it should be noted that, for larger momentum transfer values (e.g. $q = 2$ a.u.), the second-order effects should be even more pronounced [5, 11].

3.2. *Unequal energy sharing*

In comparison with the studies on equal energy configurations [2, 3, 5, 10, 15, 16, 44], insufficient attention has been devoted to the understanding of the FDCS for asymmetric energy sharing [6, 17, 20]. However, from the inspection of the U-shaped single differential cross sections [13, 27], the contributions from asymmetrical energy sharing configurations appear to be even more significant. This should, therefore, merit at least the same extent of attention as the equal-energy-sharing scenario.

We proceed again from lower to higher momentum transfers. We attribute the energy E_2 and the angle θ_2 to the slower of the two ejected electrons, with the exception of Figs. 8(b) where the opposite is true.

We start by analyzing the FDCS for q = 0.24 a.u. with 20 eV excess energy as per the kinematics considered in the Orsay experiment [3]. However, we consider the two energy sharings 8+12 eV and 5+15 eV, as done by Silenou Mengoue [20]. No experimental data are available for these cases, and only our FDCS are shown in Fig. 8.

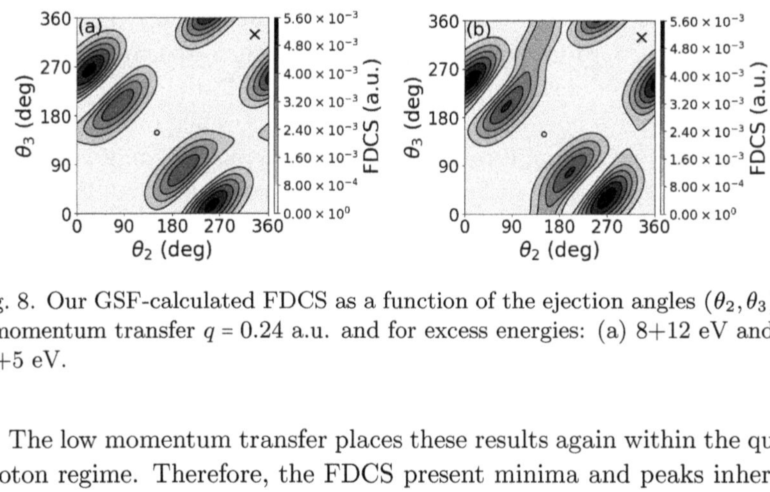

Fig. 8. Our GSF-calculated FDCS as a function of the ejection angles (θ_2, θ_3) for a momentum transfer q = 0.24 a.u. and for excess energies: (a) 8+12 eV and (b) 15+5 eV.

The low momentum transfer places these results again within the quasi-photon regime. Therefore, the FDCS present minima and peaks inherited from the dipolar regime [2, 16]. At the mentioned excess energies, the electrons are fast enough for the binary peak to be higher than the recoil peak, as was the case for the equal energy sharing scenario. A comparison of the back-to-back peaks C and C' reveals a more important emission with the fast electron ejected in the $-\boldsymbol{q}$ direction. This is compatible with the corresponding mechanism from Fig. 2.

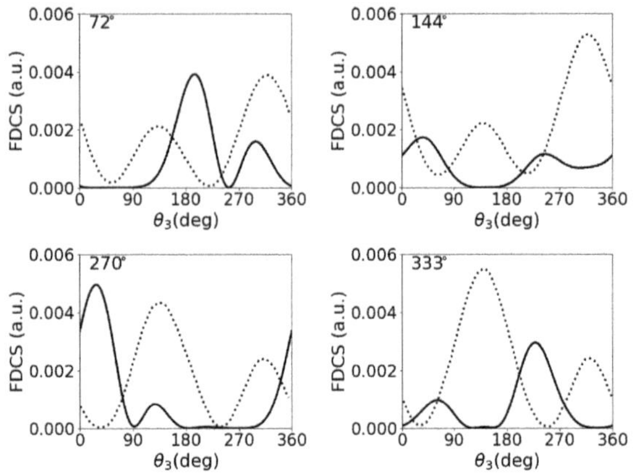

Fig. 9. Specific cuts from Fig. 8(a). The FDCS as a function of θ_3 for four fixed θ_2 values: our GSF (solid line) and the J-matrix results [20] divided by 20 (dotted line).

Specific cuts from the contour plots 8 are presented in Figs. 9 and 10. We compare our GSF-calculated FDCS with Silenou Mengoue's J-Matrix results [20]. This comparison suggests that the J-Matrix calculations tend to predict excessive emission in the $\pm \boldsymbol{q}$ directions, while our results show a more detailed structure compliant with the dipole-limit minima [2]. As an example, for $\theta_2 = 144°$ in Fig. 9, the FDCS should be small for $\theta_3 = \theta_2$, because the ejection of the two electrons in the same direction (with slightly different energies) should not be favored. We should also emphasize that the differences between the J-matrix and our results are not only on the shapes, but also on the magnitudes, as scaling factors (20) were used to put both results in the same plots.

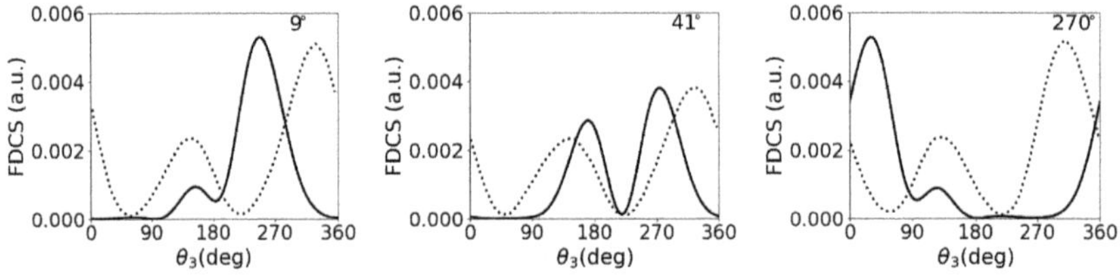

Fig. 10. Specific cuts from Fig. 8(b). The FDCS as a function of θ_3 for three fixed θ_2 values: our GSF (solid line) and the J-matrix results divided by 11 (dotted line).

We turn now to momentum transfer $q = 0.5$ a.u. for excess energies E_a of 5+10 eV and 5+25 eV as in the measurement at 2 keV incident energy performed by the Heidelberg group. Our theoretical FDCS are compared with the corresponding experimental data in Fig. 11 [17]. The observed binary-peak is in clear agreement with our first-order Born result for both excess energies. On the other hand, the experimental recoil peaks appear to be shifted with respect to the first Born theory, similarly to the discussed equal-energy scenario. The difference can be ascribed to the second-order projectile-target interactions, since our model solves the first order exactly. In the more asymmetric configuration from Fig. 11(b), there is an appreciable back-to-back structure with the fast electron emitted towards $-\boldsymbol{q}$. This is consistent with the back-to-back mechanism depicted in Fig. 2. An alternative leading to back-to-back emission is the so-called *shake-off* process, but in this case, the fast electron is ejected towards $\boldsymbol{q}$. This process was, however, disregarded in Ref. [6] on the argument that more drastic energy asymmetry would be required for it to be relevant. The small, but still visible back-to-back emission with the fast electron parallel to $\boldsymbol{q}$ (absent in the 5+10 eV case, and only very weak in the 5+25 eV emission) would require a series of hard intra-target collisions [17].

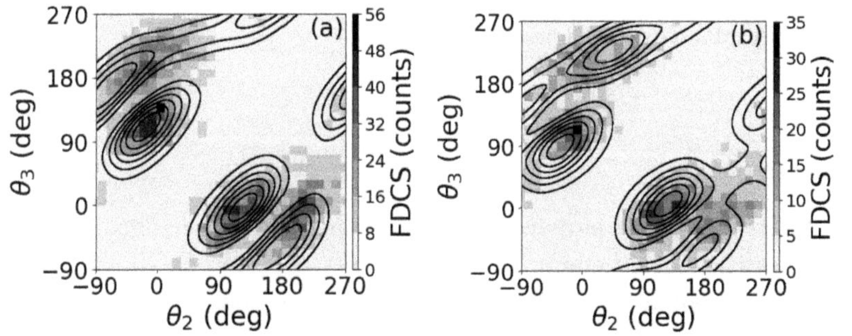

Fig. 11. The FDCS dependence on both ejection angles (θ_2, θ_3) for a momentum transfer $q = 0.5$ a.u. for excess energies (a) 5+10 eV and (b) 5+25 eV as in Ref. [17]. Here, for comparison, the theoretical results have been superimposed as contour levels over the experimental data.

In Ref. [17], the impulsive regime ($q = 2$ a.u.) was also investigated. Figure 12 shows that the GSF calculation predicts a dip in the back-to-back structure for emission of the fast electron in the $\boldsymbol{q}$ direction. This result is consistent with the back-to-back emission produced by the mechanism shown in the bottom scheme of Fig. 2. In that regard, our calculation differs from the CCC result reported in [6]. Measured FDCS are shown in Fig. 12(b). Unfortunately, the experimental set up lacked detector efficiency in that region so it is not possible to tell which of the two calculations reveals more accurate results on the back-to-back structure. Moreover, for this impulsive regime, the second-order effects are expected to be relevant. This would further complicate comparisons among different theories, even if the experiment did not have the dead zone lying around the mentioned peaks.

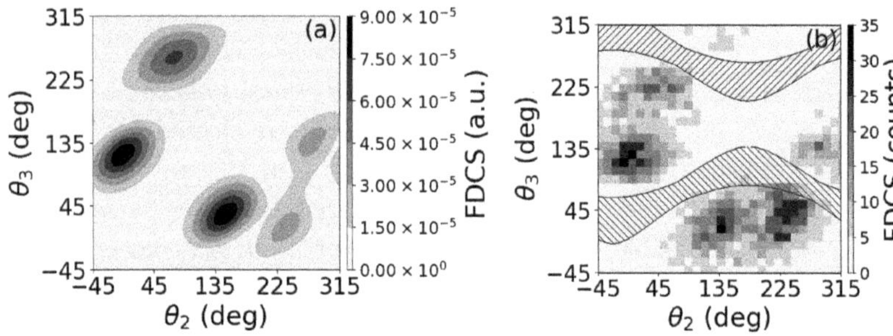

Fig. 12. The FDCS dependence on both ejection angles (θ_2, θ_3) for a momentum transfer $q = 2.0$ a.u. and for an excess energy 5+25 eV as per Ref. [17]: (a) the present GSF results, (b) experimental data.

4. Summary

In this chapter, we have surveyed fully differential cross sections, FDCS, for double ionization of helium by fast electrons, focusing on coplanar geometries with small momentum transfers. Some of the older results are completed in both symmetric and asymmetric energy sharing kinematics. Treating the projectile-target interaction to first order, we are faced with a three-body Coulomb problem. The corresponding driven equation is presently solved using the GSF method. As opposed to the other theoretical methods used within the standard first Born approximation, we extract the differential cross section from the asymptotic behavior of the double continuum scattering solution.

We first consider the equal energy sharing configurations. For small momentum transfers, and by reference to a single set of absolute experimental data, there is a long-standing controversy among theories themselves. It was recently shown that the two different *ab initio* methods, the GSF and the CCC, are in very good agreement for an excess energy of 10+10 eV, and even better for the present new results in the case of the 4+4 eV sharing. As the momentum transfer is augmented, the GSF results reproduce satisfactorily the relative experimental features. Here, however, the first sign of the second-order Born effects clearly appears. We then turned our attention to unequal energy sharing, ranging from momentum transfers of 0.24 a.u. to 0.5 a.u. In a theoretical comparison, for 0.24 a.u. our new results are in disagreement (both in the shape and magnitude) with the recently presented J-matrix. For q = 0.5 a.u. the relative scale is used for the comparison with the experimental data and the CCC results. In this case too, the signs of the second-order Born effects appear. This is further evidenced in the case with q = 2.0 a.u.

While theory and experiments have provided some key understanding of double ionization by fast electrons, there is still room for new learning. Theoretically, even within a first-order treatment, there remain some unexplained discrepancies between *ab initio* methods, which are possibly related to intrinsic difficulties of the three-body Coulomb problem. Experimentally, the measurements of small cross sections through coincidence techniques are clearly difficult. However, new absolute scale data would be most welcome to illuminate the reasons hidden behind some of the persisting differences between theory and measurements.

References

[1] I. Bray, D. V. Fursa, A. S. Kadyrov, A. T. Stelbovics, A. S. Kheifets, and A. M. Mukhamedzhanov, Electron- and photon-impact atomic ionisation, *Phys. Rep.* **520**, 135–174 (2012).

[2] A. Lahmam-Bennani, I. Taouil, A. Duguet, M. Lecas, L. Avaldi, and J. Berakdar, Origin of dips and peaks in the absolute fully resolved cross sections for the electron-impact double ionization of He, *Phys. Rev. A* **59**, 3548–3555 (1999).

[3] A. Kheifets, I. Bray, A. Lahmam-Bennani, A. Duguet, and I. Taouil, A comparative experimental and theoretical investigation of the electron-impact double ionization of He in the keV regime, *J. Phys. B* **32**, 5047 (1999).

[4] A. Dorn, R. Moshammer, C. D. Schröter, T. J. M. Zouros, W. Schmitt, H. Kollmus, R. Mann, and J. Ullrich, Double Ionization of Helium by Fast Electron Impact, *Phys. Rev. Lett.* **82**, 2496–2499 (1999).

[5] A. Dorn, A. Kheifets, C. D. Schröter, B. Najjari, C. Höhr, R. Moshammer, and J. Ullrich, Double ionization of helium by electron-impact: Complete Pictures of the Four-Body Breakup Dynamics, *Phys. Rev. Lett.* **86**, 3755–3758 (2001).

[6] A. Dorn, A. Kheifets, C. Schröter, B. Najjari, C. Höhr, R. Moshammer, and J. Ullrich, Double Ionization of Helium by Electron-Impact in the impulsive regime, *Phys. Rev. A* **65**, 032709 (2002).
[7] A. Dorn, G. Sakhelashvili, C. Höhr, A. Kheifets, J. Lower, B. Najjari, C. D. Schröter, R. Moshammer, and J. Ullrich, (e,3e) and (e,γ2e) experiments: investigations of fundamental four-body Coulomb reactions, *IOP Conf. Proc.* **177**, 41–50 (2003).
[8] A. Dorn, M. Dürr, B. Najjari, N. Haag, C. Dimopoulou, D. Nandi, and J. Ullrich, Identification of higher order contributions in three-dimensional (e,2e) cross-sections for helium, *J. Electron Spectrosc.* **161**, 2–5 (2007).
[9] J. Berakdar, A. Lahmam-Bennani, and C. Dal Cappello, The electron-impact double ionization of atoms: an insight into the four-body Coulomb scattering dynamics, *Phy. Rep.* **374**, 91–164 (2003).
[10] V. A. Knyr, V. V. Nasyrov, and Y. V. Popov, Description of (e , 2e) and (e , 3e) processes with the Jmatrix method, *AIP Conf. Proc.* **697**, 76–81 (2003).
[11] M. F. Ciappina, M. Schulz, and T. Kirchner, Reaction dynamics in double ionization of helium by electron impact, *Phys. Rev. A* **82**, 062701 (2010).
[12] M. J. Ambrosio, G. Gasaneo, and F. D. Colavecchia, Insights from the zero-angular-momentum wave in single and double ionization of He by fast electrons, *Phys. Rev. A* **89**, 012713 (2014).
[13] M. J. Ambrosio, L. U. Ancarani, D. M. Mitnik, F. D. Colavecchia, and G. Gasaneo, A Generalized Sturmian Treatment of (e,3e) Processes Described as a Three-Body Coulomb Problem, *Few-Body Systems.* **55**, 825–829 (2014).
[14] M. J. Ambrosio, F. D. Colavecchia, D. M. Mitnik, G. Gasaneo, and L. U. Ancarani, Generalized Sturmian Functions applied to double continuum problems, *J. Phys. Conf. Ser.* **601**, 012004 (2015).
[15] M. J. Ambrosio, F. D. Colavecchia, D. M. Mitnik, and G. Gasaneo, Discrepancy between theory and experiment in double ionization of helium by fast electrons, *Phys. Rev. A* **91**, 012704 (2015).
[16] M. J. Ambrosio, F. D. Colavecchia, G. Gasaneo, D. M. Mitnik, and L. U. Ancarani, Double ionization of helium by fast electrons with the Generalized Sturmian Functions method, *J. Phys. B* **48**, 055204 (2015).
[17] M. J. Ambrosio, D. M. Mitnik, A. Dorn, L. U. Ancarani, and G. Gasaneo, Double ionization of helium by 2-kev electrons in equal- and unequal-energy configurations, *Phys. Rev. A* **93**, 032705 (2016).
[18] M. Silenou Mengoue, M. G. Kwato Njock, B. Piraux, Y. V. Popov, and S. A. Zaytsev, Electron-impact double ionization of He by applying the Jacobi matrix approach to the Faddeev-Merkuriev equations, *Phys. Rev. A* **83**, 052708 (2011).
[19] M. Silenou Mengoue, Three-body-continuum Coulomb problem using a compact-kernel-integral-equation approach, *Phys. Rev. A* **87**, 022701 (2013).
[20] M. Silenou Mengoue and H. M. Tetchou Nganso, Asymmetric electron energy sharing in electron-impact double ionization of helium, *Phys. Rev. A* **94**, 062705 (2016).

[21] D. Fischer, R. Moshammer, A. Dorn, J. R. Crespo López-Urrutia, B. Feuerstein, C. Höhr, C. D. Schröter, S. Hagmann, H. Kollmus, R. Mann, B. Bapat, and J. Ullrich, Projectile-charge Sign Dependence of Four-Particle Dynamics in Helium Double Ionization, *Phys. Rev. Lett.* **90**, 243201 (2003).

[22] S. D. López, C. R. Garibotti, and S. Otranto, Double ionization of helium by bare ions: Theoretical study of the fully differential cross sections, *Phys. Rev. A* **83**, 062702 (2011).

[23] S. D. López, C. R. Garibotti, and S. Otranto, Theoretical study of the double ionization of the he isoelectronic sequence by protons and antiprotons, *Nucl. Instr. Meth. B* **283**, 63–70 (2012).

[24] S. D. López, S. Otranto, and C. R. Garibotti, Effective-momenta approach for the four-body Coulomb problem, *Phys. Rev. A* **87**, 022705 (2013).

[25] M. J. Ambrosio, D. M. Mitnik, L. U. Ancarani, G. Gasaneo, and E. L. Gaggioli, Double ionization of helium by proton impact: A generalized-Sturmian approach, *Phys. Rev. A* **92**, 042704 (2015).

[26] M. J. Ambrosio, L. U. Ancarani, A. I. Gómez, E. L. Gaggioli, D. M. Mitnik, and G. Gasaneo, Double ionization of helium by proton impact: from intermediate to high momentum transfer, *Eur. Phys. J. D* **71**, 127 (2017).

[27] G. Gasaneo, D. M. Mitnik, J. M. Randazzo, L. U. Ancarani, and F. D. Colavecchia, S-model calculations for high-energy-electron-impact double ionization of helium, *Phys. Rev. A* **87**, 042707 (2013).

[28] A. S. Kheifets, I. Bray, J. Berakdar, and C. Dal Cappello, Comparative theoretical study of (e , 3e) on helium : Coulomb-waves versus close-coupling approach, *J. Phys. B* **35**, L15 (2002).

[29] A. S. Kadyrov, A. M. Mukhamedzhanov, A. T. Stelbovics, I. Bray, and F. Pirlepesov, Asymptotic behavior of the Coulomb three-body scattered wave, *Phys. Rev. A* **68**, 022703 (2003).

[30] E. O. Alt and A. M. Mukhamedzhanov, Asymptotic solution of the Schrödinger equation for three charged particles, *Phys. Rev. A* **47**, 2004–2022 (1993).

[31] G. Gasaneo, L. U. Ancarani, D. M. Mitnik, J. M. Randazzo, A. L. Frapiccini, and F. D. Colavecchia, Three-Body Coulomb Problems with Generalized Sturmian Functions, *Adv. Quant. Chem.* **67**, 153–216 (2013).

[32] J. M. Randazzo, L. U. Ancarani, G. Gasaneo, A. L. Frapiccini, and F. D. Colavecchia, Generating optimal Sturmian basis functions for atomic problems, *Phys. Rev. A* **81**, 042520 (2010).

[33] J. M. Randazzo, A. L. Frapiccini, F. D. Colavecchia, and G. Gasaneo, Discrete sets of Sturmian functions applied to two-electron atoms, *Phys. Rev. A* **79**, 022507 (2009).

[34] J. M. Randazzo, A. L. Frapiccini, F. D. Colavecchia, and G. Gasaneo, Discrete Sets of Many-Body Sturmians, *Int. J. Quant. Chem.* **109**, 125–134 (2009).

[35] V. V. Serov, V. L. Derbov, B. B. Joulakian, and S. I. Vinitsky, Wave-packet-evolution approach for single and double ionization of two-electron systems by fast electrons, *Phys. Rev. A* **75**, 012715 (2007).

[36] L. U. Ancarani, C. D. Cappello, and G. Gasaneo, Double ionization of two-electron systems, *J. Phys.: Conf. Ser.* **212**, 012025 (2010).
[37] S. Jones and D. H. Madison, Role of the Ground State in Electron-Atom Double Ionization, *Phys. Rev. Lett.* **91**, 073201 (2003).
[38] L. U. Ancarani, T. Montagnese, and C. Dal Cappello, Role of the helium ground state in $(e, 3e)$ processes, *Phys. Rev. A* **70**, 012711 (2004).
[39] L. U. Ancarani, G. Gasaneo, F. D. Colavecchia, and C. Dal Cappello, Interplay of initial and final states for $(e, 3e)$ and $(\gamma, 2e)$ processes on helium, *Phys. Rev. A* **77**, 062712 (2008).
[40] A. S. Kheifets, Second-order Born model for two-electron atomic ionization by fast charged-particle impact, *Phys. Rev. A* **69**, 032712 (2004).
[41] M. Grin, C. dal Cappello, R. el Mkhanter, and J. Rasch, Angular distribution in (e, 3e) collisions on helium, *J. Phys. B* **33**, 131-141 (2000).
[42] L. U. Ancarani, T. Montagnese, and C. Dal Cappello, *IOP Conf. Proc.* **183**, 21-31 (2005).
[43] J. Berakdar, Incremental Approach to Strongly Correlated Many-Body Finite Systems., *Phys. Rev. Lett.* **85**, 4036–9 (2000).
[44] S. A. Zaytsev, V. A. Knyr, Y. V. Popov, and A. Lahmam-Bennani, Application of the J-matrix method to Faddeev-Merkuriev equations for (e,2e) reactions: Beyond pseudostates, *Phys. Rev. A* **75**, 022718 (2007).

Chapter 7

Interaction potentials for atoms in front of a LiF(001) surface

M.S. Gravielle[1]*, J.E. Miraglia[1] and G.A. Bocan[2]

[1]*Instituto de Astronomía y Física del Espacio (IAFE, UBA-CONICET), casilla de correo 67, sucursal 28, C1428EGA, Buenos Aires, Argentina*

[2]*Consejo Nacional de Investigaciones Científicas y Técnicas (CONICET) and Centro Atómico Bariloche, Av. Bustillo 9500, 8400 S.C. de Bariloche, Argentina*

We present certain precise representations of the atom-LiF(001) interaction for both multielectronic and hydrogen atoms. For the former, we introduce a pairwise additive (PA) potential model which includes an improved description of the electron density associated with the different lattice sites, as well as non-local electron density contributions. We apply this PA model to evaluate the interaction potential between the LiF surface with the closed-shell atoms (He, Ne, Ar, Kr) as well as with the open-shell atoms (N, S, Cl). In turn, for the H projectiles, we build a high-precision interaction potential from the density functional theory calculations, that properly takes into account the reactivity of the H-LiF(001) system, including the contribution from the van der Waals (vdW) forces.

The performance of the proposed potentials is assessed by comparing angular projectile distributions produced by fast grazing incidence with available experimental data. For multi-electronic projectiles, the angular positions of rainbow and supernumerary rainbow maxima derived within the PA potential are in good agreement with the experimental data, confirming the validity of the potential model in a wide normal energy range, from 0.2 to 80 eV. In all the cases, the vdW contributions are estimated to be negligible. In contrast, for the H atoms, the comparison of our simulations with experimental grazing-incidence fast atom diffraction patterns strongly suggests a non-negligible role of the vdW interactions in the low normal energy regime when incidence is along the ⟨100⟩ channel.

*Corresponding author. Email: msilvia@iafe.uba.ar

Contents

1. Introduction

A precise description of the interaction potential for atoms placed in front of crystal surfaces is a crucial ingredient for the study of diverse processes in surface science, as well as for the development of several modern technologies, for example in the chemical industry. In the case of LiF(001), since this ionic crystal is considered as a prototype of insulator material, its interaction with different atoms has been extensively investigated over the years. The proposed potential models for LiF surfaces range from semi-empirical approaches [1] and pairwise additive (PA) formalisms [2–4] to the descriptions based on the density functional theory (DFT) [5–7]. In the present work, we make use of the two latter methodologies in order to obtain an accurate representation of the atomic interaction with LiF(001). That is, a *non-local* PA model for multi-electronic atoms, ranging from the He to the Kr projectiles, and a high-quality interaction potential built from the DFT total energy calculations to deal with the more reactive hydrogen atom.

The PA potentials are built as a sum of binary interatomic potentials that represent the interaction of the atomic projectile with individual atomic centers in the crystal. For ionic materials, like LiF, where the electron density is highly localized around the atomic nuclei, these potential models provide an adequate representation of the atom-surface interaction, becoming a reliable alternative to more complex self-consistent *ab-initio* calculations [8–13]. However, in most works the binary potentials are derived within the local density approximation (LDA) to evaluate the kinetic and exchange contributions [14]. Thus, they do not include non-local electron density terms, which account for the inhomogeneity of the electron distributions of both crystal and atomic projectile.

Our aim is to upgrade the previous PA models [12, 15] by incorporating non-local contributions of the electron density, along with a more realistic description of the electron density associated with each ionic center of the insulator [16]. To determine the electron density around a given ionic center of the LiF crystal, we consider concentric spherical shells within which the Coulomb charges of the neighboring ions are assumed to be uniformly distributed. The resulting image resembles that of an onion and thereby we refer to this model as the onion model [17]. The potential generated by the ionic grid on a given ion, i.e. the Madelung potential, is thus taken into account in the calculations by means of such an onion model [17]. From the electronic densities of these Li^+ and F^- onions, the binary interatomic potentials are then derived by using well-established non-local functionals for the kinetic, exchange and correlation terms. The model takes also into account dispersive contributions originated from the rearrangement of the atom and crystal electron densities as a consequence of their mutual interaction, and the reconstruction of the topmost atomic layer (rumpling). In the present chapter, this improved PA model is applied to two different kinds of multi-electronic atoms interacting with LiF(001): (a) rare gases (He, Ne, Ar, Kr) that are closed-shell atoms with complete valence shells, and (b) spin-polarized atoms (N, S, Cl) that are open-shell atoms with vacancies in the outer level.

Despite their satisfactory performance for multi-electronic atoms, the PA models are not fully suitable for describing the more involved chemical nature of the H-LiF(001) interaction. Thus, the latter was modeled by the DFT calculations where projector-augmented wave (PAW) pseudopotentials were used to describe the interaction of valence. Moreover, core electrons and exchange-correlation were represented by the generalized gradient approximation (GGA) using the Perdew-Burke-Ernzerhof (PBE) functional. In addition, the van der Waals (vdW) dispersion forces, the influence of which we specifically analyzed, were included following Grimme's semi-empirical approach [18].

To test the quality of the present potential models, we resort to grazing-incidence fast atom diffraction (GIFAD or FAD) [19, 20], a powerful surface analysis technique which allows one to inspect the potential energy surface (PES) with an exceptional sensitivity [21–23]. We use experimental GIFAD data available in the literature to illustrate the good performance of our potentials in the energy range from 10 meV to few eV. However, for higher energies, as the interference structures of GIFAD spectra vanish, we make use of the angular positions of the rainbow peaks, which have a classical origin and result in a useful magnitude to probe surface interactions for energies up to 80 eV [9, 24, 25]. In all the cases, we describe the scattering process by means of the surface initial-value representation (SIVR) approximation [26, 27], which is a semi-quantum method that offers a very good representation of the GIFAD distributions in the whole angular range.

The chapter is organized as follows. The surface interaction with multi-electronic projectiles, described with the PA model, is addressed in Section 2. In this latter section, we provide details about the evaluation of (i) the atom and crystal electron densities (subsection 2.1), (ii) the partial contributions to the binary interatomic potentials (subsection 2.2), and (iii) the complete atom-surface potential, including projectile polarization and rumpling (subsection 2.3). In Section 3, the proposed PA model is examined by comparing our SIVR simulations with experimental data for axial grazing scattering. In relation to the more reactive H projectile, in Section 4.1, we present a precise DFT calculation to deal with the H-LiF(001) interaction. The main features of the proposed PESs, with and without the inclusion of the vdW forces, are analyzed in Section 4.2, while in Section 5 the influence of the vdW contributions on GIFAD spectra is studied by comparing our theoretical results with available experimental data. In Section 6, we summarize our conclusions. Atomic units ($e^2 = \hbar = m_e = 1$) are used unless otherwise stated.

2. Multi-electronic projectiles: the PA model

For multi-electronic atoms interacting with a LiF(001) surface, we derive a PA model, which is expressed as a sum of binary interatomic potentials that depend on the respective electron densities of the atom and the crystal. In this section, we will summarize all the steps required in order to: (i) determine the electron densities of both the atomic projectile and the ionic centers of the LiF crystal, (ii) use these latter densities as the input data to build the binary interatomic potentials, and (iii) evaluate the atom-surface interaction.

2.1. *Electron densities*

The electron density of the atomic projectile can be straightforwardly obtained from the square modulus of the corresponding Hartree-Fock (HF) wave functions extracted from Ref. [28]. But for the evaluation of the electron density associated with an ionic center within the crystal lattice, we need to resort to an approximate model - an onion description - in order to include the effect of the Madelung potential. In this subsection, we provide details about the derivation of the individual cation and anion electron densities in the LiF crystal within the onion model [17].

The basic idea of the onion model is to approximate the Madelung potential, produced by the whole ionic lattice at a position around a given center of the LiF crystal, by the potential originated from the spherical charged shells that take into account the asymptotic Coulomb interaction of the neighboring ions. For this purpose, we start considering a perfect cubic sample of crystal centered on an ionic site, for example, on top of a Li^+ ion. This central cation will be chosen as the active one, being excluded from the crystal grid for the evaluation of the lattice screening interaction. Hence, the grid potential produced by the long distance Coulomb interaction between the grid ions and an electron bound to the active cation reads as:

$$V_{\mathrm{G}}^{-}(\boldsymbol{r}) = \sum_{\boldsymbol{j}\neq 0} e_{\boldsymbol{j}} \, \frac{(-1)^{j_x+j_y+j_z}}{|\boldsymbol{r}-\boldsymbol{R}_{\boldsymbol{j}}|}, \tag{1}$$

where the vector $\boldsymbol{r}$ denotes the electron position and

$$\boldsymbol{R}_{\boldsymbol{j}} = \frac{a}{2}\boldsymbol{j}, \quad \text{with } \boldsymbol{j} = (j_x, j_y, j_z), \tag{2}$$

are the positions of the grid ions, a being the lattice parameter ($a = 7.597$ a.u. for LiF), and j_x, j_y, and j_z are integer numbers ranging from -5 to 5. In Eq. (1), the factors $e_{\boldsymbol{j}}$ represent the Evjen caging factors [$e_{\boldsymbol{j}} = 1$ except for ions placed at the limiting surface ($e_{\boldsymbol{j}}$ =1/2), at an arista ($e_{\boldsymbol{j}}$ =1/4) or at a vertex ($e_{\boldsymbol{j}}$ =1/8)] that warranty the charge neutrality of the sample, and the sum on j_x, j_y, and j_z excludes $\boldsymbol{j} = (0,0,0)$, which corresponds to the position of the active ion. To derive the onion model, we assume that all the neighboring ions placed a distance $R_j = |\boldsymbol{R}_{\boldsymbol{j}}|$ from the active ionic center make up a spherical shell with a superficial density of charge

$\sigma_j = Q_j/(4\pi R_j^2)$, where Q_j is the net Coulomb charge of the ions spread uniformly over the shell [17]. Then, we fit this onion potential by means of an effective Madelung potential [16]:

$$V_G^-(\boldsymbol{r}) \simeq V_M^-(r) = \frac{1}{r} - \frac{e^{-r/\lambda}}{r}\left[1 + (1 - \lambda V_{M0})\frac{r}{\lambda} + \left(\frac{1}{2} - \lambda V_{M0}\right)\left(\frac{r}{\lambda}\right)^2\right]. \quad (3)$$

Here, $\lambda \cong 0.3\ a/2$ and $V_{M0} = 0.4600$ a.u. yields the proper Madelung potential at the origin. Notice that $V_M^-(r)$ yields the correct asymptotic limit at long distances, i.e., $V_M^-(r) \to 1/r$ as $r \to +\infty$.

In this review, we use the effective Madelung potential of Eq. (3) to derive a more realistic electron density around individual crystal ions. The two-electron wave function Ψ_{Li} corresponding to an active Li^+ cation immersed in the LiF crystal, denoted as the $Li^+_@$ onion, can be obtained by solving the Schrödinger equation associated with the onion Hamiltonian:

$$H_{Li} \cong \sum_{l=1}^{2}\left(-\frac{1}{2}\nabla^2_{\boldsymbol{r}_l} - \frac{Z_{Li}}{r_l} + V_M^-(r_l)\right) + \frac{1}{2}\sum_{\substack{l,m=1\\ m\neq l}}^{2}\frac{1}{r_{lm}}. \quad (4)$$

Here, $\boldsymbol{r}_l$ is the position vector of the l electron ($l = 1, 2$) with respect to the Li^+ nuclear charge $Z_{Li} = 3$ and $r_{lm} = |\boldsymbol{r}_l - \boldsymbol{r}_m|$ is the interelectronic distance. In a similar way, the multi-electronic wave function Ψ_F corresponding to the F^- anion immersed in the LiF crystal, the $F^-_@$ onion, can be derived from the approximate Hamiltonian:

$$H_F \cong \sum_{l=1}^{10}\left(-\frac{1}{2}\nabla^2_{\boldsymbol{r}_l} - \frac{Z_F}{r_l} + V_M^+(r_l)\right) + \frac{1}{2}\sum_{\substack{l,m=1\\ m\neq l}}^{10}\frac{1}{r_{lm}}, \quad (5)$$

where $Z_F = 9$ is the nuclear charge of the active anion and $V_M^+(r) = -V_M^-(r)$ is the effective Madelung potential. Again $V_M^+(r)$ yields the correct asymptotic limit, i.e., $V_M^+(r) \to -1/r$ as $r \to +\infty$, reproducing the Coulomb potential due to the unit charge which renders the Coulomb hole that the electron leaves behind when it is removed.

In Fig. 1, we plot the unperturbed electron densities associated with the individual $Li^+_@$ and $F^-_@$ onions. These onion electron densities were obtained as the square modulus of the corresponding wave functions, Ψ_{Li} or Ψ_F, derived from full HF calculations based on the Hamiltonians of Eqs. (4) and (5), respectively [16]. In both cases, the onion electron density, as a function of the radial distance to the atomic nucleus, is compared to that of the isolated ion, Li^+ or F^-, also derived from the HF wave functions [28]. For the F^- anions, the contribution of the Madelung potential noticeably affects

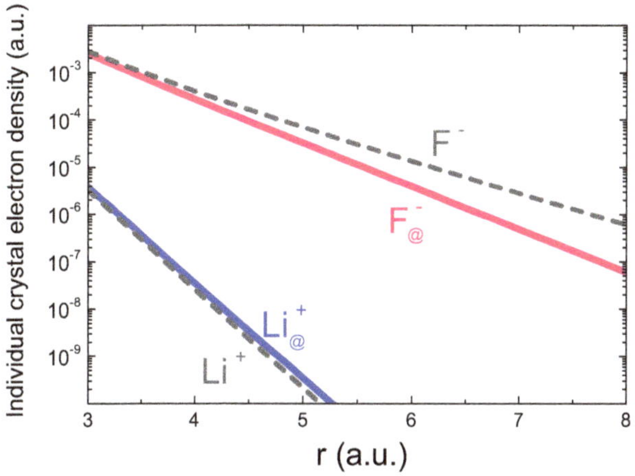

Fig. 1. Individual crystal electron density, as a function of the radial distance r to the ionic nucleus, for Li^+ and F^- ionic centers of the LiF crystal. Solid line, electron density derived from the onion model; dashed line, electron density corresponding to the isolated ion [28].

the electron density at radial distances larger than 4 a.u. Importantly, reflection in GIFAD takes place at this same distance range, thus providing one of the most useful probes for the effect of the Madelung interaction in the PA potentials. In contrast, for the Li^+ cations, the effect of the Madelung potential on the electron density is almost negligible due to the comparatively higher energy of the Li^+-bound electrons.

2.2. *Binary interatomic potentials*

The binary interatomic potentials between the neutral atom A of nuclear charge Z_{A} and electron density $n_{\mathrm{A}} = n_{\mathrm{A}}(\boldsymbol{r})$, and the crystal onions O, of nuclear charge Z_{O} and electron density $n_{\mathrm{O}} = n_{\mathrm{O}}(\boldsymbol{r})$, with $\mathrm{O} = \mathrm{Li}^+_@$ or $\mathrm{O} = \mathrm{F}^-_@$ and $Z_{\mathrm{Li}^+_@} = 3$ and $Z_{\mathrm{F}^-_@} = 9$, represent the elemental bricks used to build our PA model. These binary potentials are composed of two different contributions: a *static* potential, $V_{\mathrm{AO}}^{(\mathrm{static})}$, that describes the interaction between the atomic projectile A and the ionic center O by considering their unperturbed electron distributions as frozen, and a *dispersive* potential, $V_{\mathrm{AO}}^{(\mathrm{disper})}$, which takes into account the rearrangement of the atomic and onion electron densities as a result of their mutual interaction.

The static interatomic potential between the atom A and the onion O can be approximated as [29, 30]:

$$V_{\rm AO}^{\rm (static)}(R) = V_{\rm AO}^{\rm (e)}(R) + V_{\rm AO}^{\rm (k)}(R) + V_{\rm AO}^{\rm (x)}(R) + V_{\rm AO}^{\rm (c)}(R), \tag{6}$$

where $\boldsymbol{R}$ is the internuclear vector and the terms $V_{\rm AO}^{\rm (i)}$, with i = e,k,x,c, represent the electrostatic, kinetic, exchange and correlation potentials, respectively. All these partial contributions depend only on $R = |\boldsymbol{R}|$ as a consequence of the spherical symmetry of the interacting partners. In Eq. (6), the electrostatic interaction is expressed, as usual, in terms of $n_{\rm A}$ and $n_{\rm O}$ and reads as:

$$\begin{aligned} V_{\rm AO}^{\rm (e)}(R) = {} & \frac{Z_{\rm A} Z_{\rm O}}{R} - \int d\boldsymbol{r}\, \frac{Z_{\rm A} n_{\rm O}(\boldsymbol{r})}{|\boldsymbol{r} - \boldsymbol{R}|} - \int d\boldsymbol{r}'\, \frac{Z_{\rm O}\, n_{\rm A}(\boldsymbol{r}')}{|\boldsymbol{r}' + \boldsymbol{R}|} \\ & + \iint d\boldsymbol{r} d\boldsymbol{r}'\, \frac{n_{\rm A}(\boldsymbol{r}') n_{\rm O}(\boldsymbol{r})}{|\boldsymbol{R} + \boldsymbol{r}' - \boldsymbol{r}|}, \end{aligned} \tag{7}$$

but for the derivation of the remaining terms - $V_{\rm AO}^{\rm (k)}$, $V_{\rm AO}^{\rm (x)}$, and $V_{\rm AO}^{\rm (c)}$ - we assume that the total electron density of the atom-onion system at a given internuclear separation $\boldsymbol{R}$ can be approximated as [30]:

$$n_{\rm tot}(\boldsymbol{R}) = n_{\rm O}(\boldsymbol{r}) + n_{\rm A}(\boldsymbol{r} - \boldsymbol{R}). \tag{8}$$

Hence, the kinetic, exchange and correlation potentials can be obtained as [30]:

$$V_{\rm AO}^{\rm (i)}(R) = E_{\rm i}[n_{\rm tot}(\boldsymbol{R})] - E_{\rm i}[n_{\rm O}] - E_{\rm i}[n_{\rm A}], \quad \text{for } \ {\rm i = k, x, c}, \tag{9}$$

where the functionals $E_{\rm i}[n]$ for i = k, x, c represent the kinetic, exchange and correlation energies, respectively.

In this work, in contrast with previous models [12, 15] that involve the LDA calculations, the kinetic, exchange and correlation energies are evaluated from the well-established functionals depending not only on the local electron density $n(\boldsymbol{r})$, but also on non-local magnitudes, like $\nabla n(\boldsymbol{r})$ and $\nabla^2 n(\boldsymbol{r})$ that take into account the non-uniform behavior of the electron density. For the kinetic term, i = k, we use the Lee-Lee-Parr (LLP) approach [31], which is given by:

$$E_{\rm k}[n] = c_{\rm k} \int d\boldsymbol{r}\; n(\boldsymbol{r})^{5/3} \left(1 + \beta_{\rm k} G(\boldsymbol{r})\right), \tag{10}$$

with

$$G(\boldsymbol{r}) = \frac{g(\boldsymbol{r})^2}{1 + \gamma g(\boldsymbol{r}) \sinh^{-1}[g(\boldsymbol{r})]}, \; g(\boldsymbol{r}) = |\nabla_{\boldsymbol{r}} n(\boldsymbol{r})| / n(\boldsymbol{r})^{4/3}, \tag{11}$$

where c_{k}, β_{k} and γ are the constants. For the exchange energy, i = x, we employ the spin-dependent Becke (B) approximation [32]:

$$E_{\mathrm{x}}[n] = c_{\mathrm{x}} \int d\boldsymbol{r}\; n(\boldsymbol{r})^{4/3} \left(1 + \beta_{\mathrm{x}} G(\boldsymbol{r})\right), \tag{12}$$

where c_{x} and β_{x} are the constants. For the correlation energy, i = c, we use the Lee-Yang-Parr (LYP) approach, given by Eqs. (21) and (22) of Ref. [33], which are valid for closed- and open- shell atoms, respectively (including also $\nabla^2 n$). Then, according to the common notation for the DFT functionals [34], our static potential model should be denoted by the acronym LLPB3LYP, i.e., Lee-Lee-Parr+Becke+3-coefficient-Lee-Yang-Parr approximations for kinetic, exchange and correlation terms.

In Figs. 2 and 3, we display the electrostatic, kinetic, exchange and correlation potentials, along with the total *static* contribution given by Eq. (6), for closed- and open- shell atoms, respectively, interacting with $F^-_{@}$ (left panels) and $Li^+_{@}$ (right panels) onions. While the closed-shell atoms - He, Ne, Ar, Kr - correspond to rare gases, with fully occupied valence levels, the considered open-shell atoms - $N(^4S)$, $S(^3P)$ and $Cl(^2P)$ - present a ferromagnetic character, with partially occupied outer shells and different spin-up and spin-down electron densities. Note that for most of these projectiles, there are experimental data of the rainbow and/or GIFAD distributions available in the literature [5, 12, 13], which makes it possible to test the proposed potential model by means of grazing scattering experiments.

With the aim of analyzing the asymptotic limits of the static potential, in Figs. 2 and 3, the different terms of Eq. (6) are displayed by means of the function:

$$F^{(\mathrm{i})}_{\mathrm{AO}}(R) = R(1 + 2R^3) V^{(\mathrm{i})}_{\mathrm{AO}}(R), \quad \text{with } \mathrm{i} = \mathrm{static}, \mathrm{e}, \mathrm{k}, \mathrm{x}, \mathrm{c}. \tag{13}$$

This can be straightforwardly compared with both the internuclear and the projectile polarization contributions; the latter being ruled by the atomic polarizability α_{A}, as will be explained below. At the origin, the internuclear Coulomb interaction, corresponding to the first term of Eq. (7), dominates the static binary potential, producing $F^{(\mathrm{static})}_{\mathrm{AO}}(R) \to Z_{\mathrm{A}} Z_{\mathrm{O}}$ as $R \to 0$. As expected, the repulsive behavior of $V^{(\mathrm{e})}_{\mathrm{AO}}(R)$ prevails at small internuclear distances, but when R increases, first the kinetic and then the exchange terms gain importance. At this point, since Figs. 2 and 3 show absolute values, it is important to indicate the sign of each contribution. While the electrostatic potential for the $F^-_{@}$ onion switches its sign, becoming attractive (negative) for $R \gtrsim 1$ a.u. as a consequence of

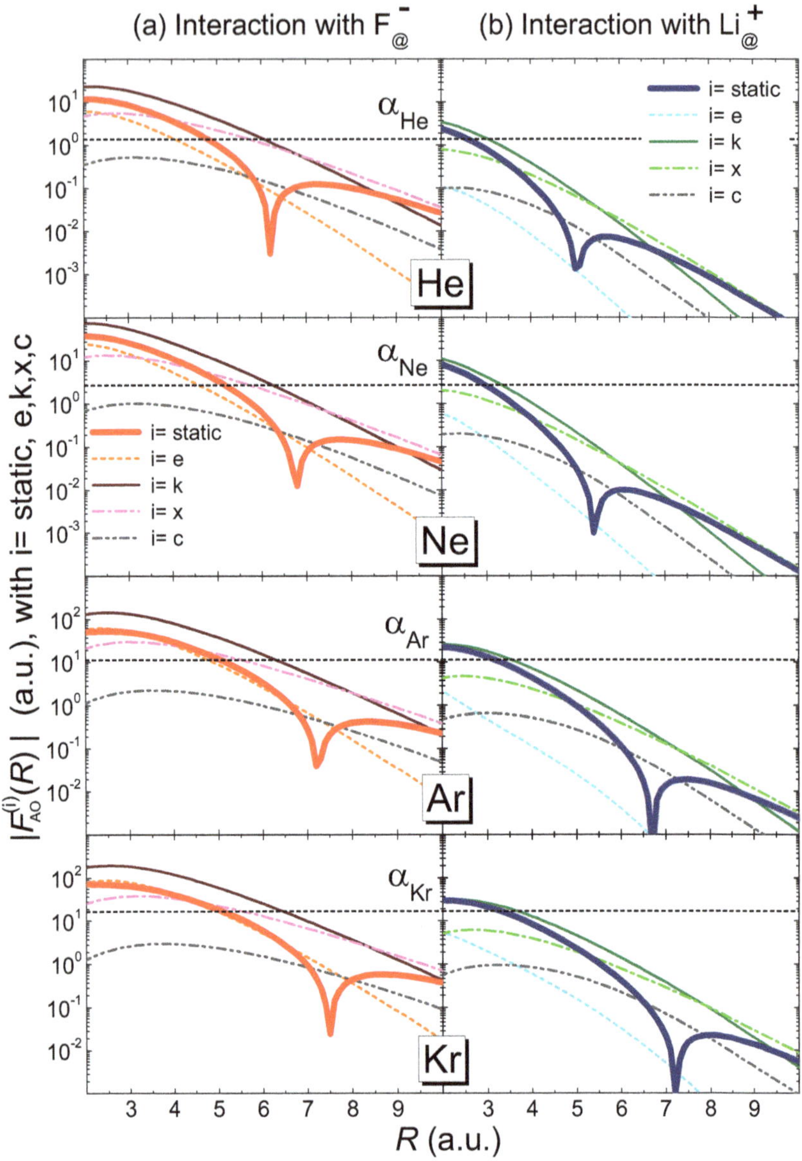

Fig. 2. Partial static contributions to the binary atom-onion potential for different closed-shell atoms. Absolute value of $F_{\rm AO}^{({\rm i})}(R)$, i = static, e, k, x, c, given by Eq. (13), as a function of the internuclear distance R, for the interaction with (a) $\rm F_{@}^{-}$ (left panels) and (b) $\rm Li_{@}^{+}$ (right panels) onions. For comparison, the horizontal dashed lines indicate the values of the corresponding projectile polarizabilities, $\alpha_{\rm A}$.

its own electronic screening, the other terms keep their signs along the whole R-range. That is, the kinetic term displays a repulsive behavior, i.e., $F_{\rm AO}^{(k)}(R) > 0$, while the exchange and correlation contributions are always attractive, with $F_{\rm AO}^{(x)}(R), F_{\rm AO}^{(c)}(R) < 0$. Regarding their relative absolute values, when R increases and consequently, the electronic density $n_{\rm tot}$ at intermediate positions between the A and O partners decreases, each term of Eq. (6) presents a different long-range dependence on the Seitz radius [35], given by $r_s = \sqrt[3]{3/(4\pi n_{\rm tot})}$. In particular, for large r_s (low $n_{\rm tot}$) values, the kinetic contribution becomes proportional to r_s^{-2} and decreases faster than the exchange term [30], which is proportional to r_s^{-1}. Hence, at distances longer than 6 a.u., the kinetic and exchange terms cancel each other, making the correlation contribution relevant. Therefore, this interplay of the different interaction mechanisms causes $F_{\rm AO}^{(\rm static)}(R)$ to change from positive to negative as R increases, giving rise to a potential well which is a characteristic of the interatomic potentials [36]. The dip of the curve corresponding to $F_{\rm AO}^{(\rm static)}(R)$, shown in Figs. 2 and 3, indicates the precise position where the sign change occurs, and this feature plays an important role within atom-surface potential models applied to the GIFAD processes. In addition, since the long-range behavior of the different terms of Eq. (6) is governed by the electron density, in Figs. 2 and 3, the R value corresponding to the dip position increases as the size of the atomic projectile augments, being larger for heavier projectiles than for lighter ones.

Concerning the dispersive potential $V_{\rm AO}^{(\rm disper)}$, since it dominates at long distances between the interacting partners, its contribution can be evaluated within a perturbative formalism. The dispersive binary potential for a given atom-onion pair is approximated, for large R values, as [37]:

$$V_{\rm AO}^{(\rm disper)}(R) \rightarrow -\frac{\alpha_{\rm A}}{2R^4} - \frac{C_{\rm AO}^{(6)}}{R^6} - \dots, \tag{14}$$

where, as before, R is the internuclear distance. In Eq. (14), $\alpha_{\rm A}$ denotes the static polarizability of the atom A, while the parameter $C_{\rm AO}^{(6)}$ is related to the vdW forces, the values of both coefficients being listed in Table III of Ref. [16]. The first term of Eq. (14) is associated with the dipole polarization of the atomic projectile induced by the long-range Coulomb interaction with the net charge of the target onion O. As observed from Figs. 2 and 3 for closed- and open- shell atoms, respectively, the projectile polarization term is comparable to the static potential at internuclear distances between approximately 4 and 5 a.u., where the absolute value

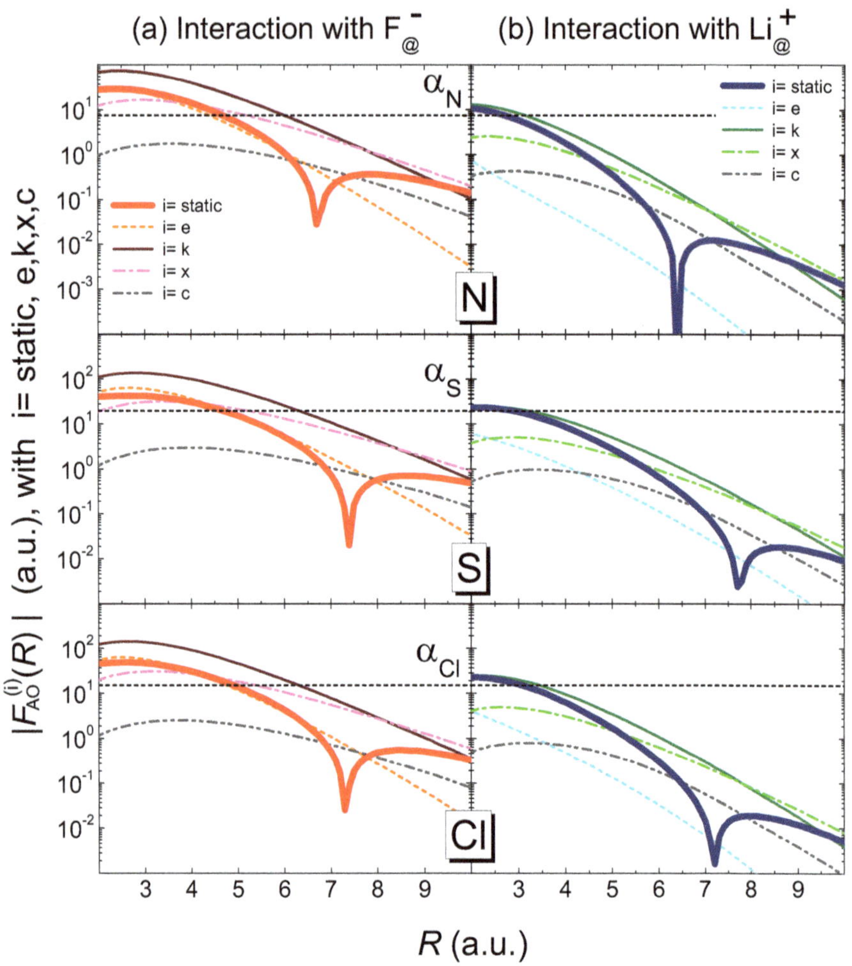

Fig. 3. Analogous to Fig. 2 for the static binary interaction of open-shell atoms - N(^4S), S(^3P), and Cl(^2P) - with: (a) $F_@^-$ and (b) $Li_@^+$ onions.

of $F_{AO}^{(static)}(R) \approx 2R^4 V_{AO}^{(static)}(R)$ is similar to α_A. Instead, for the considered multi-electronic projectiles the vdW term was found to prevail only at much longer internuclear distances [16] that are not efficiently probed by fast grazing projectiles.

2.3. *Atom-surface potential*

Within a PA scheme, the total atom-surface potential U for an atom A interacting with a LiF(001) surface can be derived from the static and dispersive potentials, given by Eqs. (6) and (14), respectively, as:

$$U(\boldsymbol{R}_{\rm A}) = U^{(\rm static)}(\boldsymbol{R}_{\rm A}) + U^{(\rm disper)}(\boldsymbol{R}_{\rm A}), \tag{15}$$

where $\boldsymbol{R}_{\rm A}$ denotes the position of the atomic projectile A with respect to the origin of the frame of reference, placed on a given ionic center of the topmost atomic layer. The first term, $U^{(\rm static)}(\boldsymbol{R}_{\rm A})$, represents the static interaction produced by the unperturbed atomic and surface electron densities, which reads as:

$$U^{(\rm static)}(\boldsymbol{R}_{\rm A}) = \sum_{\boldsymbol{j}} e_{\boldsymbol{j}} V^{(\rm static)}_{\rm AO(\boldsymbol{j})}(\boldsymbol{\rho}_{\boldsymbol{j}}), \tag{16}$$

where the index vector $\boldsymbol{j} = (j_x, j_y, j_z)$ identifies the onion $\mathrm{O}(\boldsymbol{j})$ placed at the position $\boldsymbol{R}_{\boldsymbol{j}}$ given by Eq. (2), with $\mathrm{O}(\boldsymbol{j}) = \mathrm{Li}^+_@$ or $\mathrm{F}^-_@$ according to the lattice site. In Eq. (16) $\boldsymbol{\rho}_{\boldsymbol{j}} = \boldsymbol{R}_{\rm A} - \boldsymbol{R}_{\boldsymbol{j}}$ is the position vector of the projectile A with respect to $\mathrm{O}(\boldsymbol{j})$, the factors $e_{\boldsymbol{j}}$ are the Evjen caging defined in Sec. 2.1, and the sum over $\boldsymbol{j}$ extends over $11 \times 11 \times 4 = 484$ crystal sites, including also $\boldsymbol{j} = (0,0,0)$. Furthermore, we introduce the surface rumpling $\delta_{\rm r} = 0.046$ a.u.= 0.024 Å [5] which shifts the positions of the topmost onions $\mathrm{F}^-_@$ and the $\mathrm{Li}^+_@$ by a distance $\delta_{\rm r}$ outward and inward with respect to the unreconstructed surface, respectively.

In contrast to the static contribution, to evaluate the dispersive potential $U^{(\rm disper)}$, we need to take into account for the fact that the dispersive binary terms given by Eq. (14) cannot be directly added. The dispersive potential which describes the projectile polarization effects is:

$$U^{(\rm disper)}(\boldsymbol{R}_{\rm A}) = -\frac{\alpha_{\rm A}}{2}\left|\sum_{\boldsymbol{j}} \boldsymbol{E}_{\boldsymbol{j}}(\boldsymbol{R}_{\rm A})\right|^2, \tag{17}$$

where

$$\boldsymbol{E}_{\boldsymbol{j}}(\boldsymbol{R}_{\rm A}) = f_{\boldsymbol{j}}(\rho_{\boldsymbol{j}}) Z^{(\infty)}_{\boldsymbol{j}} \frac{\boldsymbol{\rho}_{\boldsymbol{j}}}{\rho^3_{\boldsymbol{j}}} \tag{18}$$

is the electric field due to the asymptotic charge of the onion $\mathrm{O}(\boldsymbol{j})$, with $Z^{(\infty)}_{\boldsymbol{j}} = -1$ (+1) for $\mathrm{O}(\boldsymbol{j}) = \mathrm{F}^-_@$ ($\mathrm{Li}^+_@$), and $f_{\boldsymbol{j}}(\rho_{\boldsymbol{j}})$ is the screening factor given by Eq. (B1) of Ref. [16], which avoids the divergence of the field at the origin.

Notice that in Eq. (17), we have included the projectile polarization contribution only, while the dispersive vdW forces associated with the second term in Eq. (14) were left aside. This is due to the fact that for

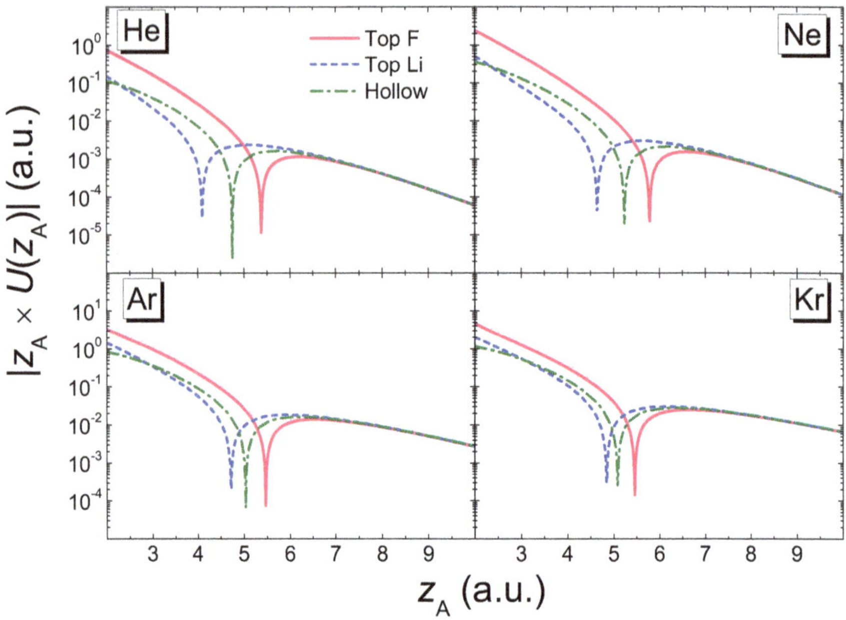

Fig. 4. Absolute value of the PA atom-surface potential $U(z_A\hat{z})$, given by Eq. (15), for closed-shell projectiles, as a function of the atom-surface distance z_A. High-symmetry positions on top of F^-, Li^- and hollow sites are considered.

multi-electronic atoms interacting with LiF(001), the contribution of the vdW forces starts to be relevant in our model at distances farther than 8 a.u. [16], i.e. away from the region probed by the GIFAD experiments. However, this feature changes drastically for the H projectiles as a consequence of the relatively high $C^{(6)}_{AO}$ values for hydrogen. In fact, for the H/LiF(001) system, the GIFAD distributions in the low normal energy regime might be used to study the vdW effects [38], as it will be discussed in Section 5.

Figures 4 and 5 show the atom-surface potential, as a function of the distance to the LiF(001) surface, for the studied closed- and open- shell atoms. In these figures, the total potential $U(\boldsymbol{R}_A)$, including the static and dispersive contributions as given by Eq. (15), is plotted for the positions on top of different high-symmetry sites.

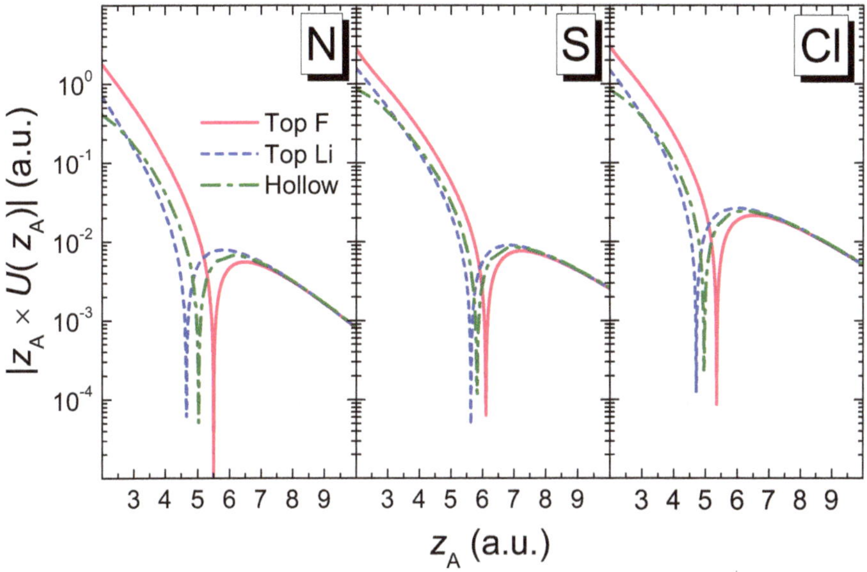

Fig. 5. Similar to Fig. 4, except for the open-shell atoms.

3. Experimental examination of the PA model for multi-electronic projectiles

In this Section, the performance of the PA potentials introduced in Section 2, is investigated by comparing the theoretical angular distributions of fast multi-electronic atoms scattered off a LiF(001) surface (under the axial grazing incidence conditions [21]) with the available experimental data. The projectile spectra are obtained by using the proposed potential, given by Eq. (15), within the SIVR approximation [26, 27]. The SIVR approach uses the method of Miller [39] to determine the semi-quantum time evolution operator from the Feynman path integral formulation, incorporating this operator within a time-dependent distorted wave formalism. In this way, the SIVR approximation accounts for quantum effects, like interferences or classical forbidden transitions, without losing a clear representation of the main mechanisms in terms of classical dynamics.

For the LiF surfaces, the angular distributions of scattered projectiles are completely governed by the normal energy $E_\perp$, associated with the motion in the plane perpendicular to the incidence channel, when they are

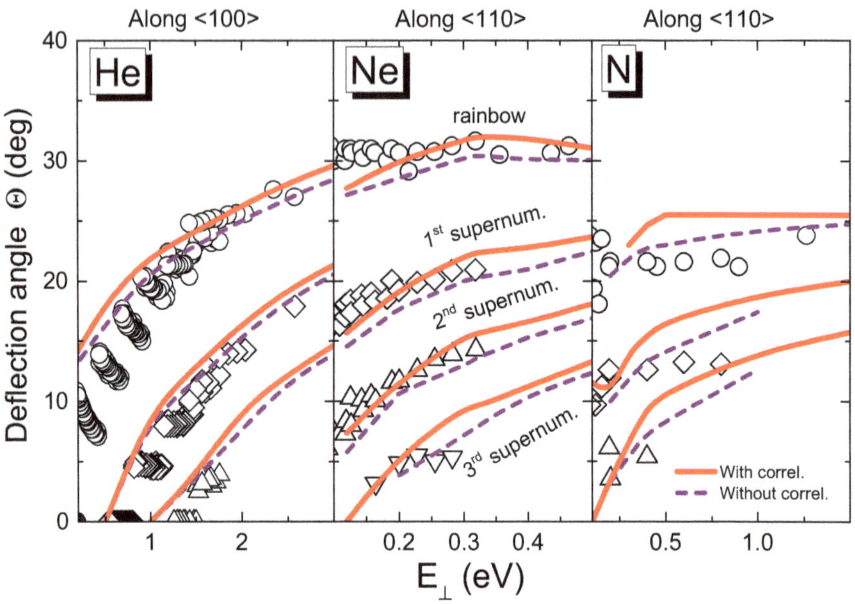

Fig. 6. Deflection angles Θ corresponding to the GIFAD maxima, as a function of the normal energy $E_\perp$, for the He^3 atoms scattered along ⟨110⟩ (left), and for Ne (middle) and N (right) atoms impinging along the ⟨110⟩ direction. In all the panels, solid (dashed) line, SIVR rainbow and supernumerary rainbow angles derived with (without) the inclusion of correlations in the PA model. Symbols: experimental data extracted from Refs. [5, 12, 16].

plotted as a function of the deflection angle Θ, with $\Theta = \arctan(\varphi_f\,/\theta_f)$ measured on the detection plane, θ_f (φ_f) being the polar (azimuthal) exit angle. For low normal energies, of the order of a few eV, the GIFAD spectra present noticeable interference structures, which come from the mix of *inter-* and *intra-* channel interferences. The *inter*-channel interference is produced by the periodic array of channels and gives rise to equally spaced Bragg diffraction peaks, while the *intra*-channel interference is due to interference inside one channel, which generates the rainbow and supernumerary rainbow maxima. Hence, the typical GIFAD patterns display Bragg peaks symmetrically placed with respect to the specular reflection direction, whose intensities become modulated by the underlying intra-channel mechanism. Since rainbow and supernumerary rainbow maxima are extremely sensitive to the shape of the PES across the incidence channel [6, 21], in the present work we use their angular positions to probe the potential model for different distances from the surface by varying the $E_\perp$ value.

In Fig. 6, angular positions of rainbow and supernumerary rainbow maxima for the He atoms impinging along the ⟨100⟩ direction, as well as for Ne and N incidence along the ⟨110⟩ channel, are compared with the GIFAD experiments as a function of $E_\perp$ [5, 12, 16]. Notice that, in contrast to rainbow peaks, supernumerary maxima originate from quantum interference, such that their positions depend on the projectile mass through the de Broglie wavelength [5]. Thus, the mass of helium projectiles - He^3 - must be explicitly indicated. Furthermore, the $E_\perp$-range where the GIFAD patterns are visible, reduces as the projectile mass increases, being smaller for the Ne than for the He impact. For the closed-shell atoms - He and Ne - the angular positions of rainbow and supernumerary rainbows obtained from the SIVR approximation by using the proposed PA potential are in good agreement with the experimental data. In relation to the correlation term of Eq. (6), its contribution is more important for the Ne and N projectiles than for the He impact, as expected. In particular, we found that for the Ne projectiles, the agreement with the experiment deteriorates when the correlation term is not included in the binary potentials, making the results run slightly below the experimental data. On the contrary, for the open-shell N atoms, the experimental values are fairly well reproduced when $V_{\mathrm{AO}}^{(c)}$ is neglected, while the inclusion of the correlation term in the potential model leads to a weak overestimation of the experimental angles.

In order to extend the experimental examination of our PA potential to high normal energies, for which the GIFAD patterns blur, we resort to the rainbow peaks, which correspond to the outermost maxima of the angular spectrum and have a classical origin. The angular positions of the rainbow maxima, Θ_{rb}, depend on the corrugation of the potential across the incidence direction, giving a stringent test for the potential model [9, 25]. In Fig. 7, we compare Θ_{rb} values (derived from the SIVR simulations based on the present PA potential) with the experimental data [12, 13] for the closed-shell - Kr and Ne - as well as the open-shell - Cl and S - atoms colliding along the ⟨100⟩ and ⟨110⟩ directions, respectively. For the closed-shell atoms and different incidence directions, the theoretical rainbow angles run along the experiments in the whole energy range. But for the considered open-shell atoms, this good accord is observed up to the intermediate $E_\perp$-values, while

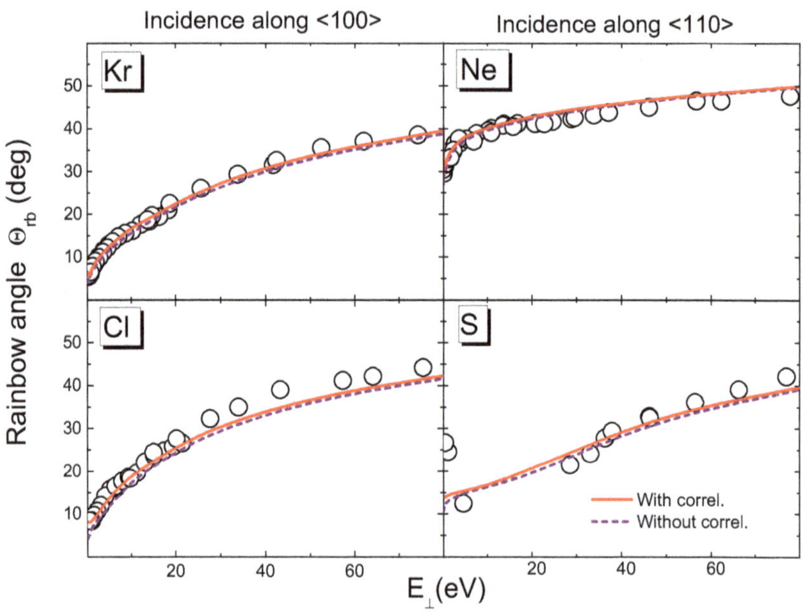

Fig. 7. Rainbow angle $\Theta_{\rm rb}$, as a function of the normal energy $E_\perp$, for closed-shell Kr and Ne (open-shell Cl and S) atoms impinging along the $\langle 110\rangle$ and $\langle 100\rangle$ channels, respectively. Symbols, experimental data from Refs. [12, 13]; lines, analogous to Fig. 6.

in the high normal energy region, the theoretical curves slightly underestimate the experimental data. Furthermore, in all the cases, the theoretical $\Theta_{\rm rb}$-curves derived with and without the inclusion of the correlation term in Eq. (6) are very alike, indicating that the correlation contribution plays a minor role for the determination of rainbow angles.

4. Hydrogen projectile: the DFT model

4.1. *The H-surface potential*

In this Section, we address the interaction potential for hydrogen atoms placed in front of a LiF(001) surface, for which the PA descriptions might be unsuitable due to the high reactivity of the interacting partners. Consequently, for the H-LiF(001) system, we build the potential from the high precision DFT calculations, performed with the VASP code [40–43], combined with a sophisticated multi-dimensional interpolation technique [44].

For the DFT calculations, we use pseudopotentials for each species (H, Li, F) where the electron-core interaction is described by means of the PAW pseudopotentials [45], while the exchange-correlation term is calculated within a semi-local GGA level, with the well-accepted PBE functional [46]. We will refer to the resulting PES as PAW-PBE. In order to assess the role of the vdW interactions, both on the PES and on a GIFAD pattern, we follow Grimme's approach [18] and introduce a semi-empirical correction to the self-consistent energy obtained with the PAW-PBE calculation. We will refer to the resulting PES as PAW-PBE-vdW. All the calculations include spin-polarization of the H atom.

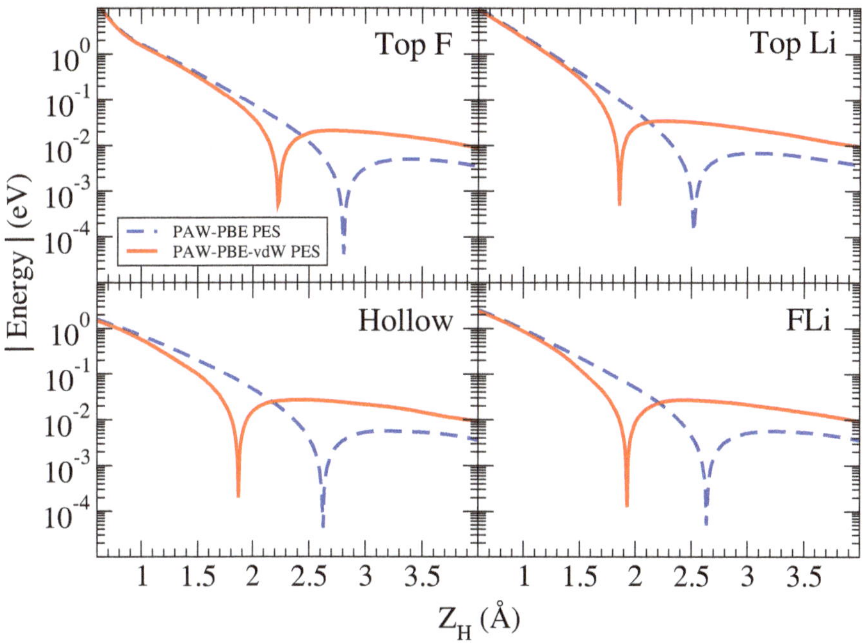

Fig. 8. Potential energy (absolute value) for the H-LiF(001) system obtained with the PAW-PBE PES (dashed line) and the PAW-PBE-vdW PES (full line). The H atom is on top of a) an F site, b) a Li site, c) a hollow site (mid-distance between neighboring F sites), d) an FLi site (mid-distance between neighboring F and Li sites)

The DFT calculations are performed for a selection of 6 high-symmetry (X_i, Y_i) configurations and 24 Z_i values. A three-dimensional (3D) potential energy grid $E(\boldsymbol{R}_{\mathrm{H}})$ is then built, where $\boldsymbol{R}_{\mathrm{H}} = (X_i, Y_i, Z_i)$ is the position of the H atom relative to the LiF(001) surface. The values of the relevant parameters for the calculation are an energy cutoff in the plane-wave expansion of 650 eV, a $5 \times 5 \times 1$ Monkhorst-Pack grid of special k-points for the Brioullin-zone integration, a supercell-slab scheme to represent the surface, with a 2×2 surface cell, a five-layer slab, and a vacuum distance of approximately 40Å. The obtained LiF lattice constant is 4.062Å for the PAW-PBE-vdW PES and 4.066Å for the PAW-PBE, slightly higher than the experimental value of 4.02Å [47]. The respective rumplings, obtained from the relaxed surface equilibrium geometry, are $\delta_r = 0.044$Å (PAW-PBE-vdW) and $\delta_r = 0.034$Å (PAW-PBE). The rumpling δ_r is here defined as the half-distance between relaxed Li and F planes. The values we obtain are consistent with low-energy electron diffraction (LEED) experiments [48] that yield a rumpling of 0.036 ± 0.1 Å and are in very good agreement with Vogt's finding 0.034Å [49], obtained with a PAW-PBE calculation. It is important to mark that, in contrast to the PA potentials discussed in the previous sections, the geometrical parameters of a DFT potential are not enforced; rather they are obtained from the well-converged system relaxations.

In Fig. 8, the potential energy (in absolute value) obtained for this system is depicted as a function of the H-surface distance Z_H for the H atom on the selected high-symmetry sites, for both PESs considered.

4.2. *The PES analysis for GIFAD along the channels ⟨110⟩ and ⟨100⟩*

The GIFAD method is extremely sensitive to the projectile-surface interaction, particularly to the profile and corrugation of the PES near the reflection region. Our PESs are 3D and no dimension reduction is made during the dynamics. However, the fast motion of the projectile along the channel is, in fact, mainly ruled by the average interaction in this direction and thus we will discuss the PESs features in these terms.

In Fig. 9, we consider the energy averages respectively along the ⟨100⟩ and ⟨110⟩ channels, and depict equipotential contours across them for both the PAW-PBE and the PAW-PBE-vdW PESs. Across a ⟨100⟩ channel, the equipotential curves have only one maximum at the border of the channel, corresponding to the rows of alternating Li^+ and F^- ions. In contrast, across a ⟨110⟩ channel the equipotential curves have local maxima both at the border and at the middle of the channel, respectively, corresponding to the rows of F^- and Li^+ ions.

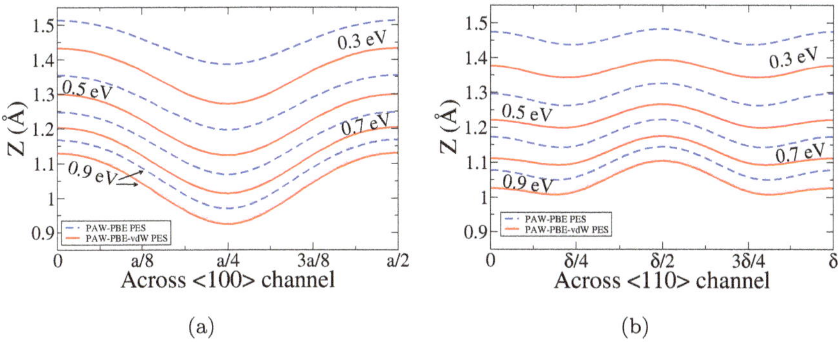

Fig. 9. Equipotential contours averaged along the incidence channel. Dashed (solid) lines for the PAW-PBE (PAW-PBE-vdW) PES. (a) Incidence along the ⟨100⟩ direction. (b) Incidence along the ⟨110⟩ direction. Here, $a/2$ and $\delta = a/\sqrt{2}$ correspond to the respective channel widths.

The corrugation $\Delta Z = Z_{\max} - Z_{\min}$ of an equipotential curve for a given energy E_0 is, in a first approach, the corrugation at the reflection region probed by an atomic projectile impinging onto the surface with a perpendicular energy $E_\perp = E_0$. Hence, its magnitude as well as the way it evolves with $E_\perp$ are key features of the PES, with marked effects on the GIFAD pattern. From a close inspection of Fig. 9, we observe that (i) the corrugation across a ⟨100⟩ channel is much higher than that across a ⟨110⟩ channel, and it also increases more rapidly; (ii) the inclusion of the vdW interactions results in a less repulsive and higher corrugated PES (these effects become stronger for lower energies); and (iii) the vdW interactions seem to play a more important role for the ⟨100⟩ channel than for the ⟨110⟩ channel. This latter feature deserves a further discussion. Along the ⟨110⟩ channel, in Ref. [11] it was found that for He/LiF(001), the presence of cationic and anionic rows induces polarization within the projectile, resulting in marked effects on the potential, and thus on the GIFAD patterns. Polarization is implicitly included in a standard DFT calculation and, therefore, in our PAW-PBE PES too. It mainly affects the projectile-surface interaction in the medium to large distance range, where it might compete with the vdW interactions, probably overshadowing them. In contrast, along the ⟨100⟩ channel, neutrally charged rows of alternating F and Li ions result in no polarization effects, and hence in the more visible vdW interactions.

5. Experimental examination of the DFT potential for the H projectiles

In this section, we present our simulated GIFAD patterns for the H projectiles and compare them to the reported experiments. We assess the performance of the PAW-PBE and PAW-PBE-vdW PESs for incidence along the ⟨100⟩ and ⟨110⟩ directions. In each case, we evaluate the role of the vdW interactions and the effect of the particular features of the PES on the intensity profiles. Analogous to the case of multi-electronic projectiles (Section 3), the GIFAD distributions are here evaluated within the SIVR approximation.

We first present and discuss the case of incidence along the ⟨100⟩ direction, with normal energy $E_{\perp} = 0.29$ eV. According to our results from Section 4.2, this is a favorable setting for studying the possible vdW effects, due to the absence of polarization effects and also to the relatively low projectile energy, which keeps the scattering dynamics restricted to certain regions, particularly the reflection region, where the vdW contributions could be relevant. In the left panel of Fig. 10, we show the projected intensities for this case, as a function of the azimuthal angle φ_f. Both the PAW-PBE and PAW-PBE-vdW simulated patterns (normalized to the Bragg peak of order $m = 1$) are displayed, together with the experiments from Ref. [7]. Remarkably, the PAW-PBE results are in a rather poor agreement with the experimental pattern, particularly regarding the relative intensity of the central peak. This is opposed to the markedly improved accord between the PAW-PBE-vdW pattern and the measurement. A key ingredient of this better performance is the increased corrugation of PAW-PBE-vdW near the reflection region (see Fig. 9(a)).

The second case, which we will address, is that of incidence along the ⟨110⟩ direction, with $E_{\perp} = 0.56$ eV, for which the vdW interactions are not expected to be of importance. In the right panel of Fig. 10, the simulated intensity distributions (normalized to the $m = 2$ Bragg maximum) are presented and compared with the experimental, projected intensity profile reported by Rousseau *et al.* [50]. We obtain very similar patterns with PAW-PBE and PAW-PBE-vdW, which satisfactorily reproduce the overall characteristics of the experiment, showing a very intense central peak accompanied by two much lower maxima associated with the $m = \pm 2$ orders. At the positions corresponding to the $m = \pm 1$ peaks, the higher intensity for the experimental $m = \pm 1$ peaks, relative to our calculations, might be attributed to experimental limitations that do not allow one to distinguish

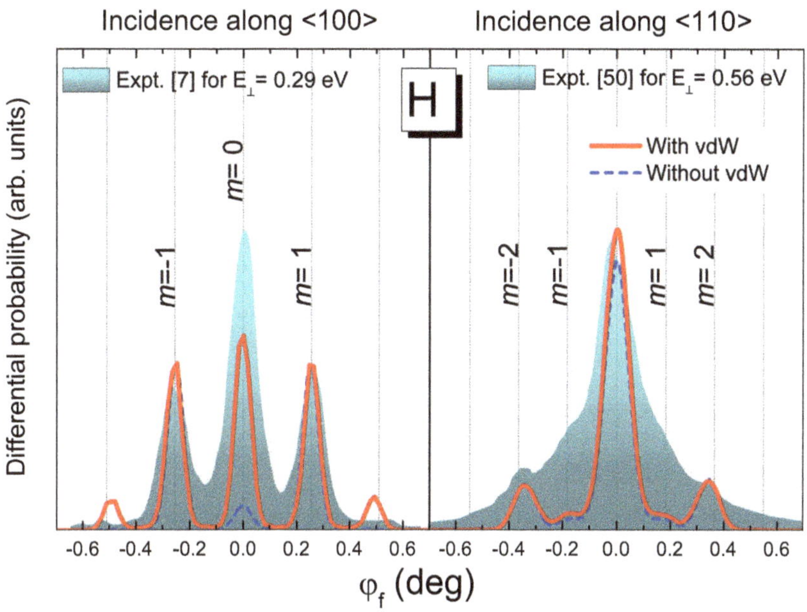

Fig. 10. The GIFAD projected patterns for incidence along ⟨100⟩ (left panel) and ⟨110⟩ (right panel) directions. The perpendicular energies are $E_{\perp} = 0.29$ and 0.56 eV, respectively. Solid (dashed) curves, the SIVR simulations using the PAW-PBE-vdW (PAW-PBE) PES. Vertical lines indicate the positions of the Bragg peaks of order m, with $m = \pm 1, \pm 2, ..$

the different Bragg orders. At such angular positions, the experimental profile seems to include contributions from the broad central maximum, which hinders an exhaustive comparison with the theoretical curves.

6. Conclusions

We have constructed a class of high-precision potentials to model the interaction of atoms approaching an alkali-halide crystal surface. The case of multi-electronic projectiles is addressed by means of a PA potential that incorporates non-local modern methods to describe the binary interactions in terms of the electron density. The contributions of the Madelung potential on the electron density around the individual ionic sites (onions) of the material, as well as non-local electron density terms, provide an improved and more realistic description, compared with the previous PA models. The present PA potential is applied to the closed-shell (He, Ne, Ar, Kr) and open-shell (N, S, Cl) atoms interacting with LiF(001), analyzing the relative importance of the different interaction mechanisms. The case of the H atoms is addressed with a refined DFT potential - PAW-PBE-VdW - which describes the relevant interactions in the system, including the vdW dispersion forces through Grimme's approach.

In both cases, the degree of accuracy of the proposed atom-surface potentials was illustrated by comparing the theoretical results for GIFAD with the available experimental data. For multi-electronic atoms, the PA potential was thoroughly probed through the highly sensitive angular positions of the rainbow and supernumerary rainbow maxima (low $E_\perp$) or just the former in the case of high $E_\perp$, where the supernumerary rainbows blur. For closed-shell atoms, we obtain a very good accord with the experiments in the whole energy range, from 0.2 to 80 eV, while for open-shell atoms, we observe a very good accord for low $E_\perp$ and a slight departure from the experimental rainbow angles for high $E_\perp$. These results are indicative of the reliability of our PA potential, particularly for the low normal energy region.

Concerning the DFT potential for the H-LiF(001) system, we have investigated the relevance of the vdW interactions by comparing the diffraction patterns obtained from the PAW-PBE and PAW-PBE-VdW PESs with the experimental GIFAD patterns for incidence along the $\langle 110 \rangle$ and $\langle 100 \rangle$ channels. While for multi-electronic projectiles with $E_\perp \geq 0.2$ eV, the vdW contributions were found negligible, they become relevant for the H impact in the low-$E_\perp$ regime, where our simulations show a marked influence of the vdW corrections for incidence along the $\langle 100 \rangle$ channel.

Acknowledgments

The authors thank financial support from CONICET, UBA, and ANPCyT of Argentina.

References

[1] D. J. Riley, A. P. Jardine, S. Dworski, G. Alexandrowicz, P. Fouquet, J. Ellis, and W. Allison, A refined He-LiF(001) potential from selective adsorption resonances measured with high-resolution helium spin-echo spectroscopy, *J. Chem. Phys.* **126**, 104702 (2007).

[2] V. Celli, D. Eichenauer, A. Kaufhold, and J. P. Toennies, Pairwise additive semi *ab initio* potential for the elastic scattering of He atoms from the LiF(001) crystal surface, *J. Chem. Phys.* **83**, 2504–2521 (1985).

[3] P. W. Fowler and J. M. Hutson, Pairwise-additive models for atom-surface interaction potentials: An *ab initio* study of He-LiF, *Phys. Rev. B.* **33**, 3724–37351 (1986).

[4] D. Eichenauer and J. P. Toennies, Pairwise additive potential models for the interaction of He atoms with the (001) surfaces of LiF, NaF, NaCl and LiCl, *Surf. Sci.* **197**, 267–276 (1988).

[5] A. Schüller, H. Winter, M. S. Gravielle, J. M. Pruneda, and J. E. Miraglia, He-LiF surface interaction potential from fast atom diffraction, *Phys. Rev. A.* **80**, 062903 (2009).

[6] A. Schüller, S. Wethekam, D. Blauth, H. Winter, F. Aigner, N. Simonović, B. Solleder, J. Burgdörfer, and L. Wirtz, Rumpling of LiF(001) surface from fast atom diffraction, *Phys. Rev. A.* **82**, 062902 (2010).

[7] A. S. Muzas, F. Gatti, F. Martín, and C. Díaz, Diffraction of H from LiF(001): From slow normal incidence to fast grazing incidence, *Nucl. Instr. Methods Phys. Res. B.* **382**, 49–53 (2016).

[8] A. Schüller and H. Winter, Supernumerary Rainbows in the Angular Distribution of Scattered Projectiles for Grazing Collisions of Fast Atoms with a LiF(001) Surface, *Phys. Rev. Lett.* **100**, 097602 (2008).

[9] A. Schüller, K. Gärtner, and H. Winter, Interaction potential for fast Ne atoms in front of LiF(001) surface, *Eur. Phys. Lett.* **81**, 37007 (2008).

[10] A. Schüller, M. Busch, J. Seifert, S. Wethekam, H. Winter, and K. Gärtner, Superstructures of oxygen and sulphur on a Fe(110) surface via fast atom diffraction, *Phys. Rev. B.* **79**, 235425 (2009).

[11] M. S. Gravielle and J. E. Miraglia, Influence of the polarization in grazing scattering of fast helium atoms from LiF(001) surfaces, *Phys. Rev. A.* **78**, 022901 (2008).

[12] M. S. Gravielle, A. Schüller, H. Winter, and J. Miraglia, Fast atom diffraction for grazing scattering of Ne atoms from a LiF(001) surface, *Nucl. Instr. Methods Phys. Res. B.* **269**, 1208–1211 (2011).

[13] M. S. Gravielle, J. Miraglia, A. Schüller, and H. Winter, Interaction potentials for multi-electron atoms in front of a LiF (001) surface from rainbow scattering, *Nucl. Instr. Methods Phys. Res. B.* **317**, 77–82 (2013).

[14] A. A. Abrahamson, Repulsive Interaction Potentials between Rare-Gas Atoms. Heteronuclear Two-Center Systems, *Phys. Rev.* **133**, A990–A1004 (1964).

[15] M. S. Gravielle and J. Miraglia, Quantum interference in grazing scattering of swift He atoms from LiF(001) surfaces: Surface eikonal approximation, *Nucl. Instr. Methods Phys. Res. B.* **267**, 610–614 (2009).

[16] J. E. Miraglia and M. S. Gravielle, Reexamination of the interaction of atoms with a LiF(001) surface, *Phys. Rev. A.* **95**, 022710 (2017).

[17] J. E. Miraglia and M. S. Gravielle, Production of excitons in grazing collisions of protons with LiF surfaces: An onion model, *Phys. Rev. A.* **84**, 062901 (2011).

[18] S. Grimme, Semiempirical GGA-Type Density Functional Constructed with a Long-Range Dispersion Correction, *J. Comput. Chem.* **27**, 1787–1799 (2006).

[19] A. Schüller, S. Wethekam, and H. Winter, Diffraction of Fast Atomic Projectiles during Grazing Scattering from a LiF(001) Surface, *Phys. Rev. Lett.* **98**, 016103 (2007).
[20] P. Rousseau, H. Khemliche, A. G. Borisov, and P. Roncin, Quantum Scattering of Fast Atoms and Molecules on Surfaces, *Phys. Rev. Lett.* **98**, 016104 (2007).
[21] H. Winter and A. Schüller, Fast atom diffraction during grazing scattering from surfaces, *Prog. Surf. Sci.* **86**, 169–221 (2011).
[22] P. Atkinson, M. Eddrief, V. H. Etgens, H. Khemliche, M. Debiossac, A. Momeni, M. Mulier, B. Lalmi, and P. Roncin, Dynamic grazing incidence fast atom diffraction during molecular beam epitaxial growth of GaAs, *Appl. Phys. Lett.* **105**, 021602 (2014).
[23] J. Seifert and H. Winter, Quantitative structure determination using grazing scattering of fast atoms: Oxygen-induced missing-row reconstruction of Mo(112), *Phys. Rev. B.* **93**, 205417 (2016).
[24] P. Tiwald, A. Schüller, H. Winter, K. Tökesi, F. Aigner, S. Gräfe, C. Lemell, and J. Burgdörfer, Interaction potentials for fast atoms in front of Al surfaces probed by rainbow scattering, *Phys. Rev. B.* **82**, 125453 (2010).
[25] U. Specht, M. Busch, J. Seifert, A. Schüller, H. Winter, K. Gärtner, R. Włodarczyk, M. Sierka, and J. Sauer, Rainbow scattering under axial surface channeling from a KCl(001) surface, *Phys. Rev. B.* **84**, 125440 (2011).
[26] M. S. Gravielle and J. E. Miraglia, Semiquantum approach for fast atom diffraction: Solving the rainbow divergence, *Phys. Rev. A.* **90**, 052718 (2014).
[27] M. S. Gravielle and J. E. Miraglia, Influence of beam collimation on fast-atom diffraction studied via a semiquantum approach, *Phys. Rev. A.* **92**, 062709 (2015).
[28] E. Clementi and C. Roetti, Roothaan-Hartree-Fock atomic wave functions, *At. Data Nucl. Data Tables.* **14**, 177–478 (1974).
[29] Y. S. Kim and R. G. Gordon, Ion-rare gas interactions on the repulsive part of the potential curves, *J Chem. Phys.* **60**, 4323–4331 (1974).
[30] R. G. Parr and W. Yang, *Density-Functional Theory of Atoms and Molecules.* Oxford University Press, New York (1989).
[31] H. Lee, C. Lee, and R. G. Parr, Conjoint gradient correction to the Hartree-Fock kinetic- and exchange-energy density functionals, *Phys. Rev. A.* **44**, 768–771 (1991).
[32] A. D. Becke, Density-functional exchange-energy approximation with correct asymptotic behavior, *Phys. Rev. A.* **38**, 3098–3100 (1988).
[33] C. Lee, W. Yang, and R. G. Parr, Development of the Colle-Salvetti correlation-energy formula into a functional of the electron density, *Phys. Rev. B.* **37**, 785–789 (1988).
[34] K. Burke, Perspective on density functional theory, *J. Chem. Phys.* **136**, 150901 (2012).
[35] N. W. Ashcroft and N. D. Mermin, *Solid State Physics* (Brooks/Cole, Cengage Learning, Belmont, USA), ch. 1 (1976).
[36] Y. S. Kim and R. G. Gordon, Unified theory for intermolecular forces between closed shell atoms and ions, *J Chem. Phys.* **61**, 1–16 (1974).

[37] K. T. Tang and J. P. Toennies, An improved simple model for the van der Waals potential based on universal damping functions for the dispersion coefficients, *J. Chem. Phys.* **80**, 3726–3741 (1984).
[38] G. A. Bocan, J. D. Fuhr, and M. S. Gravielle, van derWaals effects on grazing-incidence fast-atom diffraction for H on LiF(001), *Phys. Rev. A.* **94**, 022711 (2016).
[39] W. H. Miller, The Semiclassical Initial Value Representation: A Potentially Practical Way for Adding Quantum Effects to Classical Molecular Dynamics Simulations, *J. Phys. Chem A.* **105**, 2942–2955 (2001).
[40] G. Kresse and J. Hafner, *Ab initio* molecular dynamics for liquid metals, *Phys. Rev. B.* **47**, 558–561 (1993).
[41] G. Kresse and J. Hafner, *Ab initio* molecular dynamics for open-shell transition metals, *Phys. Rev. B.* **48**, 13115–13118 (1993).
[42] G. Kresse and J. Furthmüller, Efficiency of *ab initio* total energy calculations for metals and semiconductors using a plane-wave basis set, *Comput. Mater. Sci.* **6**, 15–50 (1996).
[43] G. Kresse and J. Furthmüller, Efficient iterative schemes for *ab initio* total-energy calculations using a plane-wave basis set, *Phys. Rev. B.* **54**, 11169–11186 (1996).
[44] H. F. Busnengo, A. Salin, and W. Dong, Representiation of the 6D Potential Energy Surface for a Diatomic Molecule near a Solid Surface, *J. Chem. Phys.* **112**, 7641–7651 (2000).
[45] G. Kresse and D. Joubert, From Ultrasoft Pseudopotentials to the Projector Augmented-Wave Method, *Phys. Rev. B.* **59**, 1758–1775 (1999).
[46] J. P. Perdew, K. Burke, and M. Ernzerhof, Generalized Gradient Approximation Made Simple, *Phys. Rev. Lett.* **77**, 3865–3868 (1996).
[47] Y. Ekinci and J. P. Toennies, Thermal Expansion of the LiF(001) Surface, *Surf. Sci.* **563**, 127–134 (2004).
[48] J. Vogt and H. Weiss, The Structure of KBr(100) and LiF(100) Single Crystal Surfaces: A Tensor Low Energy Electron Diffraction Analysis, *Surf. Sci.* **501**, 203–213 (2002).
[49] J. Vogt. *Ultradünne Molekulare Filme auf Isolatoreinkristalloberflächen: Strukturelle und Infrarotoptische Eigenschaften.* PhD thesis, Otto-von-Guericke-Universität Magdeburg, Germany (2008).
[50] P. Rousseau, H. Khemliche, N. Bundaleski, P. Soulisse, A. Momeni, and P. Roncin, Surface analysis with grazing incidence fast atom diffraction (GIFAD), *J. Phys. Conf. Ser.* **133**, 012013 (2008).

Chapter 8

A bottom-up approach for ion-water interactions: From clusters to bulk

R. Rodríguez-Segundo, D.J. Arismendi-Arrieta and R. Prosmiti*

Institute of Fundamental Physics,
Serrano 123, 28006 Madrid, Spain

Ion specific effects in aqueous environments are of fundamental interest across interdisciplinary fields, with modeling of ion-pair/ion-water/water-water interactions being one of the main research topics. So far, the water-water interactions have been widely studied yielding a variety of water potential models of increasing complexity. Although extensive experimental and theoretical studies have been reported, questions on how ions change the water network and the properties of their aqueous solutions are still open, demanding an accurate description of the underlying intermolecular forces. In this work, the first principle approaches are employed to develop generalized representations for such interactions coupled to available and reliable water models for studying ion-hydrates from finite-size clusters to bulk solutions.

Contents

*Corresponding Author. Email: rita@iff.csic.es

1. Introduction

Description of ion-water, and water-water interactions is challenging. It represents a topic of fundamental concern in several fields ranging from chemistry, biochemistry to industry [1]. For studying any physicochemical process, such as dissociation, ionization, adsorption, diffusion, collision, or energy transfer, the representation of the underlying interactions are of central importance. There are many approaches, starting from semi-empirical models [2] to *ab initio* based [3–5] methods. The selection of one of these methodologies depends on the current problem of interest, as well as on the computational power available. The perfect potential model should be as accurate as possible, and at the same time computationally fast enough for calculations in multi-dimensional systems. Nowadays, much progress has been made, although computer simulations for large systems within reasonable time is still be challenging.

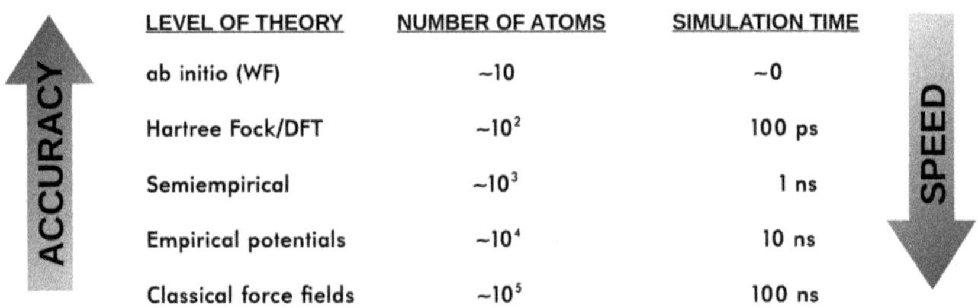

Fig. 1. Dependence of theoretical simulations on the accuracy of the underlying potential

Indeed, computer simulations have become a strong instrument for studying molecular systems. As can be seen in Fig. 1, increasing the number of atoms in a system intrinsically restricts the level of theory to be used for describing the underlying interactions. If one chooses to use very accurate methods, such as those based upon the Hartree-Fock wave function [6], simulations become computationally slow, and can be performed only for systems with few atoms. For moderate size systems, it is possible to use electronic structure methods "on-the-fly" to carry out nuclear dynamics calculations. In such computations, density functional theory (DFT) [7] is the dominant electronic structure method. However, the accessible simulation times are still not large enough to deal with many realistic systems.

Therefore, for determining the energies and forces in large-scale simulations, direct on-the-fly computationally demanding evaluations of electronic structure are commonly replaced by more efficient potentials. *Ab initio*, first-principles methods only permit the calculation of a few selected points on the potential energy surface (PES), while a full representation of the PES is usually defined by sophisticated analytical functions of the atomic coordinates. Such expressions include a limited number of parameters, determined from adjustment of the model potential to *ab initio* data. Therefore, the values of the constructed potential can be evaluated several orders of magnitude faster, and for very large numbers of configurations.

A key point in the development of PES (or a force field) is the representation and characterization of the PESs to which we adjust the function. Such functions (assuming the validity of the Born-Oppenheimer approximation [8]) depend only on the atomic positions without an explicit consideration of the electronic degrees of freedom. The representation of the PES in a closed analytical form, serves to perform computationally very efficient simulations, without solving the electronic structure problem in each simulation step. Unfortunately, the topology of the multi-dimensional PESs is extremely complicated for even simple molecular systems, since it is determined by the high complexity of the underlying quantum mechanical many-electron problem. Deriving analytical functional forms that provide reliable potential energies comparable to those of accurate electronic structure methods is a very challenging task.

Analytical representations provide a large amount of information and serve to characterize the topology of a complicated potential energy surface. The important aspects of potential energy surface are its stationary points: minima, maxima, and saddle points (first, second and higher order). Figure 2 shows a simplistic representation of a PES in two variables. Each of such points represents different configurations of the molecule, and can be characterized by its first and second derivatives (harmonic frequency analysis).

Much effort has been devoted over the last decades in devising and improving methods to construct model potentials for all imaginable applications [9]. Without a doubt, the construction of reliable potentials is still far from being a straightforward and easy task. Very often, experience and chemical intuition are required to arrive at an acceptable output. In principle, potential models are built for certain applications and, therefore, cannot be universal or even transferable for a wide range of applications.

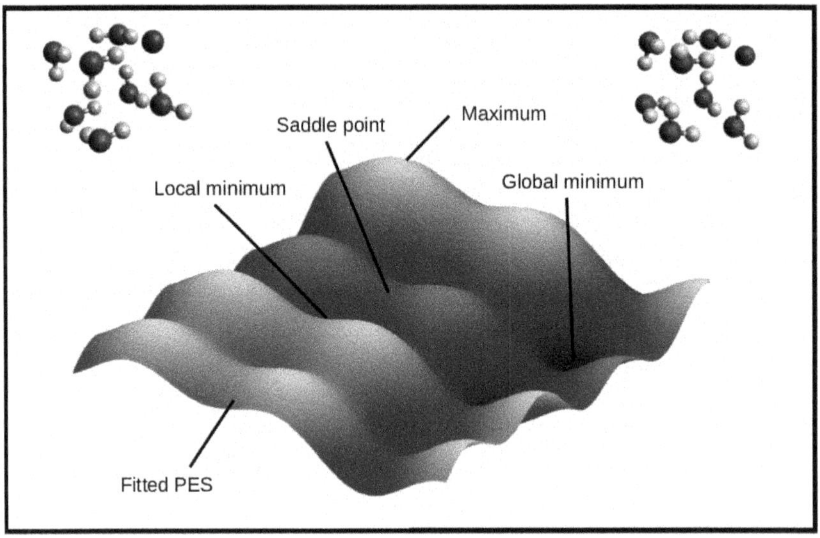

Fig. 2. Schematic representation of topological surface of potential energy surface of a molecular cluster.

A wide variety of approaches has been reported ranging from simple classical force fields to very sophisticated and highly system-specific potentials. One can classify these approaches into two categories: empirical and *ab initio*. Such a classification is based upon the type of information employed to construct these model potentials, *e.g.* making use of experimental data combined occasionally with additional theoretical results, or based exclusively on *ab initio* data from electronic structure calculations. More recently, some different first-principles approaches have been developed to provide insights into the nature of such interactions. However, investigations are still constrained by the limited predictive abilities of the empirical models, the high computational cost for correlated *ab initio* methods, and the intrinsic deficiencies of common density functionals in describing weak interactions. At first glance, a fully *ab initio* approach appears daunting, although the most advanced *ab initio*-based treatments clearly are a step forward combining computational efficiency and fitting accuracy of empirical approaches.

Thus, we adopted a bottom-up approach starting from small to medium size clusters, where valuable information has been acquired by infrared (IR) experimental studies and *ab initio* computations. As clusters bridge the gap between gas and condensed phases, they are ideal models for evaluating the intermolecular forces. The present study investigates the halide-water interactions from monohydrate to polyhydrate systems, corresponding to ion-water clusters. Based upon a bottom-up approach, by fitting the expressions for two-body short- and long-range potential to *ab initio* data for the halide-water monomer, and including polarizability (many-body terms) via the Thole-type electrostatic model, we aim to describe the ion-water interactions for larger systems up to the bulk. We focus on the representation of the interactions based on first-principle data of solvated monovalent (halide X= F^-, Cl^-, Br^-, I^-) ions, aiming at generating *ab initio*-based models for benchmarking the existing empirical/phenomenological models or energies from the DFT calculations and for investigating such ion-water processes. The plan of the present chapter is as follows: the next section briefly summarizes the theoretical methods and computational details of the electronic structure methods and functional forms used to parametrize the PESs. Section 3 presents the obtained results and their analysis, while the last section provides the summary and the conclusions of the study.

2. Methods and Computational Details

2.1. *Description of ion-water interaction potential: effective many-body contributions*

Within the many-body expansion formalism [10], the PES of a system of N monomers is represented by:

$$V_N(x_1, x_2, \cdots, x_N) = \sum_N V_{1B} + \sum_N N V_{2B} + \sum_N V_{3B} + \cdots, \tag{1}$$

starting from the terms with 1-body (1B, intramolecular), 2-bodies (2B, pair), nonadditive 3-bodies (3B), etc. The approach proposed here applies to the two first terms of the expansion that correspond to the most important components of the interaction. However, by building up a polarization potential model, higher-body effects are approximately taken into account through the induction energy part. For a system containing a halide anion X^- and n water molecules, we expressed the potential energy as a sum of the intermolecular ion-water ($V_{\mathrm{X-W}}^{\mathrm{inter}}$) as well as water-water ($V_{\mathrm{W-W}}^{\mathrm{inter}}$) interactions, and the intramolecular potential for the water molecules $V_{\mathrm{W_i}}^{\mathrm{intra}}$:

$$V_{\mathrm{X-W}} = \sum_i^n (V_{\mathrm{X-W}}^{\mathrm{inter}} + V_{\mathrm{W-W}}^{\mathrm{inter}} + V_{\mathrm{W_i}}^{\mathrm{intra}}). \tag{2}$$

For $V_{\mathrm{W_i}}^{\mathrm{intra}}$ we employ the PES given by Partridge-Schwenke (PS) [11], while for $V_{\mathrm{X-W}}^{\mathrm{inter}}$ we developed the new parametrized Thole-type model (TTM) potentials adapted to different water-water models. Thus, the $V^{\text{inter}}_{\text{W-W}}$ term is represented by polarizable TTM water-water potentials [12–16].

Furthermore, the intermolecular interaction between a halide anion and a water molecule is represented as a sum of two-body (2B) terms with short- and long-range contributions:

$$V^{\text{inter}}_{\text{X-W}} = V^{2\text{B}}_{\text{short}} + V^{2\text{B}}_{\text{long}} = V_{\text{rep}} + V_{\text{elec}} + V_{\text{disp}}. \tag{3}$$

The two-body terms are defined as the difference between the total energy of the X^-H_2O system and the one-body energies of the two isolated monomers, X^- and H_2O, with the repulsive part expressed by:

$$V_{\text{rep}} = \sum_{i,j=(\text{O/H}),\text{X}^-} A_{ij} e^{B_{ij} R_{ij}}, \tag{4}$$

where the coordinates R_{ij} are the interatomic distances, and the $\{i, j\}$ sums run over all the ion/atom systems (X^-, O or H). Here, the long-range part of the interaction is dominated by interactions between the permanent and induced moments related through the charge distributions and the dispersion forces. Within the TTM model, a negative smeared charge and a polarizable site are placed on the X^- ion. On the other hand, the water molecule positive smeared charges [11, 12, 14, 15] are placed at each hydrogen atom. Moreover, a negative smeared charge is shifted from the oxygen atom to the M-site and the polarizable sites at all the O, H atoms as seen in Fig. 3.

Therefore, the electrostatic terms are given by:

$$V_{\text{elec}} = V_{\text{TTM}} = V_{\text{TTM,elec}} + V_{\text{TTM,ind}}, \tag{5}$$

where $V_{\text{TTM,elec}}$ is the charge-charge (CC) contribution and $V_{\text{TTM,ind}} = -1/2\sum_a(\mu_a \cdot E_a)$. Here, μ_a are the induced dipoles (minimizing V_{TTM}), and E_a is the electric field at the ath interaction due to charges and dipoles. The long-range dispersion contribution is represented by the damped $1/R^6$ terms between the OX^- and HX^- ion pairs:

$$V_{\text{disp}} = \sum_{i,j=(\text{O/H}),\text{X}^-} \frac{C_6^{(ij)} f(R_{ij}, \delta_{ij})}{R_{ij}^6}, \tag{6}$$

where $f(R, \delta)$ are the Tang-Toennies damping functions [17], while $C_6^{\text{OX}^-}$ and $C_6^{\text{HX}^-}$ are the dispersion coefficients.

2.2. *The CCSD(T)-F12 and DFT electronic structure computations*

For the parametrization of the intermolecular interactions of the halide TTM model, electronic structure computations are carried out to obtain the interaction energies, anion polarizabilities, as well as dispersion energies.

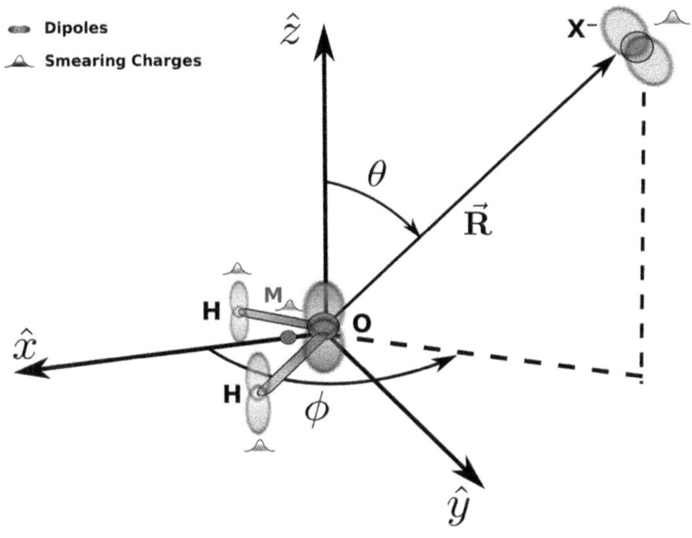

Fig. 3. Schematic representation of the positions of the charges and dipole moments in the ion TTM model, and of the internal coordinates used.

Figure 3 displays the spherical coordinates (R,θ,ϕ) used to define the X^--H_2O system, where R is the distance between the X^- anion and the oxygen atom. Further, θ and ϕ are the polar and azimuthal angles, while the H_2O molecule is fixed at its vibrationally average geometry [18] with the OH bond length r_{OH}= 0.9716257 Å and ∠ HOH= 104.69°. The interaction energies are computed at the explicitly correlated coupled cluster single double triple (CCSD(T)-F12) level of theory [19, 20], using the Molpro package [21, 22]. We found that the CCSD(T)-F12 method presents a computationally efficient way to compute energies in impressive agreement with the coupled cluster single-double and perturbative triple complete basis set (CBS) limit, CCSD(T)/CBS. In Table 1, we present the effect of the method on the interaction energies. Therein, also shown is the effect of

Table 1. Interaction energies (in kcal/mol), including the basis set superposition error (BSSE) corrected energies [28] at the CCSD(T) and CCSD(T)-F12 levels of theory for the indicated X^--H_2O clusters using the correlation-consistent polarized basis sets and effective core potentials (ECP) for the Br and I atoms.

Method/Basis set[a]	Cl^-–H_2O $\Delta E_{\rm BSSE}$	I^-–H_2O $\Delta E_{\rm BSSE}$
CCSD(T)/CBS[b]	-15.6878	-10.9814
CCSD(T)-F12b/aug-cc-pVQZ(-PP)	-15.4997	-10.9744
CCSD(T)-F12b/cc-pVQZ-F12(-PP)	-15.5450	-10.9804

[a]All electrons basis sets are used for the Cl atom, while the ECP28MDF/aug-cc-pVxZ-PP or cc-pVyZ-F12-PP (x=T,Q,5 and y=T,Q) are employed for the I one.
[b]The extrapolation CBS scheme for the total energies from Ref. [29].

using different basis sets in the calculations [23–27].

Once the total energies have converged, we could establish a direct comparison with the interaction energies at the indicated levels of theory as well as with the CBS limit values. Thus, we choose the CCSD(T)-F12b level of theory, and the interaction energies are computed for each X^--H_2O cluster in a wide range of R (in total at 50 points). Herein, the polar angle θ is varied from 0° to 90° with a step of 45°, while the azimuthal angle ϕ is taken to be 0°, 52.345°, 90° and 180°.

Regarding the $V_{\rm TTM}$ contribution, the atomic polarizabilities of each anion are computed at the CCSD(T) level of theory, as it is implemented in the Molpro package [21, 22], using even-tempered sets of diffuse functions to the correlation consistent basis sets [30]. The obtained results are found to be in very good accord with those previously reported in the literature [30–32].

Finally, we treat the long-range $C_6^{(ij)}$ dispersion terms, and the exchange-hole dipole moment (XDM) model [33] is adopted to calculate the dispersion interaction between the X^- and H_2O systems [34, 35].

2.3. *Fitting procedure*

Initially, we checked two slightly different fitting protocols for parametrization of the intermolecular $V_{\rm X-W}$ models, consisting of a standard 1-step fit directly to the raw CCSD(T)-F12 data, and a 2-step fit which allows an explicit inclusion of the dispersion interaction through the physically-meaningful C_6. Although both procedures provide fits of the same quality, we choose to employ the 2-step fit for parametrization of the intermolecular $V_{\text{X-W}}$ potentials.

In the first step, the dispersion energy V_{disp} from Eq. (6) is fitted to the XDM data. In the second step, the parameters of the repulsive energy contribution V_{rep} are fitted to the CCSD(T)-F12 reference energies with the two subtracted terms: the dispersion energy and the ion TTM contributions to the electrostatic energy.

The weighted sum of the squared residuals, $\chi^2 = \sum_N w_N[V_{\text{X-W}}^{\text{inter}}(N) - \Delta E^{\text{ref}}(N)]^2$, with respect to the reference CCSD(T)-F12 set of interaction energies is minimized with different weight-factors, such as $w(E) = 1/(E - E_{\text{min}} + DE)^2$. Here, E_{min} is the energy of the corresponding $X^-(H_2O)$ global minimum, while DE corresponds to the range of favorably weighted energies, as well as "automatic" numerically-based weighting points presenting higher deviation. For calculating the optimal values of the linear parameters, the singular value decomposition is used [36], while the nonlinear parameters were obtained through the simplex algorithm [36]. The χ^2 value of the fitting was 73.4952 (kcal/mol)^2 for the $F^-(H_2O)$ and 6.3277, 5.6403, 6.15995 (kcal/mol)^2 for $Cl^-(H_2O)$, $Br^-(H_2O)$ and $I^-(H_2O)$, respectively. Regarding the root-mean-square (rms) error of the fitting, the values of 0.151225, 0.0137858, 0.0122882 and 0.0134204 kcal/mol are obtained for each PES in the case of $X^-(H_2O)$ (X=F, Cl, Br, I).

3. Results and Discussion

3.1. *Properties of halide-water potentials*

Figure 4 compares the $V_{\text{X-W}}^{\text{inter}}$ potentials with the reference CCSD(T)-F12 set of the interaction energies as a function of the R distance at nine different (θ, ϕ) orientations for each of the $X^-(H_2O)$ clusters. One can see a very good agreement, with the computed mean-absolute errors (MAEs) from the reference CCSD(T)-F12 data listed in Table 2. Further, the MAEs with an extra set of 2250 CCSD(T)-F12 *ab initio* points (not included in the fitting procedure) are also presented. One can see that larger errors are obtained in the case of the $F^-(H_2O)$ system. These differences can be explained by considering the relatively more pronounced molecular nature of HF (i.e. higher basicity of F^-), which results in stronger (and more covalent) interactions between F^- and H_2O.

These halide-water clusters show two identical global minima for planar configurations, depending on which end of the H_2O the hydrogen bond is formed. The left panel of Fig. 5 shows the minimum energy paths (for

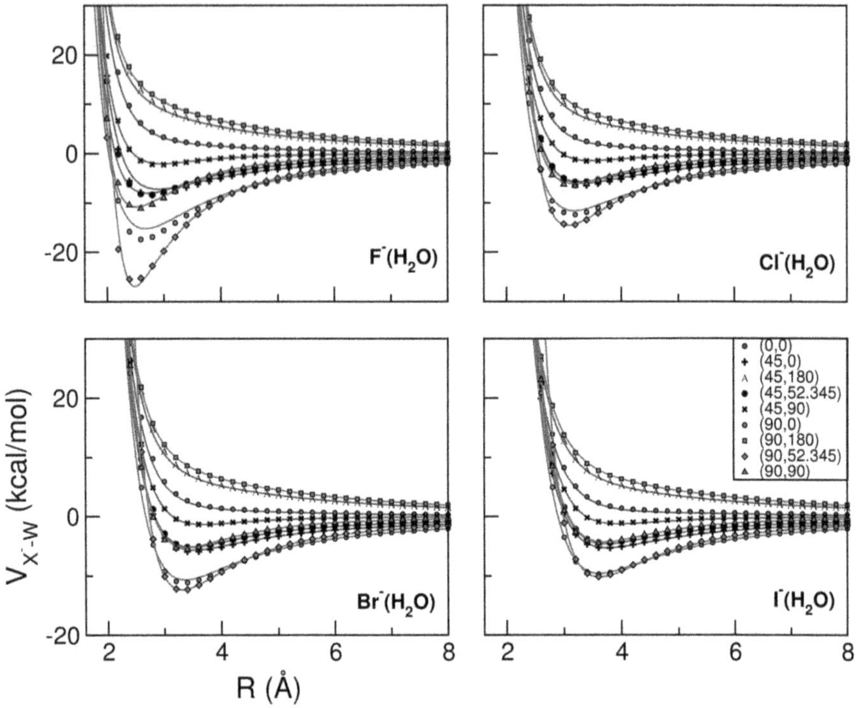

Fig. 4. The $V_{\mathrm{X-W}}$ potential curves (solid lines) and the CCSD(T)-F12 interactions energies (symbols) as a function of R distance and the angular (θ, ϕ) orientation for the indicated $X^-(H_2O)$ cluster (X=F, Cl, Br, I).

R <5 Å as a function of the ϕ angle, with θ fixed at 90° and water kept at its vibrationally average geometry) connected with the C_s minima through an isomerization barrier of the C_{2v} geometry. The $Cl^-(H_2O)$ structures at the stationary points (*e.g.* the two symmetric C_s minima as well as the two saddle-points) of the PES are also displayed. As it can be seen by comparing with the CCSD(T)-F12 values, the ion TTM potentials correctly reproduce the angular dependence, which presents minima at $\phi \approx 55°$ and $\phi \approx 305°$ due to the symmetry of the ion hydrogen bonds (IHBs). The right panel of Fig. 5 depicts the 1D-relaxed $V_{\mathrm{X-W}}$ potential curves as a function of the IHB OH distance in comparison with the corresponding CCSD(T)-F12 energies. By increasing this OH bond length, the energy values are calculated by optimizations at each step of all the remaining coordinates. Again, as expected, the distinct behavior of F^-, compared to all the other halide ions

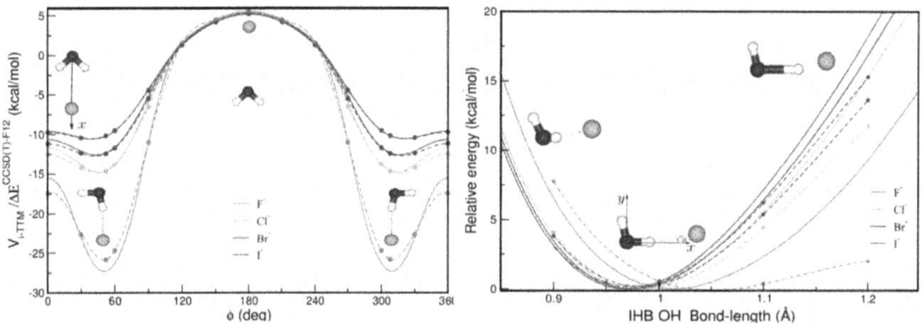

Fig. 5. The minimum energy paths for the $V_{X\text{-}W}$ PESs (solid lines) and the corresponding CCSD(T)-F12 energies (dashed lines) along the ϕ angle with H_2O molecule fixed at its vibrationally average geometry (left panel) and along the IHB OH distance with R fixed at the corresponding equilibrium value (right panel) for planar X^--H_2O.

interacting with water, becomes even more evident in these plots for the minimum energy path (MEP). This reflects the more covalent nature of the interactions between F^- and H_2O. The binding energy and the isomerization barrier of each dimer are listed in Table 2 (as per their prediction by the PES) and compared with the associated recent analytical PESs [37–39]. For comparison also included are the corresponding calculated values at the CCSD(T)-F12 level of theory as well as the earlier Møller-Plesset of second order (MP2) data [40]. Werhan *et al.* [37] have recently developed the generalized forms for potential functions, such as the extended Morse (eM) and the generalized Buckingham exponential-6 ($gBe-6$) to produce better quality fits to the Møller-Plesset of fourth order (MP4) *ab initio* data describing both the long-range and repulsive wall of the halide-water interactions. Moreover, Bowman and co-workers [38, 39] have constructed the full-dimensional PES for the $F^-(H_2O)$ and $Cl^-(H_2O)$ dimers by fitting the CCSD(T) energies to permutationally invariant polynomials. As it can be seen in Table 2, our results are close to the $eM/gBe-6$ values, especially for the heavier halide-water dimers. The observed differences can be attributed to the various sources: two different *ab initio* calculations, CCSD(T)-F12 versus MP4, the lack of angular anisotropy in the $eM/gBe-6$ functionals, particularly in the $F^-(H_2O)$ system due to the nature of this interaction as stated. By comparing with the full-dimensional PESs [38, 39], as expected, the contribution of the explicit many-body (deformation of water molecule) terms seems to be more important in the $F^-(H_2O)$ than in the $Cl^-(H_2O)$ systems. Further, we should emphasize that such a contribution decreases

Table 2. The binding energies (D_e, in kcal/mol) and the transition state barrier (ΔE^*, in kcal/mol) for the X^--H_2O clusters, as per the prediction by the CCSD(T)-F12 calculations, and by the V_{X-W}^{inter} model. Comparison with previously reported values is also presented.

Method/Model	F^-–H_2O		Cl^-–H_2O		Br^-–H_2O		I^-–H_2O	
	D_e	ΔE^*	D_e	ΔE^*	D_e	ΔE^*	D_e	ΔE^*
MP2 [40]	26.6	7.3	14.8	1.8	12.7	0.8	10.6	0.2
CCSD(T)-F12b	27.50	7.12	15.08	1.49	12.85	0.87	10.67	0.37
From Ref. [37]	26.99	>7	15.38	<2	13.38	<2	11.38	<1
From Refs. [38, 39]	27.83	7.45	<14.5	1.80	–	–	–	–
V_{X-W}^{inter}	28.15	9.68	14.90	2.23	12.64	1.25	10.65	0.53
MAE	$2.53^a/0.16^b$		$0.31^a/0.27^b$		$0.38^a/0.29^b$		$0.53^a/0.36^b$	

[a] MAE obtained from the reference data, [b] MAE for an extra set of 2250 points.

as the number of water molecules surrounding the anion increases, minimizing the distortion to each monomer.

Contour plots of the X^-H_2O ion TTM (V_{X-W}) PESs are shown in Fig. 6 in the (ϕ, R)-plane. One can see that the potential well of $F^-(H_2O)$ minima is between 2 to 3 times deeper than the corresponding ones for the other three halide-water dimers. Most importantly, the potential energy functions provide a smooth and realistic representation of the halide-water interactions including the rotation of the water monomer.

3.2. *Properties of higher-order $X^-(H_2O)_n$ potentials*

Figure 7 shows the minimum energy structures. Herein, on the left, one can see the differences between the $F^-(H_2O)_{n=1-4}$ system and the remaining clusters with $X^-=Cl^-$, Br^- and I^-. In particular, the ion TTM V_{X-W} potentials predict the C_2, C_3, and $(3+1)C_s$ for $F^-(H_2O)_n$ with n=2, 3 and 4, respectively, and C_1, C_3 and C_4 for $Cl/Br/I^-(H_2O)_n$ with n=2, 3 and 4, respectively, in accord with data from density fitting MP2 (DF-MP2) calculations.

For the $X^-(H_2O)_2$ system with X=Cl, Br and I, the C_1 structure exhibits, apart of the two IHBs, an additional weaker water-water hydrogen bond (HB). On the other hand, the C_2 structures are the 1st-order transition states, higher in energy by 0.47, 0.55 and 0.67 kcal/mol for the Cl^-, Br^- and I^- trimers, respectively, at the DF-MP2 level of theory.

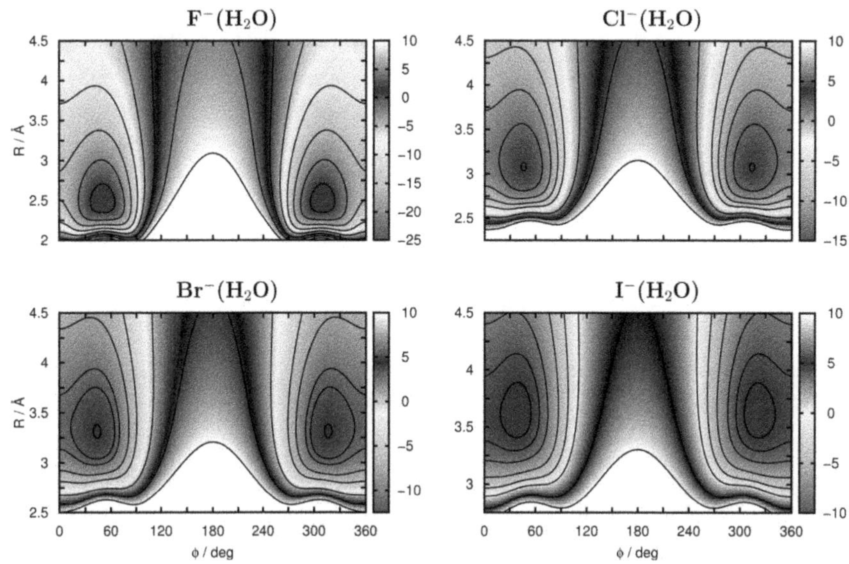

Fig. 6. Two-dimensional contour plots in the (ϕ, R)-plane calculated using the $V_{\text{X-W}}$ potentials for the $X^-(H_2O)$ dimers with X = F, Cl, Br and I. The equipotential curves are shown for the X^- anion moving around a planar H_2O molecule fixed at its vibrationally average structure.

One can see in Fig. 7 that, as the number of water molecules increases, they are directly bonded to the X^- anion via the ion-water HB thus forming single-shell structures, such as the C_3 or (3+0) and C_4 or (4+0) or multi-shell configurations, like the (3+1)C_s, with the outer shell water connected to the first hydration shell via the water-water HBs. Hereafter, we use the notation $(m + m')$, where m is the number of water molecules involved in the ion-water HBs and m' is the remaining number of water molecules, that in the multi-shell configurations are involved in the water-water HBs within each cluster. For these small halide-water clusters, we consider the anion as being on the surface of the water molecules as in the previous studies of such systems [40]. Typically, for these anions, the first solvation shell contains around six water molecules. For this reason, we focus on clusters of about that size in order to further investigate the surface or the interior ion hydration structures, and subsequently we extend the analysis to larger clusters with n up to 8.

On the right panels of Fig. 7, we presents the optimal structures from the prediction by the $V_{\text{X-W}}$ potential for the $X^-(H_2O)_n$ system with n=5-

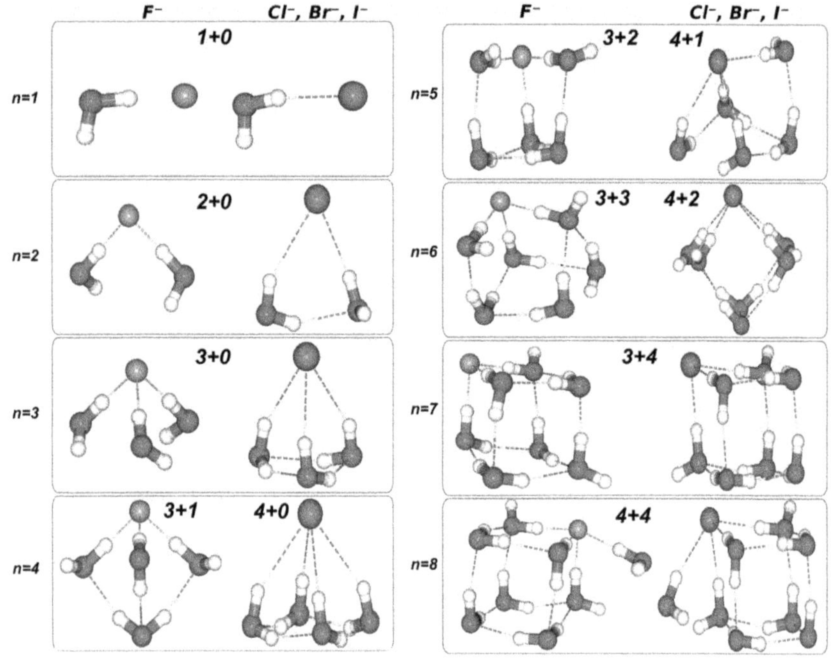

Fig. 7. Optimized structures for the indicated $X^-(H_2O)_n$ system with X = F, Cl, Br and I from the $V_{X\text{-}W}$ PESs.

8. In all the structures with n=5-8, the anion is preferentially accommodated at the surface of the water molecules, with the hydrogen atoms in the 1st hydration shell pointing at the anion. Here, again, the formation of multiple shell structures can be seen. Following the same pattern as for the n=4 clusters, the minimum energy structures of the $F^-(H_2O)_{n=5,6}$ system show the (3+2) and (3+3) configurations. On the other hand, the $Cl/Br/I^-(H_2O)_{n=5,6}$ systems correspond to the (4+1) and (4+2) configurations. For the n=7 and 8 clusters, we found that the optimal structures are the same for all the halide-water clusters, corresponding to the (3+4) and (4+4) configurations, respectively. We also found that the (3+1) and (3+2) configurations for the $F^-(H_2O)_{n=4,5}$ clusters as well as the (3+4) configurations for the $Cl/Br/I^-(H_2O)_{n=7}$ clusters seem to be slightly more stable than the other structures.

In order to carry out the direct comparisons of the $V_{X\text{-}W}$ PESs with the first-principle data for larger $X^-(H_2O)_n$ clusters, we first analyze the performance of different types of the DFT functionals. Comparisons of

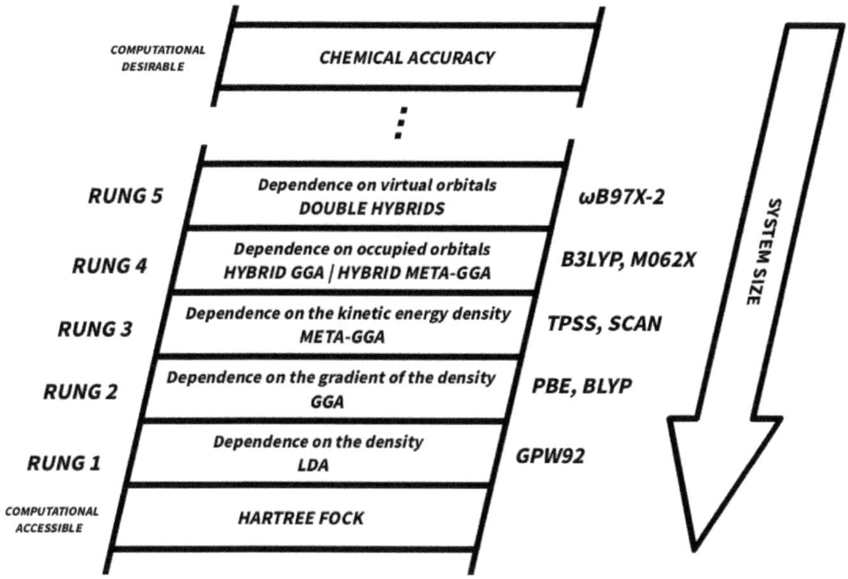

Fig. 8. Jacob's ladder of DFT

such DFT approaches with the previously employed wave function-based method, DF-MP2, will enable us to assess their quality as well as the quality of the present PESs with the increased cluster size. A previous study [3] (which includes the dispersion corrections) has checked the performance of 12 different functionals, selected from various types of functionals along various rungs (see Fig. 8), such as the generalized gradient approximation (GGA), Meta-GGA, hybrid-GGA, hybrid-meta-GGA, double-hybrids and the range-separated hybrids with the D3 dispersion [34, 41].

In general, it has been found that the functionals including dispersion perform better than those without dispersion. Further, for all the four halide-water cluster sizes, the DFT functionals TPSS-D3BJ, B3LYP-D3BJ, LC-ωPBE and ωB97XD are in closer accord with the reference DF-MP2 energies than the other functionals. Also the B3LYP-D3BJ functional clearly has the best performance for the F^-, Cl^-, and Br^--water clusters compared to the DF-MP2 energies, while for the I^--water cluster, the four of them show very similar behavior. Recently, a new Meta-GGA functional, SCAN [42], has become available and it was presently employed. Once the performance of the DFT functionals has been evaluated, we continue with

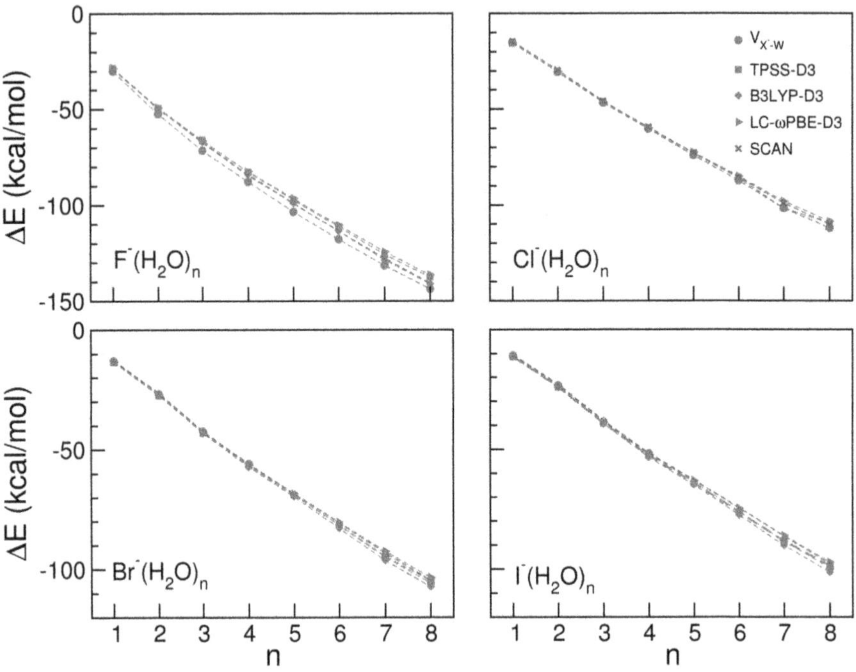

Fig. 9. The interaction energies obtained from the indicated DFT functional with the best performance as a function of the cluster size compared to the V_{X-W} values.

the evaluation of the V_{X-W} model. Moreover, the previous MP2 calculations [40] are available for halide-water clusters up to n=6 and these results compare well with the present optimized structures on the PESs as seen in Fig. 7. Further, in Fig. 9, we display the plots of the interaction energy as a function of the cluster size for the $X^-(H_2O)_{n=1-8}$ system. These plots are for the results calculated at the DFT-D3 level of theory by employing 5 functionals with the best performance, and using the V_{X-W} potentials.

The quality of the V_{X-W} model and its performance compared to computationally expensive DFT-D evaluations was checked. As the DFT simulations are becoming applicable in aqueous solutions of simple ions, the present model potentials offer an alternative for an accurate description of ionic interactions. However, such DFT approaches must be benchmarked against highly accurate electronic structure calculations, and the *ab initio* based PESs constructed for small model systems. For example, the polarizable TTM models for ion-water interactions could serve as reference for judging the performance of the DFT functionals [43] for more complex aqueous systems.

4. Summary and conclusions

We developed the full dimensional, polarizable *ab initio* based potential models for describing the halide-water interactions. The present TTM models include anisotropy in both short- and long-range interactions. The parameters of their functional forms were determined by a two-step fitting procedure to electronic structure calculations for the X^-H_2O complex. In particular, the dispersion coefficients were evaluated first from the DFT computations by employing the XDM model with the imposed correct asymptotic behavior at large distances. For the electrostatic part, the Thole model was used, with the anion polarizabilites and the interaction energies calculated at the CCSD(T) and CCSD(T)-F12 level of the theory, respectively. The quality of the analytical PESs was assessed by direct comparisons with the results from the DF-MP2 and various DFT/DFT-D approaches, showing a high fidelity of the model potentials for larger halide-water clusters. We note that the problems in the $F^-(H_2O)$ system should be attributed to the larger distortion of the water molecule in this cluster. Thus, for representing the strongly hydrogen-bonded stretches for the smallest $F^-(H_2O)$ and $F^-(H_2O)_2$ clusters, such water-deformed configurations should be added.

The results presented in this chapter are obtained entirely from a first-principle treatment, following a bottom-up approach, and are focused on the energetics and structural characterization of poly-hydrated halide systems. The obtained data suggest that the polarizable two-body halide-water potentials can indeed be used to describe the ion-water gas phase interactions and, moreover, in conjunction with various TTM water-water models, can serve for benchmarking the DFT approaches and/or the existing empirical/phenomenological models. The present approach can be also used in molecular simulations for studying collisional phenomena and other ion-water processes, e.g. transport and hydration/solvation processes.

Acknowledgments

The authors thank to Centro de Calculo del IFF, SGAI (CSIC) and CESGA for allocation of computer time. This work has been supported by MINECO grant No. FIS2014-51933-P, and by COST Action CM1405(MOLIM).

References

[1] P. Ball, Life's Matrix: A Biography of Water, *University of California Press* (2001).

[2] C. Vega and J. L. F. Abascal, Simulating water with rigid non-polarizable models: a general perspective, *Phys. Chem. Chem. Phys.* **13**, 19663–19688 (2011).

[3] D. J. Arismendi-Arrieta, M. Riera, P. Bajaj, R. Prosmiti, and F. Paesani, I-TTM model for ab initio-based ion–water interaction potentials. 1. halide–water potential energy functions, *J. Phys. Chem. B*, **120**, 1822–1832 (2016).

[4] G. A. Cisneros, K. T. Wikfeldt, L. Ojamë, J. Lu, Y. Xu, H. Torabifard, A. P. Bartók, G. Csányi, V. Molinero, and F. Paesani, Modeling molecular interactions in water: From pairwise to many-body potential energy functions, *Chem. Rev.*, **116**, 7501–7528 (2016).

[5] L.-P. Wang, K. A. McKiernan, J. Gomes, K. A. Beauchamp, T. Head-Gordon, J. E. Rice, W. C. Swope, T. J. Martínez, and V. S. Pande, Building a more predictive protein force field: A systematic and reproducible route to AMBER-FB15, *J. Phys. Chem. B*, **121**, 4023–4039 (2017).

[6] T. Helgaker, P. Jorgensen, and J. Olsen, *Molecular Electronic Structure Theory*, *Wiley* (2008).

[7] R. Parr and Y. Weitao, Density-Functional Theory of Atoms and Molecules. International Series of Monographs on Chemistry, *Oxford University Press*, (1994).

[8] M. Born and R. Oppenheimer, Zur quantentheorie der molekeln, *Ann. Phys.* **389**, 457–484 (1927).

[9] D. J. Wales, J. P. K. Doye, M. A. Miller, P. N. Mortenson, and T. R. Walsh, Energy Landscapes: From Clusters to Biomolecules, *Adv. Chem. Phys.* pp. 1–111. (2007).

[10] J. N. Murrell, S. Carter, S. C. Farantos, P. Huxley, and A. Varandas, *Molecular Potential Energy Function*, *Wiley* (1984).

[11] H. Partridge and D. W. Schwenke, The determination of an accurate isotope dependent potential energy surface for water from extensive ab initio calculations and experimental data, *J. Chem. Phys.*, **106**, 4618–4639 (1997).

[12] C. J. Burnham and S. S. Xantheas, Development of transferable interaction models for water. IV. a flexible, all-atom polarizable potential (TTM2-F) based on geometry dependent charges derived from an ab initio monomer dipole moment surface, *J. Chem. Phys.* **116**, 5115–5124 (2002).

[13] G. S. Fanourgakis and S. S. Xantheas, The flexible, polarizable, thole-type interaction potential for water (TTM2-F) revisited, *J. Phys. Chem. A*, **110**, 4100–4106 (2006).

[14] G. S. Fanourgakis and S. S. Xantheas, Development of transferable interaction potentials for water. V. extension of the flexible, polarizable, thole-type model potential (TTM3-F, v. 3.0) to describe the vibrational spectra of water clusters and liquid water, *J. Chem. Phys.*, **128**, 074506 (2008).

[15] C. J. Burnham, D. J. Anick, P. K. Mankoo, and G. F. Reiter, The vibrational proton potential in bulk liquid water and ice, *J. Chem. Phys.* 128,154519 (2008).
[16] V. Babin, C. Leforestier, and F. Paesani, Development of a "first principles" water potential with flexible monomers: Dimer potential energy surface, vrt spectrum, and second virial coefficient, *J. Chem. Theory Comput.* **9**, 5395–5403 (2013).
[17] K. T. Tang and J. P. Toennies, An improved simple model for the van der waals potential based on universal damping functions for the dispersion coefficients, *J. Chem. Phys.* **80**, 3726–3741 (1984).
[18] E. M. Mas, K. Szalewicz, R. Bukowski, and B. Jeziorski, Pair potential for water from symmetry-adapted perturbation theory, *J. Chem. Phys.* **107**, 4207–4218 (1997).
[19] T. B. Adler, G. Knizia, and H.-J. Werner, A simple and efficient CCSD(T)-F12 approximation, *J. Chem. Phys.* **127**,221106 (2007).
[20] G. Knizia, T. B. Adler, and H.-J. Werner, Simplified CCSD(T)-F12 methods: Theory and benchmarks, *J. Chem. Phys.* **130**,054104 (2009).
[21] H.-J. Werner, P. Knowles, G. Knizia, R. Manby, M. Schütz Molpro, version 2012.1, a package of ab initio programs (2012).
[22] H.-J. Werner, P. J. Knowles, G. Knizia, F. R. Manby, and M. Schütz, Molpro: a general-purpose quantum chemistry program package, *WIREs: Comput. Mol. Sci.* **2**, 242–253 (2012).
[23] T. H. Dunning, Gaussian basis sets for use in correlated molecular calculations. i. the atoms boron through neon and hydrogen, *J. Chem. Phys.* **90**, 1007–1023 (1989).
[24] K. A. Peterson, T. B. Adler, and H.-J. Werner, Systematically convergent basis sets for explicitly correlated wavefunctions: The atoms H, He, B-Ne, and Al-Ar, *J. Chem. Phys.* **128**, 084102 (2008).
[25] J. G. Hill and K. A. Peterson, Correlation consistent basis sets for explicitly correlated wavefunctions: Pseudopotential-based basis sets for the post-d main group elements Ga-Rn, *J. Chem. Phys.* **141**,094106 (2014).
[26] K. A. Peterson, D. Figgen, E. Goll, H. Stoll, and M. Dolg, Systematically convergent basis sets with relativistic pseudopotentials. II. small-core pseudopotentials and correlation consistent basis sets for the post-d group 16-18 elements, *J. Chem. Phys.* **119**, 11113–11123 (2003).
[27] K. A. Peterson, B. C. Shepler, D. Figgen, and H. Stoll, On the spectroscopic and thermochemical properties of ClO, BrO, IO, and their anions, *J. Phys. Chem. A*, **110**, 13877–13883 (2006).
[28] S. Boys and F. Bernardi, The calculation of small molecular interactions by the differences of separate total energies. some procedures with reduced errors, *Mol. Phys.* **19**, 553–566 (1970).
[29] K. A. Peterson, D. E. Woon, and T. H. Dunning, Benchmark calculations with correlated molecular wave functions. IV. the classical barrier height of the $H+H_2 \rightarrow H_2+H$ reaction, *J. Chem. Phys.* **100**, 7410–7415 (1994).

[30] D. E. Woon and T. H. Dunning, Gaussian basis sets for use in correlated molecular calculations. IV. calculation of static electrical response properties, *J. Chem. Phys.* **100**, 2975–2988 (1994).

[31] C. Hättig and B. A. Heß, Tdmp2 calculation of dynamic multipole polarizabilities and dispersion coefficients for the halogen anions F^-,Cl^-,Br^- and I^-, *J. Chem. Phys.* **108**, 3863–3870 (1998).

[32] M. Trumm, Y. O. G. Martínez, F. Réal, M. Masella, V. Vallet, and B. Schimmelpfennig, Modeling the hydration of mono-atomic anions from the gas phase to the bulk phase: The case of the halide ions F^-, Cl^-, and Br^-, *J. Chem. Phys.* **136**, 044509 (2012).

[33] A. D. Becke and E. R. Johnson, Exchange-hole dipole moment and the dispersion interaction revisited, *J. Chem. Phys.* **127**, 154108 (2007).

[34] M. J. Frisch, G. W. Trucks, H. B. Schlegel, G. E. Scuseria, M. A. Robb, J. R. Cheeseman, G. Scalmani, V. Barone, B. Mennucci, G. A. Petersson, H. Nakatsuji, M. Caricato, X. Li, H. P. Hratchian, A. F. Izmaylov, J. Bloino, G. Zheng, J. L. Sonnenberg, M. Hada, M. Ehara, K. Toyota, R. Fukuda, J. Hasegawa, M. Ishida, T. Nakajima, Y. Honda, O. Kitao, H. Nakai, T. Vreven, J. A. Montgomery, Jr., J. E. Peralta, F. Ogliaro, M. Bearpark, J. J. Heyd, E. Brothers, K. N. Kudin, V. N. Staroverov, R. Kobayashi, J. Normand, K. Raghavachari, A. Rendell, J. C. Burant, S. S. Iyengar, J. Tomasi, M. Cossi, N. Rega, J. M. Millam, M. Klene, J. E. Knox, J. B. Cross, V. Bakken, C. Adamo, J. Jaramillo, R. Gomperts, R. E. Stratmann, O. Yazyev, A. J. Austin, R. Cammi, C. Pomelli, J. W. Ochterski, R. L. Martin, K. Morokuma, V. G. Zakrzewski, G. A. Voth, P. Salvador, J. J. Dannenberg, S. Dapprich, A. D. Daniels, C. Farkas, J. B. Foresman, J. V. Ortiz, J. Cioslowski, and D. J. Fox. Gaussian 09 Revision d.01 Gaussian Inc. Wallingford CT (2009).

[35] A. Otero-de-la Roza and E. R. Johnson, Non-covalent interactions and thermochemistry using XDM-corrected hybrid and range-separated hybrid density functionals, *J. Chem. Phys.* **138**, 204109 (2013).

[36] W. H. Press, S. A. Teukolsky, W. T. Vetterling, and B. P. Flannery, Numerical Recipes in Fortran 90: The Art of Parallel Scientific Computing, *Cambridge University Press* (1996).

[37] J. C. Werhahn, D. Akase, and S. S. Xantheas, Universal scaling of potential energy functions describing intermolecular interactions. II. the halide-water and alkali metal-water interactions, *J. Chem. Phys.* **141**,064118 (2014).

[38] J. Rheinecker and J. M. Bowman, The calculated infrared spectrum of Cl-H_2O using a new full dimensional ab initio potential surface and dipole moment surface, *J. Chem. Phys.* **125**, 133206 (2006).

[39] E. Kamarchik, D. Toffoli, O. Christiansen, and J. M. Bowman, Ab initio potential energy and dipole moment surfaces of the F-(H_2O) complex, *Spectrochim. Acta A*, **119**, 59 – 62 (2014).

[40] J. Kim, H. M. Lee, S. B. Suh, D. Majumdar, and K. S. Kim, Comparative ab initio study of the structures, energetics and spectra of $X^-(H_2O)_{n=1--4}$ [X=F,Cl,Br,I] clusters, *J. Chem. Phys.* **113**, 5259–5272 (2000).

[41] S. Grimme, S. Ehrlich, and L. Goerigk, Effect of the damping function in dispersion corrected density functional theory, *J. Comput. Chem.* **32**, 1456–1465 (2011).

[42] J. Sun, A. Ruzsinszky, and J. P. Perdew, Strongly constrained and appropriately normed semilocal density functional, *Phys. Rev. Lett.* **115**, 036402 (2015).

[43] M. Kohagen, E. Pluhařová, P. E. Mason, and P. Jungwirth, Exploring ion–ion interactions in aqueous solutions by a combination of molecular dynamics and neutron scattering, *J. Phys. Chem. Lett.* **6**, 1563–1567 (2015).

Chapter 9

The stopping power of hydrogen for protons and antiprotons

J.J. Bailey*, I.B. Abdurakhmanov, A.S. Kadyrov and I. Bray

Curtin Institute for Computation and Department of Physics and Astronomy, Curtin University, GPO Box U1987, Perth, WA 6845, Australia

Calculations of the stopping cross sections for antiprotons and protons incident on hydrogen are presented. Antiproton collisions are modeled using a single-center convergent close-coupling method. For proton-hydrogen collisions, a two-center convergent close-coupling method is used. The latter approach explicitly models electron capture processes via the inclusion of bound states of the projectile in the expansion of the scattering wave function. This accounts for both positive and neutral charge states of the projectile in calculation of the proton-hydrogen stopping cross section. For hydrogen-hydrogen collisions a single-center convergent close-coupling method is used to model one-electron processes, while the first Born approximation is used to model two-electron processes.

Contents

*Corresponding author. Email: jackson.bailey@postgrad.curtin.edu.au

1. Introduction

Stopping power data are fundamentally important in a great number of fields that rely upon particle transport in matter, including astrophysics, radiation protection, fusion research, material analysis, space exploration, cancer therapy, etc. In particular, stopping power plays a crucial role in the effectiveness of proton therapy, as it forms the main input for radiation dose simulations [1] performed with Monte Carlo codes such as GEANT4 [2, 3]. In fact, there are databases that aim to provide accurate stopping powers for various applications [4, 5].

In this chapter we explore the topic of stopping powers from a theoretical perspective. Specifically, we will be focusing on the simplest target, namely atomic hydrogen. We will look at how the stopping power of this atom is calculated for proton and antiproton projectiles. Calculating the stopping power of hydrogen for an antiproton projectile is relatively simple in comparison to proton, as energy losses are due to electronic excitation and ionization processes only. On the other hand, collisions involving protons open up the possibility for electron capture (charge exchange). Therein, the proton captures a target electron forming a hydrogen atom. This process plays an important role in the calculation of the stopping power and, therefore, must be included if one wishes to obtain accurate results. However, including electron capture processes significantly increases the complexity of calculations as the problem requires a two-center approach. Additionally, the electron capture process in proton-hydrogen collisions further complicates calculations as the newly formed hydrogen atom will continue to interact with the stopping medium. Therefore, when calculating the stopping power of hydrogen for protons, one must also consider the collisions of hydrogen with hydrogen. Not only does the aforementioned collision system have a possibility of excitation and ionization of the target, but now excitation and ionization of the projectile are also possible. These processes can occur individually, resulting in single excitation or ionization, or simultaneously, resulting in double excitation, double ionization, or ionization with excitation. All these must be included in the calculation of the stopping power. The total stopping power of hydrogen for protons then becomes the sum of the proton-hydrogen and hydrogen-hydrogen stopping powers weighted by their corresponding charge-state fractions. In this chapter, we will present our approach to calculate the stopping powers of hydrogen for protons as well as antiprotons, and compare the results with the other theoretical approaches. Currently there are no experimental data for either system.

The first theoretical study of proton stopping in hydrogen was performed by Dalgarno and Griffing [6]. They applied the first Born approximation (FBA) to calculate the proton-hydrogen and hydrogen-hydrogen stopping cross sections. Rearrangement processes in the case of proton-hydrogen scattering and two-electron processes in the case of hydrogen-hydrogen scattering were included. The results were combined by weighting each contribution by its charge-state fraction to obtain the total stopping cross section.

Later, Schiwietz [7] performed single-center coupled-channel atomic-orbital (AO) calculations for the proton fraction of the beam, where wave packets were used to model the continuum. Furthermore, they used the FBA calculations for the hydrogen fraction (including only single-excitation and single-ionization processes), and the experimental H_2 charge-state fractions of Allison [8] to obtain the total stopping cross section. Agreement with the calculations of Dalgarno and Griffing [6] above 125 keV was obtained. Schiwietz and Grande [9] further developed the aforementioned work of Schiwietz [7] by replacing the single-center AO results below 30 keV with the two-center (AO+) ones, which included electron capture. The result was a significant reduction in the proton-hydrogen stopping cross section in this region. Additionally, a screened potential was used to perform the AO calculations for hydrogen-hydrogen collisions, including only single excitation and single ionization. The results were again combined by weighting each contribution with the experimental H_2 charge-state fractions to obtain the total stopping cross section. Agreement with the previous calculations of Schiwietz [7] was achieved above 100 keV and below 15 keV. However, in the intermediate energy region, a larger stopping cross section was obtained.

Fainstein *et al.* [10] used the continuum distorted wave – eikonal initial state (CDW-EIS) model to calculate the stopping cross section for protons impinging on atomic hydrogen. The total stopping cross section was obtained by combining their calculations with the FBA hydrogen-hydrogen results of Dalgarno and Griffing [6]. Agreement with all the previous calculations above 125 keV was obtained, but below this energy, different results were obtained.

Moving to antiproton projectiles, Schiwietz *et al.* [11, 12] performed the first theoretical calculations of the stopping cross section for antiprotons in hydrogen. They carried out calculations using atomic-orbital (AO) close coupling, distorted wave (DW) Born, and generalized adiabatic-ionization (AI) methods. It was found that the first order contribution to the stopping cross section dominates at high velocities with higher order effects becoming important at intermediate velocities. At the same time, near-adiabatic dynamics prevailed in the low velocity limit. The AO and DW calculations were in mutual agreement above the stopping maximum. Cabrera-Trujillo *et al.* [13] were the next to contribute from a theoretical perspective. They used the electron nuclear dynamics (END) formalism to calculate the antiproton-hydrogen stopping cross section up to 300 keV. The results showed reasonable agreement with the AO method of Schiwietz *et al.* [11, 12]. The latest development in solving the problem comes from Lühr and Saenz [14] who calculated the stopping cross sections for antiprotons in hydrogen between 1 keV and 6.4 MeV. They used a semiclassical close-coupling approach to the solution of the time-dependent Schrödinger equation with the radial wave function expanded in a B-spline basis. Lühr and Saenz [14] obtained good agreement with the calculations of Schiwietz *et al.* [11, 12], and there was reasonable agreement with the calculations of Cabrera-Trujillo *et al.* [13] as well.

Our approach to calculating the antiproton-hydrogen stopping cross sections is described in Ref. [15]. We use a single-center semiclassical close-coupling approach, where the scattering wave function is expanded in a basis of target states built from the orthogonal Laguerre functions. The number of basis functions is increased until a convergent result is obtained. The method is referred to as the single-center convergent close-coupling (CCC) approach. To model proton collisions with hydrogen, we use a two-center CCC approach. In this approach electron capture channels are explicitly included by expanding the scattering wave function in a basis of both target and projectile states. This is important for calculating both the stopping cross sections and charge-state fractions. To model hydrogen collisions with hydrogen, we use a combination of the single-center CCC approach and the FBA. In the single-center CCC model, one atom remains fixed in the ground state. Therefore, only single excitation and single ionization can be taken into account, and coupling between channels is included. Subsequently, the FBA is used to account for the two-electron processes of double excitation, double ionization, and ionization with excitation.

The single-center CCC approach has previously been employed to calculate stopping cross sections for antiproton collisions with atoms and molecules [15–17], as well as scattering cross sections for antiproton-hydrogen collisions [18, 19]. Additionally, the two-center CCC approach has been used to calculate scattering cross sections for proton-hydrogen collisions [20–22]. Preliminary results of the proton-hydrogen stopping cross section with the two-center CCC approach have been published in Ref. [23].

In what follows, we will first present theoretical details of our approach to modeling the collisions of protons, antiprotons and hydrogen atoms with hydrogen. We will then describe our target structure calculations, followed by the procedure for calculating the stopping powers. Lastly, we present the results of our calculations compared to the theories described above.

2. Description of the approach

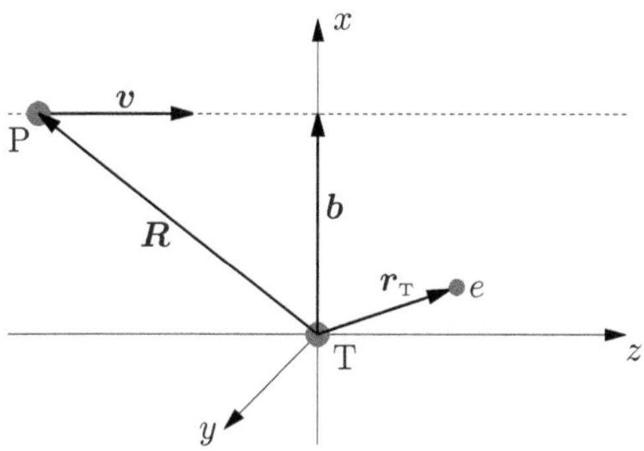

Fig. 1. Coordinates used for the collision of a projectile (P) with a target atom. The target nucleus (T) is at the origin, with the projectile's velocity directed along the z-axis.

For all the CCC calculations, we use the semiclassical approximation to formulate a set of coupled-channel differential equations that describe the scattering system. In the semiclassical approximation, the target electron is treated fully quantum-mechanically, while the motion of the projectile is treated classically. In the laboratory frame, the projectile is assumed to be moving with velocity $\boldsymbol{v}$, along a straight line toward the target at an impact parameter $\boldsymbol{b}$, as illustrated in Fig. 1. The position of the projectile with respect to the target nucleus is then given by $\boldsymbol{R}(t) = \boldsymbol{b} + \boldsymbol{v}t$, where t is time and $t = 0$ corresponds to the distance of closest approach. The velocity of the projectile is taken to be along the z-axis and the impact parameter is taken to be along the x-axis. Hence, the position of the projectile along the z-axis is $z = vt$.

2.1. *Proton-hydrogen collisions*

The CCC approach is based upon the exact time-independent Schrödinger equation and uses an expansion for the total scattering wave function which correctly represents both the target and projectile centers. Here, we describe the main steps (more details are given in Ref. [22]).

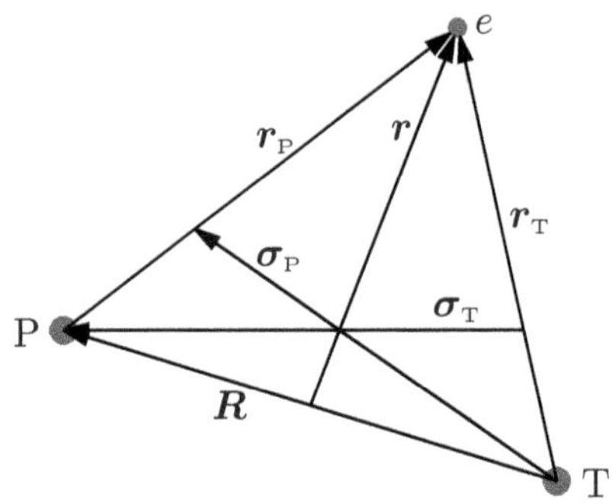

Fig. 2. Sketch of the Jacobi coordinates used for the proton-hydrogen system. The target nucleus is denoted as T, while P indicates the projectile proton.

To describe the proton-hydrogen system, we utilize the Jacobi coordinates illustrated in Fig. 2. Here, $\boldsymbol{r}_\mathrm{T}$ is the position of the electron relative to the target proton, whereas $\boldsymbol{r}_\mathrm{P}$ is the position of the electron relative to the projectile proton. Similarly, $\boldsymbol{\sigma}_\mathrm{T}$ is the position of the projectile proton relative to the center-of-mass of the target proton-electron system, while $\boldsymbol{\sigma}_\mathrm{P}$ is the position of the center-of-mass of the projectile proton-electron system relative to the target proton. Finally, $\boldsymbol{r}$ is the position of the electron relative to the center-of-mass of the two-proton system.

The exact non-relativistic Schrödinger equation for the total scattering wave function Ψ is:

$$H\Psi = E\Psi, \tag{1}$$

where E is the total energy of the system and H is the full three-body Hamiltonian. The Hamiltonian can be written in the following equivalent forms:

$$H = -\frac{1}{2\mu}\nabla^2_{\boldsymbol{\sigma}_\mathrm{T}} + H_\mathrm{T} + V_\mathrm{P} = -\frac{1}{2\mu}\nabla^2_{\boldsymbol{\sigma}_\mathrm{P}} + H_\mathrm{P} + V_\mathrm{T}. \tag{2}$$

Here, μ is the reduced mass of the proton-hydrogen system, H_T and H_P are the target and projectile atom Hamiltonians, V_T is the interaction of the target proton with the projectile atom, and V_P is the interaction of the projectile proton with the target atom. Hamiltonians H_T and H_P are given by:

$$H_\mathrm{T} = -\frac{1}{2}\nabla^2_{\boldsymbol{r}_\mathrm{T}} - \frac{1}{r_\mathrm{T}}, \qquad H_\mathrm{P} = -\frac{1}{2}\nabla^2_{\boldsymbol{r}_\mathrm{P}} - \frac{1}{r_\mathrm{P}}, \tag{3}$$

while the interactions $V_{\rm T}$ and $V_{\rm P}$ read as

$$V_{\rm T} = \frac{1}{R} - \frac{1}{r_{\rm T}}, \qquad V_{\rm P} = \frac{1}{R} - \frac{1}{r_{\rm P}}. \tag{4}$$

The total scattering wave function is expanded in terms of a set of $N_{\rm T}$ target pseudostates ψ_α and $N_{\rm P}$ projectile pseudostates ψ_β according to:

$$\Psi = \sum_{\alpha=1}^{N_{\rm T}} A_\alpha(\boldsymbol{\sigma}_{\rm T})\psi_\alpha(\boldsymbol{r}_{\rm T})e^{i\boldsymbol{k}_\alpha\cdot\boldsymbol{\sigma}_{\rm T}} + \sum_{\beta=1}^{N_{\rm P}} B_\beta(\boldsymbol{\sigma}_{\rm P})\psi_\beta(\boldsymbol{r}_{\rm P})e^{i\boldsymbol{k}_\beta\cdot\boldsymbol{\sigma}_{\rm P}}, \tag{5}$$

where $\boldsymbol{k}_\alpha$ is the relative momentum of the projectile proton and the target atom in channel α. Similarly, $\boldsymbol{k}_\beta$ is the relative momentum of the target proton and the projectile atom in channel β. The total energy of the system E is:

$$E = \frac{k_\alpha^2}{2\mu} + \epsilon_\alpha = \frac{k_\beta^2}{2\mu} + \epsilon_\beta. \tag{6}$$

Furthermore, the pseudostates ψ_α and ψ_β represent both bound and continuum states, and are constructed to satisfy the conditions:

$$\langle\psi_{\gamma'}|H_{\rm T}|\psi_\gamma\rangle = \epsilon_\gamma\delta_{\gamma'\gamma}, \qquad \langle\psi_{\gamma'}|\psi_\gamma\rangle = \delta_{\gamma'\gamma}. \tag{7}$$

It must be emphasized that although the pseudostates within each set are orthogonal to each other, a pseudostate from one set is not orthogonal to a pseudostate from the other set. Details of the target pseudostates are given in Section 3.

The scattering wave function (5) is substituted into the Schrödinger equation (1) and the result is projected onto the conjugate of each term in the expansion, i.e. $\psi^*_{\alpha'}(\boldsymbol{r}_{\rm T})e^{-i\boldsymbol{k}_{\alpha'}\cdot\boldsymbol{\sigma}_{\rm T}}$ and $\psi^*_{\beta'}(\boldsymbol{r}_{\rm P})e^{-i\boldsymbol{k}_{\beta'}\cdot\boldsymbol{\sigma}_{\rm P}}$. After some lengthy algebra, we arrive at the final set of two-center coupled-channel differential equations that describes proton scattering from hydrogen:

$$\begin{cases} i\dot{A}_{\alpha'} + i\sum\limits_{\beta=1}^{N_{\rm P}} \dot{B}_\beta\mathcal{K}_{\alpha'\beta} = \sum\limits_{\alpha=1}^{N_{\rm T}} A_\alpha\mathcal{D}_{\alpha'\alpha} + \sum\limits_{\beta=1}^{N_{\rm P}} B_\beta\mathcal{Q}_{\alpha'\beta}, \\ i\sum\limits_{\alpha=1}^{N_{\rm T}} \dot{A}_\alpha\tilde{\mathcal{K}}_{\beta'\alpha} + i\dot{B}_{\beta'} = \sum\limits_{\alpha=1}^{N_{\rm T}} A_\alpha\tilde{\mathcal{Q}}_{\beta'\alpha} + \sum\limits_{\beta=1}^{N_{\rm P}} B_\beta\tilde{\mathcal{D}}_{\beta'\beta}, \\ \alpha' = 1,\ldots,N_{\rm T}, \quad \beta' = 1,\ldots,N_{\rm P}, \end{cases} \tag{8}$$

where the dots over A and B denote the time derivative. In Eq. (8), the direct scattering matrix elements $\mathcal{D}_{\alpha'\alpha}$ and $\tilde{\mathcal{D}}_{\beta'\beta}$ are:

$$\mathcal{D}_{\alpha'\alpha} = e^{i(\epsilon_{\alpha'}-\epsilon_\alpha)t}\int d\boldsymbol{r}_{\rm T}\psi^*_{\alpha'}(\boldsymbol{r}_{\rm T})V_{\rm P}\psi_\alpha(\boldsymbol{r}_{\rm T}), \tag{9}$$

and

$$\tilde{\mathcal{D}}_{\beta'\beta} = e^{i(\epsilon_{\beta'}-\epsilon_\beta)t} \int d\boldsymbol{r}_{\rm P}\psi^*_{\beta'}(\boldsymbol{r}_{\rm P})V_{\rm T}\psi_\beta(\boldsymbol{r}_{\rm P}). \tag{10}$$

The overlap matrix elements $\mathcal{K}_{\alpha'\beta}$ and $\tilde{\mathcal{K}}_{\beta'\alpha}$ are:

$$\mathcal{K}_{\alpha'\beta} = e^{i(-v^2t/2+(\epsilon_{\alpha'}-\epsilon_\beta)t)} \int d\boldsymbol{r}_{\rm T}\psi^*_{\alpha'}(\boldsymbol{r}_{\rm T})e^{i\boldsymbol{v}\cdot\boldsymbol{r}_{\rm T}}\psi_\beta(\boldsymbol{r}_{\rm P}), \tag{11}$$

and

$$\tilde{\mathcal{K}}_{\beta'\alpha} = e^{i(-v^2t/2+(\epsilon_{\beta'}-\epsilon_\alpha)t)} \int d\boldsymbol{r}_{\rm P}\psi^*_{\beta'}(\boldsymbol{r}_{\rm P})e^{-i\boldsymbol{v}\cdot\boldsymbol{r}_{\rm P}}\psi_\alpha(\boldsymbol{r}_{\rm T}). \tag{12}$$

The electron transfer matrix elements $\mathcal{Q}_{\alpha'\beta}$ and $\tilde{\mathcal{Q}}_{\beta'\alpha}$ are:

$$\mathcal{Q}_{\alpha'\beta} = e^{i(-v^2t/2+(\epsilon_{\alpha'}-\epsilon_\beta)t)} \int d\boldsymbol{r}_{\rm T}\psi^*_{\alpha'}(\boldsymbol{r}_{\rm T})e^{i\boldsymbol{v}\cdot\boldsymbol{r}_{\rm T}}(H_{\rm P}+V_{\rm T}-\epsilon_\beta)\psi_\beta(\boldsymbol{r}_{\rm P}), \tag{13}$$

and

$$\tilde{\mathcal{Q}}_{\beta'\alpha} = e^{i(-v^2t/2+(\epsilon_{\beta'}-\epsilon_\alpha)t)} \int d\boldsymbol{r}_{\rm P}\psi^*_{\beta'}(\boldsymbol{r}_{\rm P})e^{-i\boldsymbol{v}\cdot\boldsymbol{r}_{\rm P}}(H_{\rm T}+V_{\rm P}-\epsilon_\alpha)\psi_\alpha(\boldsymbol{r}_{\rm T}). \tag{14}$$

In Eqs. (8)-(14), the exponential factors arise naturally and not from the introduction of electron translation factors (for details, see Ref. [22]).

The system of differential equations (8) is solved with the initial conditions $A_{\alpha'}(t=-\infty,\boldsymbol{b}) = \delta_{\alpha' i}$ and $B_{\beta'}(t=-\infty,\boldsymbol{b}) = 0$. This implies that the target is in the initial state ψ_i. For all the calculations, we take $i = 1s$, i.e. the target is initially in the ground state. The dependence of $A_{\alpha'}$ and $B_{\beta'}$ on the orientation of $\boldsymbol{b}$ can be factored out such that the probability for transition from some initial state of the target i into any final target state f or any final projectile state k is:

$$p_f(b) = |A_f(t=+\infty,b) - \delta_{fi}|^2, \qquad p_k(b) = |B_k(t=+\infty,b)|^2, \tag{15}$$

where $A_f(t=+\infty,b)$ and $B_k(t=+\infty,b)$ are the probability amplitudes. The set of equations (8) is solved within the region $[-z_{\rm max}, z_{\rm max}]$, where parameter $z_{\rm max}$ is increased until convergent results are obtained. The direct scattering matrix elements are evaluated in spherical coordinates. The overlap and electron transfer matrix elements are evaluated in prolate spheroidal coordinates [20].

2.2. *Antiproton-hydrogen collisions*

Due to the absence of electron capture channels, the single-center scattering equations for the antiproton-hydrogen system are trivial in comparison to the two-center proton-hydrogen ones. To obtain the antiproton-hydrogen scattering equations, we simply change the sign of the projectile interaction (4) and drop all terms relating to the second center. With this, the expansion of the total scattering wave function becomes:

$$\Psi = \sum_{\alpha=1}^{N_{\mathrm{T}}} A_\alpha(\boldsymbol{\sigma}_{\mathrm{T}})\psi_\alpha(\boldsymbol{r}_{\mathrm{T}})e^{i\boldsymbol{k}_\alpha\cdot\boldsymbol{\sigma}_{\mathrm{T}}}. \tag{16}$$

The final set of coupled-channel differential equations reads as:

$$i\dot{A}_{\alpha'} = \sum_{\alpha=1}^{N_{\mathrm{T}}} A_\alpha \mathcal{D}_{\alpha'\alpha};\ \alpha' = 1,\dots,N_{\mathrm{T}}. \tag{17}$$

Here, the matrix elements $\mathcal{D}_{\alpha'\alpha}$ are given by Eq. (9), but with the opposite sign. Equation (17) is solved in a similar manner described for proton-hydrogen.

2.3. *Hydrogen-hydrogen collisions*

For the collisions of hydrogen with hydrogen, we must introduce the coordinate $\boldsymbol{r}'_{\mathrm{P}}$, which is the position of the projectile electron relative to the projectile nucleus. As stated earlier, we use a single-center CCC approach for the calculation of one-electron processes and the FBA for the calculation of two-electron processes. The details of both methods are given here. Taken together, we refer to this procedure as the "CCC+B2e" approach.

First, we consider the single-center CCC approach. This approach closely resembles the one described for antiproton-hydrogen in Section 2.2. With the projectile atom fixed in the ground state, the total scattering wave function is expanded in terms of a complete set of N_{T} target pseudostates ψ_α according to:

$$\Psi = \sum_{\alpha=1}^{N_{\mathrm{T}}} A_\alpha(\boldsymbol{\sigma}_{\mathrm{T}})\psi_\alpha(\boldsymbol{r}_{\mathrm{T}})\psi_{1s}(\boldsymbol{r}'_{\mathrm{P}})e^{i\boldsymbol{k}_\alpha\cdot\boldsymbol{\sigma}_{\mathrm{T}}}. \tag{18}$$

Details of the target pseudostates are given in Section 3. Additionally, the total energy of the system E is:

$$E = \frac{k_\alpha^2}{2\mu} + \epsilon_\alpha + \epsilon_{1s}. \tag{19}$$

The total Hamiltonian H is written as:

$$H = -\frac{1}{2\mu}\nabla^2_{\boldsymbol{\sigma}_{\mathrm{T}}} + H_{\mathrm{T}} + H'_{\mathrm{P}} + V_{\mathrm{P}}, \tag{20}$$

where μ is the reduced mass of the hydrogen-hydrogen system. Here, H_{T} is the target atom Hamiltonian defined earlier in Eq. (3) and the projectile atom Hamiltonian H'_{P} is:

$$H'_{\mathrm{P}} = -\frac{1}{2}\nabla^2_{\boldsymbol{r}'_{\mathrm{P}}} - \frac{1}{r'_{\mathrm{P}}}. \tag{21}$$

Also, V_P is the projectile-target interaction, which is given by:

$$V_P = \frac{1}{R} - \frac{1}{|\boldsymbol{R} - \boldsymbol{r}_T|} - \frac{1}{|\boldsymbol{R} + \boldsymbol{r}'_P|} + \frac{1}{|\boldsymbol{R} + \boldsymbol{r}'_P - \boldsymbol{r}_T|}. \tag{22}$$

Substituting the scattering wave function (18) into the Schrödinger equation (1), and following Sections 2.1 and 2.2, we obtain the final set of coupled-channel differential equations:

$$i\dot{A}_{\alpha'} = \sum_{\alpha=1}^{N_T} A_\alpha \mathcal{H}_{\alpha'\alpha}; \ \alpha' = 1, \ldots, N_T, \tag{23}$$

where $\mathcal{H}_{\alpha'\alpha}$ are the direct-scattering matrix elements

$$\mathcal{H}_{\alpha'\alpha} = e^{i(\epsilon_{\alpha'} - \epsilon_\alpha)t} \int d\boldsymbol{r}_T d\boldsymbol{r}'_P \psi^*_{\alpha'}(\boldsymbol{r}_T)\psi_{1s}(\boldsymbol{r}'_P) V_P \psi_{1s}(\boldsymbol{r}'_P)\psi_\alpha(\boldsymbol{r}_T). \tag{24}$$

Again, Eq. (23) is solved in a similar manner described for the proton-hydrogen system.

To model two-electron processes, i.e. double excitation, double ionization, and ionization with excitation, we use the FBA. In the FBA, the transition amplitude for the scattering process:

$$\mathrm{H}(1s) + \mathrm{H}(1s) \to \mathrm{H}(\alpha) + \mathrm{H}(\beta)$$

is given by

$$T_{\alpha,\beta} = \iiint d\boldsymbol{r}_T d\boldsymbol{r}'_P d\boldsymbol{R}\, \psi_{1s}(\boldsymbol{r}_T)\psi^*_\alpha(\boldsymbol{r}_T)\psi_{1s}(\boldsymbol{r}'_P)\psi^*_\beta(\boldsymbol{r}'_P) e^{i\boldsymbol{K}\cdot\boldsymbol{R}}$$
$$\times \left(\frac{1}{R} - \frac{1}{|\boldsymbol{R} + \boldsymbol{r}'_P|} - \frac{1}{|\boldsymbol{R} - \boldsymbol{r}_T|} + \frac{1}{|\boldsymbol{R} + \boldsymbol{r}'_P - \boldsymbol{r}_T|} \right). \tag{25}$$

Here, $\boldsymbol{K} = \boldsymbol{k}_i - \boldsymbol{k}_f$ is the momentum transfer, where $\boldsymbol{k}_i$ and $\boldsymbol{k}_f$ are the initial and final momenta of the projectile, respectively. For the FBA calculations, we will choose $\boldsymbol{K}$ to be along the z-axis. Additionally, α and β represent the final states of the target and projectile atoms, respectively. If the final state of the target is a bound state, then $\alpha = n_\alpha l_\alpha m_\alpha$, where n, l and m are the principal, orbital and magnetic quantum numbers, and $\psi_{n_\alpha l_\alpha m_\alpha}$ is given by the known wave functions of the hydrogen atom. On the other hand, if the final state of the target is a continuum state, then $\alpha = \boldsymbol{k}_e$, where $\boldsymbol{k}_e$ is the momentum of the ejected electron, and $\psi^-_{\boldsymbol{k}_e}$ is given by the two-body Coulomb wave function (see below). Similarly, if the final state of the projectile is a bound state, then $\beta = n_\beta l_\beta m_\beta$, and if it is a continuum state then $\beta = \boldsymbol{k}'_e$.

One benefit of the FBA is that Eq. (25) can be evaluated completely analytically. We start by writing the Coulomb potential $1/r$ as the Fourier transform of its momentum-space representation according to:

$$\frac{1}{r} = (2\pi)^{-3} \int d\boldsymbol{k} \frac{4\pi}{k^2} e^{-i\boldsymbol{k}\cdot\boldsymbol{r}}. \tag{26}$$

Now substituting Eq. (26) into (25) and using the relations:

$$\int d\boldsymbol{r} e^{i(\boldsymbol{k}-\boldsymbol{k}')\cdot\boldsymbol{r}} = (2\pi)^3\delta(\boldsymbol{k}-\boldsymbol{k}'), \qquad \int d\boldsymbol{k} f(\boldsymbol{k})\delta(\boldsymbol{k}-\boldsymbol{k}') = f(\boldsymbol{k}'), \tag{27}$$

we obtain the expression

$$T_{\alpha,\beta} = \frac{4\pi}{K^2}\big(\delta_{\alpha,1s} - \mathcal{F}_\alpha(\boldsymbol{K})\big)\big(\delta_{\beta,1s} - \mathcal{F}_\beta(-\boldsymbol{K})\big), \tag{28}$$

where

$$\mathcal{F}_\gamma(\boldsymbol{K}) = \int d\boldsymbol{r}\psi_{1s}(\boldsymbol{r})\psi^*_\gamma(\boldsymbol{r})e^{i\boldsymbol{K}\cdot\boldsymbol{r}}. \tag{29}$$

The function $\mathcal{F}_\gamma$ will be referred to as the individual-atom amplitude. It can be obtained analytically for both excitation and ionization processes. By definition, $\mathcal{F}_\gamma$ is the Fourier transform of $(2\pi)^{3/2}\psi_{1s}(\boldsymbol{r})\psi^*_\gamma(\boldsymbol{r})$.

For excitation transitions, both the initial- and final-state wave functions are obtained from the general expression for the wave functions of the hydrogen atom:

$$\psi_{nlm}(\boldsymbol{r}) = R_{nl}(r)Y_{lm}(\hat{\boldsymbol{r}}), \tag{30}$$

where R_{nl} are the radial functions and Y_{lm} are the spherical harmonics. Utilizing the above expressions, Eq. (29) can be evaluated for any excitation transition. Since we choose $\boldsymbol{K}$ to be aligned along the z-axis, the resulting amplitudes are non-zero only when the change in magnetic quantum number is equal to zero. As both atoms are initially in the ground state, only final states with $m = 0$ need to be considered.

For ionization transitions, the final-state wave function is taken to be the two-body Coulomb wave function:

$$\psi^-_{\boldsymbol{k}_e}(\boldsymbol{r}) = (2\pi)^{-3/2}e^{-\pi\eta/2}\Gamma(1-i\eta)e^{i\boldsymbol{k}_e\cdot\boldsymbol{r}}\,{}_1F_1(i\eta;1;-ik_e r - i\boldsymbol{k}_e\cdot\boldsymbol{r}), \tag{31}$$

where Γ is the Gamma function, ${}_1F_1$ is the confluent hypergeometric function, and $\eta = -1/k_e$ is the Sommerfeld parameter. Following the ideas of Guth and Mullin [24], who calculated the Fourier transform of $\psi^-_{\boldsymbol{k}_e}(\boldsymbol{r})$, it can be shown that $\mathcal{F}_{\boldsymbol{k}_e}$ is:

$$\begin{aligned}\mathcal{F}_{\boldsymbol{k}_e}(\boldsymbol{K}) &= -\frac{(2\pi)^{3/2}}{\sqrt{\pi}}\frac{\partial}{\partial\nu}\left(\frac{4\pi}{(2\pi)^3}e^{-\pi\eta/2}\Gamma(1+i\eta)\frac{\left[K^2-(k_e+i\nu)^2\right]^{i\eta}}{\left[(\boldsymbol{k}_e+\boldsymbol{K})^2+\nu^2\right]^{1+i\eta}}\right)\Bigg|_{\nu=1}\\ &= \frac{2\sqrt{2}}{\pi}e^{-\pi\eta/2}\Gamma(1+i\eta)\frac{\left[K^2-(k_e+i)^2\right]^{i\eta}}{\left[(\boldsymbol{k}_e+\boldsymbol{K})^2+1\right]^{1+i\eta}}\\ &\quad\times\left(\frac{(1+i\eta)[K^2-(k_e+i)^2]+(1-i\eta)[(\boldsymbol{k}_e+\boldsymbol{K})^2+1]}{[(\boldsymbol{k}_e+\boldsymbol{K})^2+1][K^2-(k_e+i)^2]}\right).\end{aligned} \tag{32}$$

3. Target structure

For the generation of atomic hydrogen pseudostates, we follow the ideas of Bray and Stelbovics [25]. The pseudostates ψ_α in Eqs. (5), (16) and (18) are expressed according to Eq. (30) with:

$$R_{nl}(r) = \frac{1}{r}\sum_{k=1}^{N_l} B^l_{nk}\xi_{kl}(r). \tag{33}$$

Here, ξ_{kl} is a complete set of orthonormal basis functions, N_l is the number of basis functions for a given l and B^l_{nk} are the expansion coefficients that are found by diagonalization of the target Hamiltonian in the complete basis. In other words, the pseudostates ψ_α satisfy the conditions specified in Eq. (7).

In Eq. (33), we choose the basis functions to be the Laguerre functions:

$$\xi_{kl}(r) = \left(\frac{\lambda_l(k-1)!}{(2l+1+k)!}\right)^{1/2}(\lambda_l r)^{l+1}\exp(-\lambda_l r/2)L^{2l+2}_{k-1}(\lambda_l r), \tag{34}$$

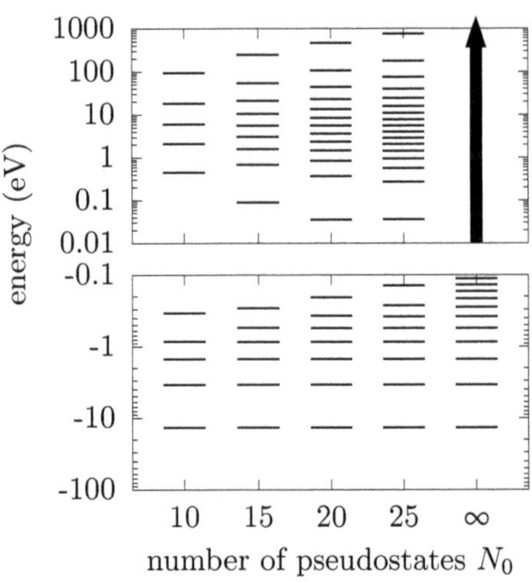

Fig. 3. Energy levels of atomic hydrogen obtained from a Laguerre basis with N_0 functions for $l = 0$. Also shown is a part of the true spectrum of the hydrogen atom, which corresponds to a Laguerre basis with $N_0 = \infty$ functions.

where $L^{2l+2}_{k-1}(\lambda_l r)$ are the associated Laguerre polynomials and λ_l is an exponential fall-off parameter. Typically, we choose λ_l to give the most accurate ground state of the target with a minimum number of basis functions. The choice of λ_l does not affect the final result, but it does affect the speed of convergence. Specific values of λ_l will be given in Section 5.

The choice of the basis as a set of orthogonal Laguerre functions allows us to model the whole spectrum of the target atom. As the size of the one-electron basis increases, the low-lying states will converge to the bound states of the target, while the remaining (pseudo) states will provide an effective representation of the high-lying bound states of the target atom and an increasingly dense square-integrable representation of the target continuum. This is illustrated in Fig. 3, where the pseudostate energies obtained from a Laguerre basis with N_0 functions for $l = 0$ are plotted against the eigenenergies of the hydrogen atom. Higher orbital angular momentum states are distributed similarly.

4. Stopping power

The stopping power is the energy loss per unit path length and in general is defined as:

$$-\frac{dE}{dx} = N_a S(E_0), \tag{35}$$

where $S(E_0)$ is referred to as the stopping cross section. It depends on the incident energy of the projectile E_0, and is related to the stopping power through the density of target atoms in the stopping medium N_a. For heavy projectiles, it can be assumed that the total stopping cross section is the sum of two contributions, the nuclear and the electronic stopping cross sections, i.e. $S = S_{\mathrm{e}} + S_{\mathrm{n}}$. In this chapter, we consider only the electronic stopping cross sections, as the nuclear stopping cross sections are negligible above 5 keV incident energy. However, calculation of the nuclear part is trivial and is detailed in Ref. [16].

First, details of the proton-hydrogen electronic stopping cross section will be discussed. Following this, the antiproton-hydrogen electronic stopping cross section will be presented, which is simpler to calculate, as stated.

Due to the possibility of charge exchange, the incident proton can capture an electron and form a hydrogen atom. This newly formed hydrogen atom will continue to interact with the stopping medium, losing energy and potentially losing and gaining electrons many times. For this reason all possible charge-states of the beam must be considered when calculating the total stopping cross section for protons incident on atomic hydrogen. Therefore, the total stopping cross section for the proton-hydrogen system is:

$$S(E_0) = f^{\mathrm{H}^+} S^{\mathrm{H}^+} + f^{\mathrm{H}^0} S^{\mathrm{H}^0}. \tag{36}$$

Here, S^{H^+} is the stopping cross sections for a beam consisting entirely of protons (positive charge), S^{H^0} is the stopping cross sections for a beam consisting entirely of hydrogen atoms (neutral charge), whereas f^{H^+} and f^{H^0} are the positive and neutral charge-state fractions of the beam, respectively. In this work, we neglect the negative charge state as the probability of H^- formation is insignificant. The charge-state fractions f^{H^+} and f^{H^0} are calculated from the total electron capture cross section σ_{c} in proton-hydrogen collisions and the total electron loss cross section σ_{l} in hydrogen-hydrogen collisions according to:

$$f^{\mathrm{H}^+} = \sigma_{\mathrm{l}}/(\sigma_{\mathrm{c}} + \sigma_{\mathrm{l}}), \qquad f^{\mathrm{H}^0} = \sigma_{\mathrm{c}}/(\sigma_{\mathrm{c}} + \sigma_{\mathrm{l}}). \tag{37}$$

This highlights the importance of having a two-center approach which can provide accurate electron capture cross sections. Further details on σ_{c} and σ_{l} are given below.

The positive-charge-state electronic stopping cross section $S_{\mathrm{e}}^{\mathrm{H}^+}$ is the result of three possible energy loss processes in the proton-hydrogen collision system. These are excitation and ionization of the target, and capture of the target electron to a bound state of the projectile. The stopping cross section is, therefore, written as:

$$S_{\mathrm{e}}^{\mathrm{H}^+}(E_0) = \sum_{f=1}^{\infty}(\epsilon_f - \epsilon_i)\sigma_{fi} + \int_0^{E_0+\epsilon_i}(\epsilon - \epsilon_i)\frac{d\sigma}{d\epsilon}d\epsilon + \sum_{k=1}^{\infty}(\epsilon_k - \epsilon_i + v^2/2)\sigma_{ki}, \tag{38}$$

where ϵ_i is the energy of the ground state of the target i. Here, σ_{fi} is the cross section for excitation to a state f of energy ϵ_f, $d\sigma/d\epsilon$ is the single differential cross section for ionization of the electron with energy ϵ, and σ_{ki} is the cross section for electron capture to a state k of energy ϵ_k. Additionally, the $v^2/2$ term represents the kinetic energy of an electron traveling with the speed of the incident proton after being captured.

To accurately model electron capture processes, we use a two-center coupled-channel approach with pseudostates centered on both the target and projectile, as described in Section 2.1. For the calculation of the stopping cross sections, we include bound and continuum pseudostates on the target center, while only negative-energy pseudostates for the projectile center are considered. With this model, the first two terms in Eq. (38) that represent excitation and ionization become a single sum over N_{T} negative- and positive-energy target-centered pseudostates, while the third term becomes a sum over N_{P} negative-energy projectile-centered pseudostates. Thus, we obtain:

$$S_{\mathrm{e}}^{\mathrm{H}^+} \approx \sum_{f=1}^{N_{\mathrm{T}}}(\epsilon_f - \epsilon_i)\sigma_{fi} + \sum_{k=1}^{N_{\mathrm{P}}}(\epsilon_k - \epsilon_i + v^2/2)\sigma_{ki}. \tag{39}$$

Here, the cross sections for the direct transitions σ_{fi} and rearrangement transitions σ_{ki} are obtained by integration of the transition probabilities (15) over the impact parameter according to:

$$\sigma_{fi} = 2\pi \int_0^\infty p_{fi}(b) b db, \qquad \sigma_{ki} = 2\pi \int_0^\infty p_{ki}(b) b db. \tag{40}$$

Furthermore, the total electron capture cross section σ_c, which is required for the calculation of the charge-state fractions (37), is the sum of all electron-capture cross sections (40), i.e.:

$$\sigma_c = \sum_{k=1}^{N_P} \sigma_{ki}. \tag{41}$$

The neutral-charge-state electronic stopping cross section $S_e^{H^0}$ is the result of many possible energy loss processes in the hydrogen-hydrogen collision system. These are excitation or ionization of either the target or projectile, simultaneous excitation or ionization of both the target and projectile, and excitation of either the target or projectile with ionization of the other. Including all these terms, the stopping cross section becomes:

$$\begin{aligned} S_e^{H^0} =& 2\sum_{f=1}^{\infty}(\epsilon_f - \epsilon_i)\sigma_{fi} + \sum_{f\neq i}^{\infty}\sum_{k\neq i}^{\infty}(\epsilon_f - \epsilon_i + \epsilon_k - \epsilon_i)\sigma_{fi,ki} \\ &+ 2\int_0^{E_0+\epsilon_i}(\epsilon - \epsilon_i)\frac{d\sigma}{d\epsilon}d\epsilon + \int_0^{E_0+\epsilon_i}\int_0^{E_0+\epsilon_i}(\epsilon - \epsilon_i + \epsilon' - \epsilon_i)\frac{d\sigma}{d\epsilon d\epsilon'}d\epsilon d\epsilon' \\ &+ 2\sum_{f\neq i}^{\infty}\int_0^{E_0+\epsilon_i}(\epsilon_f - \epsilon_i + \epsilon - \epsilon_i)\frac{d\sigma_{fi}}{d\epsilon}d\epsilon, \end{aligned} \tag{42}$$

where σ_{fi} is the cross section for excitation of one hydrogen atom to a state f of energy ϵ_f while the other hydrogen atom remains in the ground state. Further, $\sigma_{fi,ki}$ is the cross section for excitation of both hydrogen atoms, one to a state f of energy ϵ_f and the other to a state k of energy ϵ_k. Quantity $d\sigma/d\epsilon$ is the differential cross section for ionization of one hydrogen atom to an energy ϵ, while the other hydrogen atom remains in the ground state. Also $d\sigma/d\epsilon d\epsilon'$ is the differential cross section for ionization of both hydrogen atoms, one with energy ϵ and the other with energy ϵ'. Moreover, $d\sigma_{fi}/d\epsilon$ is the differential cross section for ionization of one hydrogen atom to an energy ϵ, while the other hydrogen atom is excited to a state f of energy ϵ_f. Additionally, the factor of 2 in the first, third, and last terms of Eq. (42) is due to the symmetry of the system.

To model both one- and two-electron processes, we use a combination of two approaches. The first is a single-center coupled-channel approach. This approach is used to model single excitation and single ionization processes only since one atom remains fixed in the ground state. Since the continuum is discretized in the single-center coupled-channel approach, the first and third terms of Eq. (42) that represent the stopping cross sections associated with single excitation and ionization, become a single sum over N_{T} negative- and positive-energy pseudostates. The second approach, which is used for the calculation of two-electron processes, is the FBA. In this approach we limit ourselves to a total of N_{b} bound states. Combining these two models, the neutral-charge-state electronic stopping cross section (42) becomes:

$$\begin{aligned} S_{\mathrm{e}}^{\mathrm{H}^0} =& 2\sum_{f=1}^{N_{\mathrm{T}}}(\epsilon_f - \epsilon_i)\sigma_{fi} + \sum_{f\neq i}^{N_{\mathrm{b}}}\sum_{k\neq i}^{N_{\mathrm{b}}}(\epsilon_f - \epsilon_i + \epsilon_k - \epsilon_i)\sigma_{fi,ki} \\ &+ \int_0^{E_0+\epsilon_i}\int_0^{E_0+\epsilon_i}(\epsilon - \epsilon_i + \epsilon' - \epsilon_i)\frac{d\sigma}{d\epsilon d\epsilon'}d\epsilon d\epsilon' \\ &+ 2\sum_{f\neq i}^{N_{\mathrm{b}}}\int_0^{E_0+\epsilon_i}(\epsilon_f - \epsilon_i + \epsilon - \epsilon_i)\frac{d\sigma_{fi}}{d\epsilon}d\epsilon. \end{aligned} \tag{43}$$

Here, the cross sections for one-electron transitions σ_{fi} are obtained using Eq. (40). Additionally, the cross sections for the two-electron processes of double-excitation, double-ionization, and ionization-with-excitation are calculated from the transition amplitudes in the FBA (28) according to:

$$\sigma_{fi,ki} = \frac{\mu^2}{2\pi k_i^2}\int |T_{n_f l_f, n_k l_k}|^2 K dK, \tag{44}$$

$$\frac{d\sigma}{d\epsilon d\epsilon'} = k_{\mathrm{e}}k'_{\mathrm{e}}\frac{\mu^2}{2\pi k_i^2}\iiint |T_{\boldsymbol{k}_{\mathrm{e}},\boldsymbol{k}'_{\mathrm{e}}}|^2 K dK d\Omega_{\boldsymbol{k}_{\mathrm{e}}}\Omega_{\boldsymbol{k}'_{\mathrm{e}}}, \tag{45}$$

and

$$\frac{d\sigma_{fi}}{d\epsilon} = k_{\mathrm{e}}\frac{\mu^2}{2\pi k_i^2}\iint |T_{n_f l_f,\boldsymbol{k}_{\mathrm{e}}}|^2 K dK d\Omega_{\boldsymbol{k}_{\mathrm{e}}}, \tag{46}$$

respectively. In Eqs. (44)-(46), the integrals over K are over the range from $K_{\mathrm{min}} = k_i - k_f$ to $K_{\mathrm{max}} = k_i + k_f$ and evaluated numerically, while the integrals over $\Omega_{\boldsymbol{k}_{\mathrm{e}}}$ are evaluated analytically. The final momentum of the projectile k_f is obtained from the energy conservation law and depends on the final states of the atoms, while $\boldsymbol{k}_{\mathrm{e}}$ is the momentum of the ejected electron. Furthermore, the total electron loss cross section σ_{l}, which is required for the calculation of the charge-state fractions (37), is the sum of all the cross sections corresponding to ionization of the target atom. Therefore, it is written as the sum of the total single ionization cross section σ_{si}, total double ionization cross section σ_{di} and total ionization-with-excitation cross section $\sigma_{\rm ie}$, so that $\sigma_{\rm l} = \sigma_{\rm si} + \sigma_{\rm di} + \sigma_{\rm ie}$. In the aforementioned model, the total single ionization cross section is given by the sum of all cross sections for one electron transitions to positive-energy states:

$$\sigma_{\rm si} = \sum_{f,\epsilon_f>0}^{N_{\rm T}} \sigma_{fi}. \tag{47}$$

The total double ionization and ionization-with-excitation cross sections are calculated from Eqs. (45) and (46), respectively, as:

$$\sigma_{\rm di} = \int_0^{E_0+\epsilon_{1s}} \int_0^{E_0+\epsilon_{1s}} \frac{d\sigma}{d\epsilon d\epsilon'} d\epsilon d\epsilon', \tag{48}$$

and

$$\sigma_{\rm ie} = \sum_{f\neq i}^{N_{\rm b}} \int_0^{E_0+\epsilon_{1s}} \frac{d\sigma_{fi}}{d\epsilon} d\epsilon. \tag{49}$$

For antiproton-hydrogen collisions, since there is no possibility of electron capture, the electron stopping cross section is simply given by the first term of Eq. (39).

5. Results of calculations

In this Section, the results of our calculations for the antiproton-hydrogen and proton-hydrogen electronic stopping cross sections will be presented and compared to existing theoretical calculations. When using a coupled-channel approach where the scattering wave function is expanded in a set of target pseudostates, it is important to establish convergence of the stopping cross section with increasing size of the underlying basis. Therefore, we will start by specifying the basis parameters N_l and $l_{\rm max}$ (the maximum value of orbital angular momentum included) that produced the convergent result. Specifically, we say that convergence is achieved when the stopping cross section varies no more than 2% with an increase in either N_l or $l_{\rm max}$. Therefore, our calculations are estimated to be accurate to within 2%.

5.1. *Antiproton-hydrogen stopping cross section*

The CCC results for antiproton-hydrogen have been published in Refs. [15, 17]. Convergence in the calculations of the electronic stopping cross sections for antiproton-hydrogen collisions is achieved with $l_{\rm max} = 6$ and $N_l = 30 - l$. These basis parameters result in a total of 1267 target states to be used in the solution of the coupled-channel differential equations (17). Also, the basis function exponential fall-off parameter λ_l is chosen to be 2 for all l.

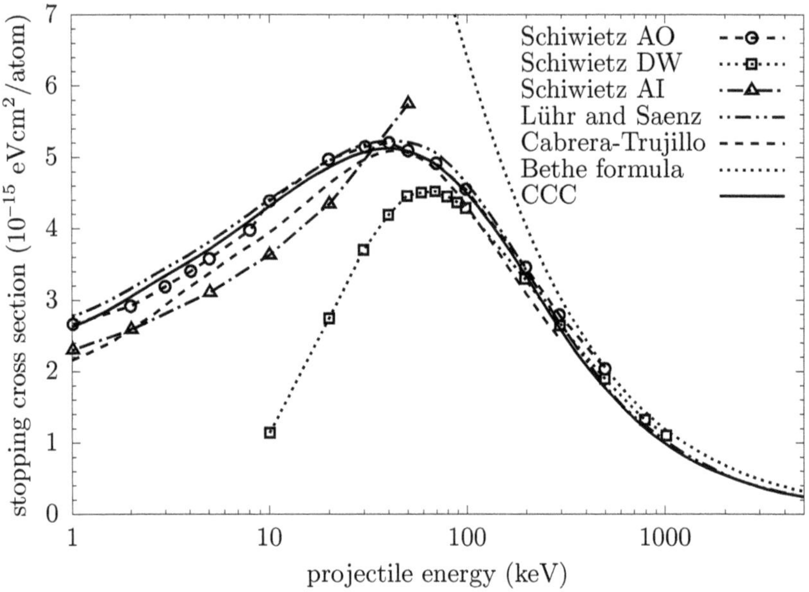

Fig. 4. Electronic stopping cross section for antiprotons incident on hydrogen. The CCC calculations (first published in Ref. [15]) are compared with calculations of Schiwietz *et al.* [11, 12], Lühr and Saenz [14], and Cabrera-Trujillo *et al.* [13].

Figure 4 shows the CCC results for the antiproton-hydrogen electronic stopping cross section together with the theoretical calculations of Schiwietz *et al.* [11, 12] (AO, DW, and AI), Lühr and Saenz [14] and Cabrera-Trujillo *et al.* [13]. Also shown is the result obtained using Bethe's formula [26]. The CCC results are in good agreement with the AO calculations of Schiwietz *et al.* [11, 12] and the results of Lühr and Saenz [14] over the whole energy range. Also, good agreement with the calculations of Cabrera-Trujillo *et al.* [13] and the DW calculations by Schiwietz *et al.* [11, 12] is seen above 30 keV and 100 keV, respectively. Additionally, the CCC calculations tend toward those from Bethe's formula at high energies whenever the latter is applicable.

5.2. *Proton-hydrogen stopping cross section*

For proton-hydrogen collisions, we will present the results for the positive and neutral charge-state contributions to the total stopping cross sections. Calculated charge-state fractions will also be presented, followed by the results for the total stopping cross sections.

5.2.1. *Positive charge state*

Convergence in calculations of the electronic stopping cross section for proton-hydrogen collisions was achieved with $l_{\max} = 8$ and $N_l = 30 - l$, where $l_{\max}$ is the maximum value of orbital angular momentum included in the expansion of the scattering wave function and N_l is the number of basis functions for a given l. These basis parameters result in a total of 1896 target states and 159 projectile states to be used in the solution of the coupled-channel differential equations (8). Also, the basis function exponential fall-off parameter λ_l is chosen to be 2 for all l.

In Fig. 5, we present our result for the proton-hydrogen electronic stopping cross section together with the calculations of Dalgarno and Griffing [6], Schiwietz [7], Schiwietz and Grande [9], and Fainstein *et al.* [10]. We use the two-center CCC approach, meaning energy losses due to electron capture processes are explicitly included, as well as energy losses due to excitation and ionization. The CCC results are in good agreement with

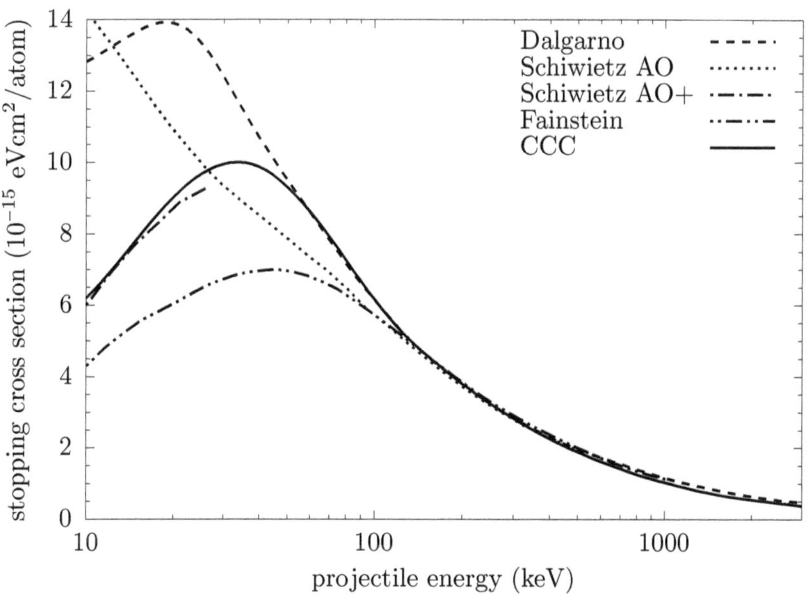

Fig. 5. Electronic stopping cross section for protons incident on hydrogen. The two-center CCC calculations (first published in Ref. [32]) are compared with the results of Dalgarno and Griffing [6], Schiwietz [7] (AO), Schiwietz and Grande [9] (AO+), and Fainstein *et al.* [10].

the FBA calculation of Dalgarno and Griffing [6] above 50 keV. They are also in good agreement with the AO calculations of Schiwietz [7] and the CDW-EIS calculations of Fainstein *et al.* [10] above 130 keV. Furthermore, in the lower energy region, we obtain reasonable agreement with the two-center AO+ calculations of Schiwietz and Grande [9]. The fact that the AO+ results are slightly lower than the CCC ones above 15 keV is likely to be due to the inclusion of more target and projectile states in our calculations compared to those of Schiwietz and Grande [9]. In addition, comparing the single-center AO calculations of Schiwietz [7] to the two-center CCC and AO+ calculations, we see that the explicit inclusion of electron capture channels results in a significant difference in the proton-hydrogen electronic stopping cross section below 100 keV.

5.2.2. *Neutral charge state*

Convergence in the CCC calculations of the electronic stopping cross section for hydrogen-hydrogen collisions was achieved with $l_{\text{max}} = 15$ and $N_l = 30 - l$. These basis parameters result in a total of 5080 target states used in the solution of the coupled-channel differential equations (23). Also, the basis function exponential fall-off parameter λ_l is chosen to be 4 for all l. It should be noted that the value for l_{max} is significantly larger than that required for proton-hydrogen collisions as a single-center approach is used. This is to generate higher-energy continuum states compared to the proton calculations. Furthermore, in the FBA calculations, we include excitations to all states for which $n \leq 8$ and $l \leq 3$.

In Fig. 6, we present our CCC+B2e results for the hydrogen-hydrogen electronic stopping cross sections together with the calculations of Dalgarno and Griffing [6], as well as Schiwietz and Grande [9]. The CCC+B2e calculations include energy losses due to single excitation and single ionization calculated in the single-center CCC approach, as well as energy losses due to double excitation, double ionization, and ionization with excitation calculated in the FBA. Also shown in Fig. 6 are the results arising solely from the CCC calculations, meaning only energy losses due to single excitation and ionization are included, as well as the results of the FBA calculations for all energy loss processes (labeled as Born) and one-electron processes only (labeled as B1e). First, we note that the CCC results are in agreement with the B1e results at high energies, where the latter is considered accurate. Specifically, good agreement is seen above 300 keV. However, at lower energies, below 200 keV, the coupling between channels in the CCC approach

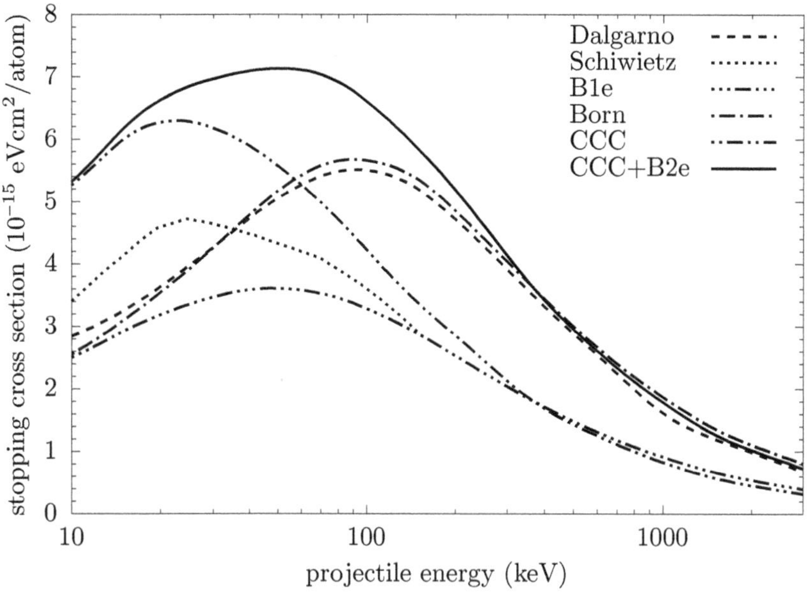

Fig. 6. Electronic stopping cross section for hydrogen incident on hydrogen. Calculations (first published in Ref. [32]) are compared with those of Dalgarno and Griffing [6], as well as Schiwietz and Grande [9]. The results labeled as CCC and B1e include one-electron processes only, while the CCC+B2e and the Born results include one- and two-electron processes (see the main text for details).

results in significantly larger stopping cross sections when compared to the B1e results. In this energy region, the results of the CCC calculations are much larger than those due to the AO calculations of Schiwietz and Grande [9], although both methods are based on a similar approach. It could be that the results of Schiwietz and Grande [9] did not have a sufficient number of states, whereas the CCC calculations include a much larger number of target states. Further, turning to the FBA calculations, we note that above 40 keV, there is a small but systematic disagreement between the present work and that of Dalgarno and Griffing [6]. This is due to the fact that we include excitation to all states with $n \leq 8$ and $l \leq 3$, whereas Dalgarno and Griffing [6] include excitations up to the $n = 3$ shell only. On the other hand, below 20 keV the FBA calculations of Dalgarno and Griffing [6] are slightly higher, as they have included an estimated contribution to the stopping cross section due to the H^- formation. Lastly, comparing our calculations that include one-electron processes (CCC & B1e) to those that include one- and two-electron processes (CCC+B2e & Born), we see that double excitation, double ionization, and ionization with excitation make a substantial contribution to the stopping cross section above 20 keV. As such, the CCC+B2e calculations are considered to be our most accurate results.

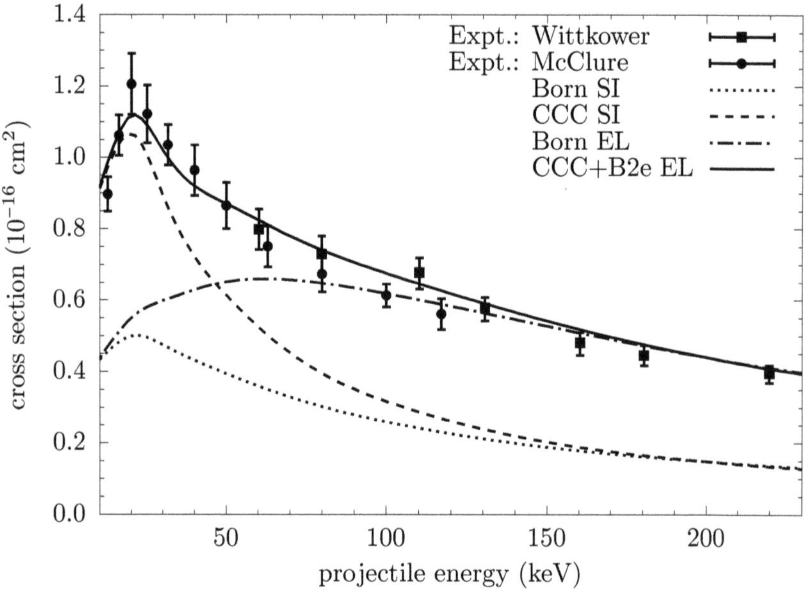

Fig. 7. Total cross section for electron loss by the projectile in hydrogen collisions with hydrogen. The CCC+B2e calculations (see the text for details) are compared to the experimental data of Wittkower *et al.* [27] and McClure [28]. Also shown are the Born calculations for total electron loss (Born), as well as the Born and CCC calculations for single ionization only, labeled as Born SI and CCC SI, respectively (first published in Ref. [32]).

The aforementioned statement can be validated by considering the total cross section for electron loss by the projectile, as there are experimental data with which to compare with the theoretical results. Since the stopping cross section is dominated by ionization processes, this may prove useful in assessing the accuracy of the hydrogen-hydrogen stopping cross section. In Fig. 7, we present our CCC+B2e calculation for the total electron loss cross section compared to the experimental data of Wittkower *et al.* [27] and McClure [28]. Also shown are the results of the FBA calculations for the total electron loss cross section (Born EL), as well as the FBA and single-center CCC calculations for the single ionization cross section (denoted as Born SI and CCC SI, respectively). The CCC+B2e results are in good agreement with the experimental data over the whole energy region considered. On the other hand, the Born EL calculations significantly underestimate experimental data below 70 keV projectile energy. This illustrates the benefit of using a coupled-channel approach for the one-electron processes, as is done presently. This becomes evident when we compare the CCC SI and Born SI calculations, where the latter significantly underestimates the former below 100 keV. Furthermore, the importance of including the two-electron processes becomes apparent when two models that include one-electron processes only, i.e. CCC SI and Born SI, are compared to those that include both one- and two-electron processes, i.e. CCC+B2e EL and Born EL. As can been seen, the CCC SI calculations underestimate experiment above 20 keV projectile energy and the Born SI calculations underestimate experiment at all projectile energies considered.

5.2.3. *Charge-state fraction*

In Fig. 8, we present our results for the positive (f^{H^+}) and neutral (f^{H^0}) charge-state fractions for beam of protons passing through hydrogen. They are displayed alongside the calculations of Dalgarno and Griffing [6] and Fainstein *et al.* [10], as well as the experimental data of Allison [8] (which were used in the calculation of the total stopping cross section by Schiwietz [7] as well as Schiwietz and Grande [9]). We obtain good agreement with Dalgarno and Griffing [6] above 40 keV projectile energy and with Fainstein *et al.* [10] above 150 keV. Furthermore, although the experimental data of Allison [8] were measured for a molecular hydrogen target, we obtain reasonable agreement with the latter over the whole energy range. Overall, the calculations by the present method describe the experimental data much better than the corresponding results from the other methods. Additionally, from Fig. 8, we can make certain statements about the composition of the beam passing through the target. Firstly, above 200 keV projectile energy, the beam is comprised entirely of protons. As the projectile energy falls, the proportion of hydrogen begins to rise, reaching 50% of the beam composition at 50 keV. Below 50 keV, hydrogen atoms make up the majority of the beam, reaching 90% of the beam composition at 10 keV.

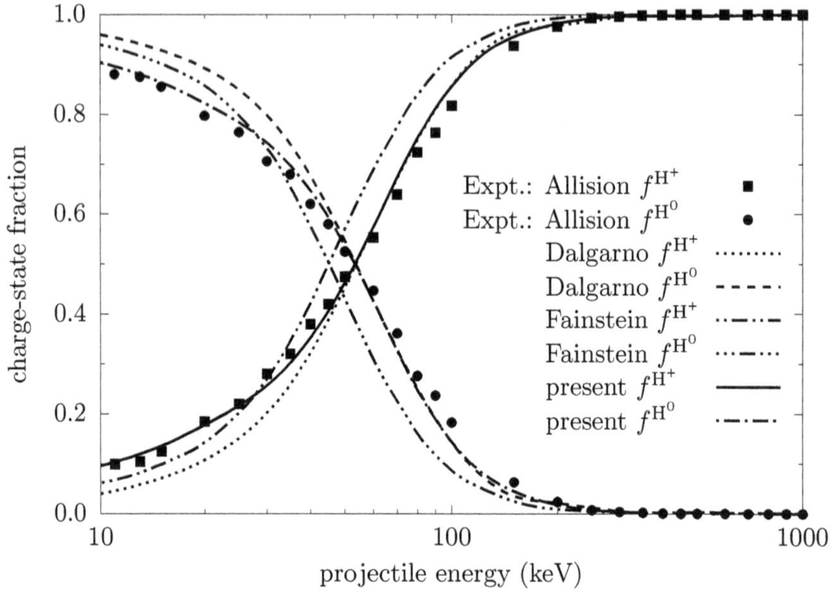

Fig. 8. Positive (f^{H^+}) and neutral (f^{H^0}) charge-state fractions for protons passing through hydrogen. The present results (first published in Ref. [32]) are compared to those of Dalgarno and Griffing [6] and Fainstein *et al.* [10]. The experimental data of Allison [8], that were measured for a molecular hydrogen target, are also shown.

5.2.4. *Total stopping cross section*

In Fig. 9, we present our results for the total electronic stopping cross sections for protons passing through hydrogen together with the theoretical calculations of Dalgarno and Griffing [6], Schiwietz [7], Schiwietz and Grande [9] and Fainstein *et al.* [10]. At present, no experimental data for this system exist. Good agreement with the calculations of Dalgarno and Griffing [6] is seen above 100 keV projectile energy, while agreement with the calculations of Schiwietz [7], Schiwietz and Grande [9] and Fainstein *et al.* [10] is obtained above 125 keV. However, below 100 keV, there are substantial deviations among all the theoretical approaches. These deviations cannot be attributed to either the positive or neutral charge-state contributions since there are large deviations between theories in both cases, as seen in Figs. 5 and 6. We can, however, make the statement that our calculations for the positive charge-state contribution are the most sophisticated

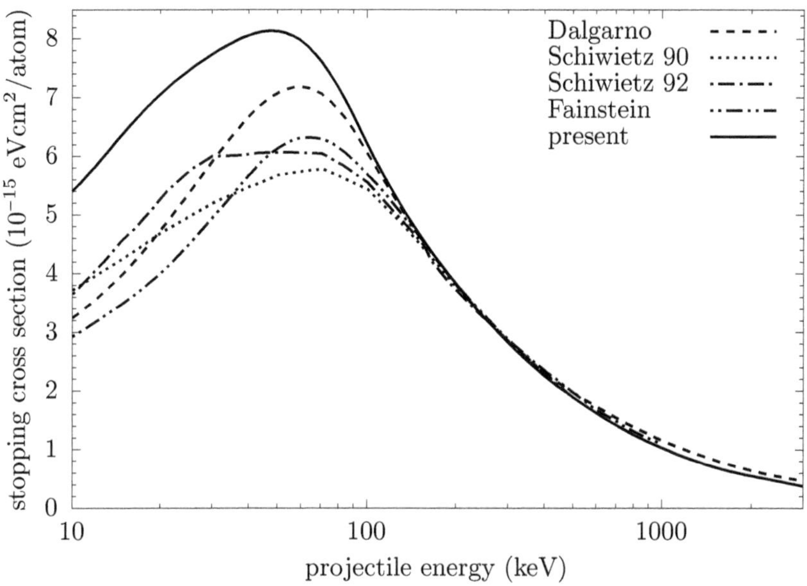

Fig. 9. Total electronic stopping cross section for protons passing through hydrogen. The present results (first published in Ref. [32]) have been obtained using the two-center CCC method for the positive charge-state contribution and the "CCC+B2e" method for the neutral charge-state contribution (see the main text for details). Also shown are the theoretical calculations of Dalgarno and Griffing [6], Schiwietz [7], Schiwietz and Grande [9], as well as Fainstein *et al.* [10]. The experimental data of Reynolds *et al.* [29], Reiter *et al.* [30] and Golser and Semrad [31] for protons passing through molecular hydrogen are shown as well.

and accurate, as we employ a large two-center expansion of the scattering wave function, which explicitly includes electron capture channels. Also, for the neutral charge-state contribution, our approach produces the most accurate total cross section for electron loss by the projectile.

6. Conclusions

In conclusion, we have investigated stopping of protons and antiprotons in a medium made of hydrogen atoms. We applied the single-center CCC approach to the calculation of the antiproton-hydrogen stopping cross section. Good agreement with existing theories is obtained over the whole energy range considered. The stopping cross section for protons passing through hydrogen has also been calculated. For the positive charge state of the projectile, we applied the two-center CCC approach. In this approach, bound states of the projectile are included in the expansion of the scattering wave function. This permits accurate modeling of electron capture processes. For the neutral charge state of the projectile (H atom), we applied a single-center CCC approach to the calculation of one electron processes and the first Born approximation to the calculation of two electron processes. Both one- and two-electron processes make a significant contribution to the stopping cross sections. The calculated total electron-capture cross sections in proton-hydrogen collisions and the total electron loss cross section in hydrogen-hydrogen collisions are used to obtain the charge-state fractions, which are, in turn, used to obtain the total stopping cross sections for protons passing through hydrogen. Good agreement with all the existing theories is obtained above 125 keV incident energy. However, below this energy, there are significant deviations among all the theoretical calculations. Our calculations for the positive charge-state contribution are the most sophisticated and accurate since we employ a large two-center expansion of the scattering wave function. Also, for the neutral charge-state contribution our approach produces the most accurate total cross section for electron loss by the projectile.

Acknowledgments

The work was supported by the Australian Research Council. We are grateful for access to the Australian National Computing Infrastructure Facility and the Pawsey Supercomputing Centre in Western Australia.

References

[1] Dž. Belkić, ed. *Theory of Heavy Ion Collision Physics in Hadron Therapy*, vol. 65, *Advances in Quantum Chemistry*, Elsevier, Amsterdam (2014).
[2] S. Chauvie, P. Nieminen, and M. G. Pia, Geant4 model for the stopping power of low energy negatively charged hadrons, *IEEE Trans. Nucl. Sci.* **54**, 578–584 (2007).
[3] S. Agostinelli *et al.* (Geant4 Collaboration), Geant4–a simulation toolkit, *Nucl. Instrum. Methods Phys. Res. A.* **506**, 250 – 303 (2003).
[4] ICRU, Report 49, Stopping powers and ranges for protons and alpha particles, International Commission on Radiation Units and Measurements, Bethesda, MD (1993).
[5] ICRU, Report 73, Stopping of ions heavier than helium, International Commission on Radiation Units and Measurements, Bethesda, MD (2005).
[6] A. Dalgarno and G. W. Griffing, Energy loss of protons passing through hydrogen, *Proc. R. Soc. A.* **232**, 423–434 (1955).

[7] G. Schiwietz, Coupled-channel calculation of stopping powers for intermediate-energy light ions penetrating atomic H and He targets, *Phys. Rev. A.* **42**, 296–306 (1990).
[8] S. K. Allison, Experimental results on charge-changing collisions of hydrogen and helium atoms and ions at kinetic energies above 0.2 keV, *Rev. Mod. Phys.* **30**, 1137–1168 (1958).
[9] G. Schiwietz and P. L. Grande, Electronic stopping of protons at intermediate velocities, *Nucl. Instrum. Methods Phys. Res. B.* **69**, 10 – 17 (1992).
[10] P. D. Fainstein, V. H. Ponce, and A. E. Martinez, Distorted-wave calculation of stopping powers for light ions traversing H targets, *Phys. Rev. A.* **47**, 3055–3061 (1993).
[11] G. Schiwietz, U. Wille, R. Díez Muiño, P. D. Fainstein, and P. L. Grande, Stopping mechanisms of negative heavy particles in gas targets, *Nucl. Instrum. Methods Phys. Res. B.* **115**, 106 – 110 (1996).
[12] G. Schiwietz, U. Wille, R. Díez Muiño, P. D. Fainstein, and P. L. Grande, Comprehensive analysis of the stopping power of antiprotons and negative muons in He and H_2 gas targets, *J. Phys. B.* **29**, 307–321 (1996).
[13] R. Cabrera-Trujillo, J. R. Sabin, Y. Öhrn, and E. Deumens, Stopping of swift antiprotons by hydrogen atoms and the Barkas correction, *Phys. Rev. A.* **71**, 012901 (2005).
[14] A. Lühr and A. Saenz, Stopping power of antiprotons in H, H_2, and He targets, *Phys. Rev. A.* **79**, 042901 (2009).
[15] J. J. Bailey, A. S. Kadyrov, I. B. Abdurakhmanov, D. V. Fursa, and I. Bray, Antiproton stopping in atomic targets, *Phys. Rev. A.* **92**, 022707 (2015).
[16] J. J. Bailey, A. S. Kadyrov, I. B. Abdurakhmanov, D. V. Fursa, and I. Bray, Antiproton stopping in H_2 and H_2O, *Phys. Rev. A.* **92**, 052711 (2015).
[17] J. Bailey, A. Kadyrov, I. Abdurakhmanov, D. Fursa, and I. Bray, Antiproton stopping power data for radiation therapy simulations, *Phys. Med.* **32**, 1827 – 1832 (2016).
[18] I. B. Abdurakhmanov, A. S. Kadyrov, I. Bray, and A. T. Stelbovics, Coupled-channel integral-equation approach to antiproton-hydrogen collisions, *J. Phys. B.* **44**, 075204 (2011).
[19] I. B. Abdurakhmanov, A. S. Kadyrov, and I. Bray, Wave-packet continuum-discretization approach to ion-atom collisions: Nonrearrangement scattering, *Phys. Rev. A.* **94**, 022703 (2016).
[20] I. B. Abdurakhmanov, A. S. Kadyrov, S. K. Avazbaev, and I. Bray, Solution of the proton-hydrogen scattering problem using a quantum-mechanical two-center convergent close-coupling method, *J. Phys. B.* **49**, 115203 (2016).
[21] S. K. Avazbaev, A. S. Kadyrov, I. B. Abdurakhmanov, D. V. Fursa, and I. Bray, Polarization of lyman-α emission in proton-hydrogen collisions studied using a semiclassical two-center convergent close-coupling approach, *Phys. Rev. A.* **93**, 022710 (2016).
[22] I. B. Abdurakhmanov, J. J. Bailey, A. S. Kadyrov, and I. Bray, Wave-packet continuum-discretization approach to ion-atom collisions including rearrangement: Application to differential ionization in proton-hydrogen scattering, *Phys. Rev. A.* **97**, 032707 (2018).

[23] J. J. Bailey, A. S. Kadyrov, I. B. Abdurakhmanov, and I. Bray, Hybrid approach to calculating proton stopping power in hydrogen, *J. Phys. Conf. Ser.* **777**, 012010 (2017).

[24] E. Guth and C. J. Mullin, Momentum representation of the coulomb scattering wave functions, *Phys. Rev.* **83**, 667–668 (1951).

[25] I. Bray and A. T. Stelbovics, Convergent close-coupling calculations of electron-hydrogen scattering, *Phys. Rev. A.* **46**, 6995–7011 (1992).

[26] H. Bethe, Zur theorie des durchgangs schneller korpuskularstrahlen durch materie, *Ann. Phys.* **397**, 325–400 (1930).

[27] A. B. Wittkower, G. Levy, and H. B. Gilbody, An experimental study of electron loss during the passage of fast hydrogen atoms through atomic hydrogen, *Proc. Phys. Soc.* **91**, 306–309 (1967).

[28] G. W. McClure, Ionization and electron transfer in collisions of two H atoms: 1.25-117 keV, *Phys. Rev.* **166**, 22–29 (1968).

[29] H. K. Reynolds, D. N. F. Dunbar, W. A. Wenzel, and W. Whaling, The stopping cross section of gases for protons, 30-600 keV, *Phys. Rev.* **92**, 742–748 (1953).

[30] G. Reiter, N. Kniest, E. Pfaff, and G. Clausnitzer, Proton and helium stopping cross sections in H_2, He, N_2, O_2, Ne, Ar, Kr, Xe, CH_4 and CO_2, *Nucl. Instrum. Methods Phys. Res. B.* **44**, 399 – 411 (1990).

[31] R. Golser and D. Semrad, Energy loss of hydrogen and helium ions in hydrogen and helium gas: looking for exceptions from velocity proportionality, *Nucl. Instrum. Methods Phys. Res. B.* **69**, 18 – 21 (1992).

[32] J.J. Bailey, I.B. Abdurakhmanov, A.S. Kadyrov, I. Bray and A.M. Mukhamedzhanov, Proton-beam stopping in hydrogen, *Phys. Rev.* **99**, 042701 (2019).

Chapter 10

Physics models for Monte Carlo simulations in carbon ion therapy

S. Guatelli[1*], D. Bolst[1], Z. Francis[2], S. Incerti[3,4], V. Ivanchenko[5,6] and A.B. Rosenfeld[1]

[1] *Centre for Medical Radiation Physics, University of Wollongong, Australia*

[2] *Saint Joseph University, Beirut, Lebanon*

[3] *CNRS/IN2P3, Centre d'Etudes Nucléaires de Bordeaux-Gradignan, France*

[4] *Université Bordeaux, Centre d'Etudes Nucléaires de Bordeaux-Gradignan, France*

[5] *The European Organization for Nuclear Research (CERN)*

[6] *Tomsk State University, Tomsk, Russia*

Monte Carlo codes are widely used in carbon ion therapy as tools to verify clinical treatment planning systems, to develop associated quality assurance detectors and also to improve the delivery of the treatment to the patient. This review illustrates the physics capability of Geant4 which is one of the most widely used Monte Carlo codes of its genre in medical physics. We focus on carbon ion therapy and on validation of Geant4 with respect to experimental measurements. Finally, the mixed radiation field of a typical clinical carbon ion beam is shown, as calculated by Geant4. Although Geant4 has an advanced overall performance for carbon ion therapy, the present chapter emphasizes that this versatile code can, nevertheless, be further improved in terms of both physics models development and validation studies.

*Corresponding author. Email: susanna@uow.edu.au

Contents

1. Introduction

Carbon ion therapy is a type of radiotherapy consisting of the use of carbon nuclei ($^{12}C^{6+}$, or ^{12}C for short) to destroy the tumor. With respect to X-ray radiotherapy and proton therapy, ^{12}C ion therapy has better conformity to the target tumor in terms of dose delivery, because the Bragg peak (BP) is coupled with a significantly smaller lateral scattering of the incident particles. This radiotherapy is more advantageous for deep-seated and radioresistant tumors, due to its higher relative biological effectiveness (RBE). Additionally, ^{12}C ion therapy is more suitable to hypofractionation, which delivers higher doses in a small number of treatment sessions for each patient. This, in turn, improves convenience for the patient as well as cost effectiveness [1].

The National Institute of Radiological Science (NIRS) in Japan was the first institute to implement clinical ^{12}C ion therapy in 1994. It was then followed by the Gesellschaft für Schwerionenforschung (GSI) in Germany in 1997, the Hyogo Ion Beam Medical Center (Japan) in 2002, the Institute of Modern Physics (Lanzou, China) in 2009, together with the University Hospital Heidelberg (Germany), Gunma University (Japan) and the Fondazione Centro Nazionale Adroterapia Oncologica (Italy). Currently, there are ten facilities in Europe and Asia performing carbon ion therapy, with more planned or under construction [1]. According to the Particle Therapy Co-Operative Group (PTCOG) [2] statistics, a total of 20000 patients have been treated with ^{12}C ions in the world from 1954 to 2015 [3].

Monte Carlo codes, modeling particle transport in matter, such as FLUKA [4, 5], Geant4 [6–8], MCNPX [9], SHIELD/SHIELD-HIT [10, 11] and PHITS [12] can all be used in ^{12}C ion therapy to characterize the mixed radiation field in the patient [13], as well as to verify clinical treatment planning systems (TPSs). They can calculate the physical quantities serving as the input data to TPSs [13]. These codes can be used to optimize the design of novel Quality Assurance detector technology [14, 15].

2. The Geant4 physics capability for ^{12}C ion therapy

Geant4 is a Monte Carlo toolkit which describes the interactions of particles with matter. It was developed as a successor to Geant3. In a CERN Research & Development project, started in 1994, Geant4 was successfully designed using the Object-Oriented (OO) Technology in the C++ programming language [6]. The first public release of Geant4 was in 1998. Currently, Geant4 is the most widely used Monte Carlo code for modeling of high energy physics (HEP) experiments [16]. The adoption of the OO technology has allowed the functionality of Geant4 to be extended to application domains other than HEP, such as space science and medical physics. Since its first release, Geant4 has regularly been updated, further developed and maintained by an international collaboration of more than 100 scientists. As a result, Geant4 became the most powerful Monte Carlo code devoted to radiation physics. This is due to an extensive functionality of Geant4, ranging from geometry and physics to particle tracking, event biasing and multi-threaded capability.

In the last ten years, Geant4 has become frequently used for physics applications in the medical field. In 2009, it was reported [17] that the first Geant4 general paper [6] had more than 6000 citations out of which ~20% were from medical physics, second to ~40% from HEP – the original application field of Geant4. Given an extensive interest in Geant4 within the medical physics community, GATE [18], GAMOS [19] and, more recently, TOPAS [20] have been developed to facilitate the use of Geant4 in imaging and in radiotherapy.

Geant4 has a versatile functionality in terms of physics, covering a wide range of particle energies, spanning from eV to TeV. Its physics processes can be subdivided into the following main components: electromagnetic physics, optical photon processes, decay of subatomic particles, radioactive decay and hadronic physics, which includes hadron/ion/gamma/lepto-nuclear interactions. The design of this toolkit allows the alternative configurations of the interaction models applicable to various energy ranges and provides different levels of physics accuracy as well as CPU performance.

Geant4 interfaces allow the substitution of detailed simulations of particle tracks with fast parametrization within the regions of interest [21].

This chapter intends to give a brief account of the Geant4 physics processes and models which are of particular relevance for ^{12}C ion therapy. The reader may refer to the Geant4 general articles [6–8] to have a full description of the evolution and current status of the Geant4 physics component.

2.1. *Geant4 electromagnetic physics*

The Geant4-EM (electromagnetic) physics component describes the following processes of interest for ^{12}C ion therapy: ionization, multiple scattering, bremsstrahlung for charged particles, Rayleigh scattering, photoelectric effect, Compton scattering and pair/triplet production. Two alternative physics approaches are provided within Geant4: the "Standard" and "Low Energy EM" sublibraries. Both of them adopt a condensed history technique, developed by Berger *et al.* in 1963 [22]. This latter technique consists of condensing a large number of interactions of one particle into a single potential [23]. In particular, the low-energy limit of the "Standard" models is 1 keV, while it is 100 eV for the "Low Energy" models. The "Low Energy" physics has two alternative approaches, one based on the Livermore data libraries and the other relying on the Penelope Monte Carlo code [24], rewritten from FORTRAN to C++.

The Geant4-DNA (deoxyribonucleic acid) sublibrary, described in more detail in section 2.3, is intended for simulation of very low energy particle transport at the cellular level [25]. It includes a full set of models for liquid water [25] and also some other models for solid state materials such as silicon [26]. The DNA models implement discrete interactions without adopting the condensed history technique.

The atomic de-excitation [27, 28] is interfaced with any EM physics library, and describes the emission of X-rays and Auger electrons. Details of the physics component of Geant4 of interest for ^{12}C ion therapy are described here as released in Geant4 10.3. Nowadays, alternative Geant4 physics lists are provided to facilitate the selection of the most appropriate set of physics models for a specific application scenario. The EM physics lists can be found in the directory geant4.10.03/source/physics_lists/constructors/electromagnetic.

For medical physics applications, G4EmStandardPhysics_option4 is suggested. This has been developed with the goal of providing the most accu-

Process	G4EmStandardPhysics_option4
Multiple scattering	For e+/e-, G4UrbanMscModel below 100 MeV and WentzelVI model, combined with G4eCoulombScatteringModel for large scattering angles, above 100 MeV. For protons the WentzelVI model is used together with the G4CoulombScattering. G4hMultipleScattering for deuteron, triton and heavier ions.
e-/e+ ionization	Penelope model
e-/e+ Bremsstrahlung	Seltzer-Berger model
Rayleigh scattering	Livermore model
Photoelectric effect	Livermore model
Compton scattering	G4KleinNishinaModel for energies above 20 MeV, G4LowEPComptonModel for energies below 20 MeV
Gamma Conversion	Penelope and Standard models below and above 1 GeV, respectively
Proton ionization	Model based on ICRU49 electronic and nuclear stopping power data
Ion ionization	Model based on ICRU73 ion electronic stopping power data and ICRU49 nuclear stopping power data

Table 1. Geant4 version 10.3: processes and physics models of the G4EmStandardPhysics_option4 list.

rate physics models of Geant4 for applications in medical physics. The alternative options are G4EmStandardPhysics_option3, G4EmPenelopePhysics and G4EmLivermorePhysics. Table 1 gives the processes and physics models adopted in G4EmStandardPhysics_option4. The mentioned alternative physics lists offer the same physics models for heavy ions such as protons and ^{12}C ions.

The EM physics processes of heavy charged particles are ionization, multiple scattering and elastic scattering with the target nuclei [29, 30]. The heavy charged particles of interest for ^{12}C ion therapy are the incident ^{12}C ions, protons (from the scattering of neutrons), nuclei (from the fragmentation process) such as α particles and heavier nuclei that have a high

RBE. The class G4hIonization describes energy loss of protons, while the class G4ionIonization manages energy loss of deuteron, triton and any ion with charge greater than 1.

The class G4ionIonization manages the configuration of models that compute electronic stopping powers, energy loss fluctuations and cross sections for production of δ-electrons [31]. Only δ-electrons with a kinetic energy above the user defined threshold of production $T_{\rm cut}$ are explicitly generated and tracked in the simulation. If $T_{\rm cut}$ is smaller than the mean excitation/ionization potential I of the target material, δ-electrons would be produced with a kinetic energy higher than I. An implication of this model is that the atomic/molecular structure of the target is not taken into account at this level. Such structures are important in nano-scale studies. If the energy of δ-electrons to be generated is below $T_{\rm cut}$ (or below I if $T_{\rm cut} < I$), they are not generated as secondaries; rather they are treated as continuous energy loss of the parent particle along a simulation step [31]. The continuous energy loss is determined by means of restricted stopping powers, with the constraint that the minimum energy transferred to δ-electrons is $T_{\rm cut}$ (or I if $T_{\rm cut} < I$). The model of the continuous energy loss depends on the kinetic energy (KE) of the projectile.

For KE > 2 MeV/u, the continuous energy loss is calculated by means of the Bethe-Bloch formula [31]. For O(keV/u) < KE< 2 MeV/u, the stopping power is modeled using the ICRU parametrization. In particular, the ICRU49 is adopted for protons and α particles, whereas the ICRU73 is used for heavier ions [32]. It is important to note that in order to use the ICRU73 stopping powers, materials must coincide with the Geant4 NIST materials or have their names matching the names of the corresponding Geant4 NIST materials. For example, the user can create water from the pertinent elemental composition, but the assigned name should be G4_WATER which is the name of the Geant4 NIST water. For materials made of only one element, this requirement does not apply, i.e. any name for that material can be used. If the ICRU73 parametrization is not used, the stopping powers for heavy ions are derived from the proton stopping powers documented in the ICRU49 report [33] using the effective charge approximation [34].

The mean ionization/excitation potential I of water, when using the ICRU73 stopping power, is 78 eV. Overall, constructing materials using the ICRU73 parametrization would lead to better agreement with experimental measurements of the Bragg curve [31]. More details are provided in Refs. [31, 35]. For energies below O (keV/u), the free electron gas model is adopted [21].

2.2. *The Geant4 hadronic physics for ^{12}C ion therapy*

This sub-section provides an overview of Geant4 hadronic physics models of interest for ^{12}C ion therapy. For a comprehensive description of this physics component, the reader is referred to the Geant4 general article [8].

The secondary radiation field produced in ^{12}C ion therapy is much more complex compared to the more conventional treatment modality of X-ray therapy. This is due to the hadronic interactions between the incident beam and the target nuclei which can cause secondary fragments to be produced. The process of fragmentation is commonly described by the abrasion-ablation model [36]. The abrasion-ablation model involves the compound of nucleons from the projectile and target nucleus. This compound is an excited pre-fragment amalgamate which de-excites into smaller nuclei. The projectile fragment continues to travel in a direction and velocity similar to those of the original projectile, while the target fragment remains stationary. Because the produced fragments are lighter than the original incident ion and have a similar velocity, their range is larger than the range of the primary ion. Additionally, the angular distributions of these produced fragments are wider than what would be produced by multiple Coulomb scattering alone [37]. The increased range and greater lateral scattering of the fragments imply that the dose is deposited away from the treatment target volume, i.e. in the tumor surroundings which may include healthy tissue and/or organs at risk. An example of the effect of fragmentation of radiotherapeutic nuclei will be shown later in sub-section 2.5.

Figure 1 shows a partial inventory of hadronic processes and physics models of interest for ^{12}C ion therapy, up to 10 GeV/u. Geant4 offers complementary and alternative physics models to describe hadronic interactions, depending on particle type and energy range.

At present, three intranuclear cascade models for hadrons, including protons and neutrons, are offered in Geant4: Bertini, Binary and INCL++ cascades. The Bertini cascade [38] is valid for projectiles with incident energies between 0 and 15 GeV. Although this model has its own pre-compound and de-excitation model, an option exists for using the native Geant4 pre-compound and de-excitation modules shown in Fig. 1. The Binary cascade [39] simulates cascades generated by the incident projectiles with energy below 10 GeV. This model relies on the native Geant4 pre-compound and de-excitation component. The Liège Intranuclear Cascade model (INCL++) [40] can be used up to 15 GeV incident energy. By default, INCL++ uses the Geant4 native de-excitation immediately after

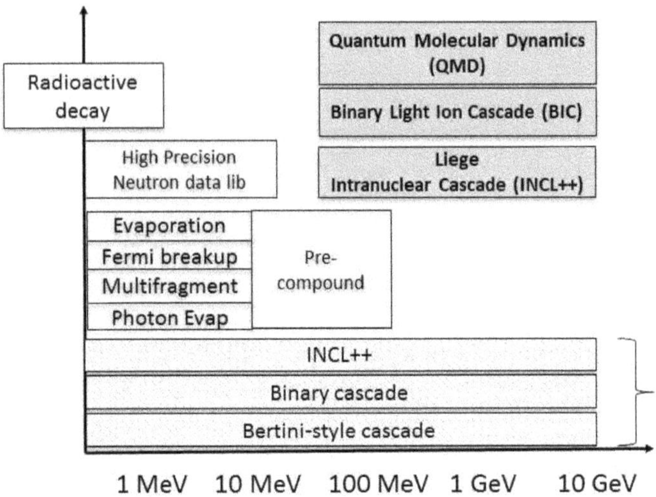

Fig. 1. Partial inventory of Geant4 version 10.3 hadronic physics models of interest for ^{12}C ion therapy.

the cascade stage, without including the pre-equilibrium phase. Currently, it is also possible to couple the ABLA V3 de-excitation model with INCL++ [41, 42].

The elastic scattering of hadrons is also modeled in Geant4. In particular, the elastic cross sections from G4GGNuclNuclCrossSection, described by Grichine *et al.* [43], is calculated based upon a simplified Glauber approach. This is coupled with G4NuclNuclDiffuseElastic [44], which is an implementation of the diffuse Fresnel model by taking into account the diffuse nuclear halo, as well as the Coulomb effects [8]. The NeutronHP models, for elastic, inelastic, capture and fission reactions, are recommended for Geant4 applications where an accurate transport of neutrons is required, including ^{12}C ion therapy. These models depend on the Geant4 Neutron Data Library (G4NDL) [8].

It is important to be able to model nuclear fragmentation accurately in a Monte Carlo code such as Geant4, because at clinical energies up to ~400 MeV/u, the incident ^{12}C ions generate a complex radiation field, with ~70% of the primary beam (for a 400 MeV/u beam) undergoing fragmentation before the Bragg peak [45]. Unfortunately, there is no unique, rigorous model which is capable of describing nuclear fragmentation in detail. Geant4 offers alternative models to describe nucleus-nucleus hadronic interactions, such as the binary light ion cascade (BIC), the quantum molecular dynamics model (QMD) and the INCL++ model [8].

The BIC model describes the interaction between a projectile and a single nucleon of the target nucleus interacting in the overlap region as Gaussian wave functions. Within Geant4, the BIC model [39] is valid above 80 MeV and below 10 GeV/u [8]. By contrast, the QMD model considers all the nucleons of the target and projectile, each with their own wave function. The QMD has an option "Frag" which changes the production yields of secondary fragments. The INCL++ model describes the nucleons as a Fermi free gas in a static potential well.

The pre-compound model, shown in Fig. 1, is the native Geant4 pre-equilibrium model, based upon a version of the semiclassical exciton model [46] and is used as the back-end stage of several cascade models, including the BIC, QMD and INCL++. It handles the de-excitation of the remnant nucleus from its formation immediately following a cascade or a high energy collision, until it reaches equilibrium.

The de-excitation of the target nucleus depends on its mass and on its residual energy. Geant4 handles the evaporation of photons, the nuclear evaporation with emission of light fragments up to α particles, based upon the Weisskopf-Ewing model [47], or up to ^{28}Mg, based upon the generalized evaporation model (GEM) [48, 49]. Geant4 also handles fission [49, 50], statistical multi-fragmentation [51] and Fermi break-up [51]. The photon evaporation handles both the emission of discrete gamma rays according to the E1, M1 and E2 transition probabilities taken from the evaluated nuclear structure data file (ENSDF) [52], and the continuous gamma rays according to the E1 giant dipole resonance strength distribution [8].

For the application of Geant4 in ^{12}C ion therapy, it is essential to accurately model the Fermi break-up (which de-excites light fragments with $Z < 9$, $A < 17$) as well as the photon evaporation. This is critical for the tracking of the Bragg curve from the emitted prompt gamma rays. The fission is applicable to high atomic number nuclei only and, therefore, it is not relevant in ^{12}C ion therapy, unless there are implants in the patient [53].

2.3. *Geant4-DNA for ^{12}C ion therapy*

The Geant4-DNA project is principally motivated by the need to include track structure physics processes in Geant4 in a way which is adequate for radiobiological studies. The first models were released within Geant4 in 2007. These models can simulate the interactions of particles (step-by-step) allowing a detailed description of track structures in liquid water at the cellular and subcellular levels.

In the case of electrons, Geant4-DNA uses models for simulation of elastic scattering, ionization, electronic excitation, vibrational excitation and molecular attachment [25]. Three alternative physics model sets are provided by Geant4 10.4 Beta (June 2017) for modeling of elastic and inelastic interactions, as shown in Table 2. Electron interactions are described in the energy range between a few eV up to 1 MeV.

Proton and neutral hydrogen atom interactions are described in the energy range from 100 MeV down to ~100 eV. The excitation process is described by the model of Miller and Green for incident energy below 500 keV and Born and Bethe theories above 500 keV [54, 55].

Ionization is modeled following the Rudd semi-empirical approach in the energy range 100 eV - 500 keV [54]. Born and Bethe theories as well as the dielectric formalism are adopted above 500 keV. The charge change and nuclear scattering are described by analytical parametrization by Dingfelder *et al.* [54] and the classical approach by Everhart *et al.* [63], respectively.

Ionization and excitation cross sections with α particles as projectiles were obtained from the protons cross sections using a velocity scaling procedure. The charge transfer processes with the resulting screening effect for neutral and single charged helium particles are described by the semiempirical models of Dingfelder *et al.* [64]. The α particle interactions are described in the energy range between ~1 keV and 400 MeV.

Francis *et al.* [65] included the ionization process for dominant ion species of interest for ^{12}C ion therapy (Li, Be, B, C, N, O, Si, Fe) in Geant4-DNA, which uses a velocity scaling and a global effective charge by Booth and Grant [66]. Excitation, elastic scattering and ion fragmentations are not taken into account. However, for ion fragmentation, Geant4-DNA physics can be activated together with the hadronic physics component of Geant4 [65].

Because the elastic scattering and excitation are not modeled, the applicability of the ionization model is restricted to energies above ~1 MeV/u. Below this threshold energy, transfer due to elastic scattering of nuclei and charge transfer processes become important and cannot be neglected. The ionization cross sections for heavy ions were calculated using the Rudd formula with the relativistic approach described by Plante and Cucinotta [67].

Process	*Physics model*
Elastic scattering	• Screened Rutherford model and Brenner-Zaider model below 200 eV [25] • Screened Rutherford model with improved screening parameter by Uehara [56] • Independent atom method (IAM) model by Mott *et al.* and very low energy data in ice, from the CPA100 track structure code [57] , [58] • Partial wave model by Champion *et al.* [25]
Ionization and excitation	• Dielectric formalism-based model of Emfietzoglou using the Heller optical data up to 1 MeV, including the low energy corrections [59] • Improved alternative version of the Emfietzoglou model [60] • Relativistic binary encounter Bethe (RBEB) model from the CPA100 track structure code [58, 61]

Table 2. Alternative Geant4-DNA models describing elastic scattering, ionization and electronic excitation of electrons in liquid water, as released in Geant4 10.4 beta [62].

Francis and collaborators [68] used this ionization model applicable to heavy ions to calculate the energy deposition with spatial distribution at nano-scale in liquid water. They focused on the ^{12}C ions (among other projectiles) with the energy of interest for clinical applications. An illustration along these lines is given in Fig. 2, which shows the number of processes occurring from a ^{12}C ion incident on liquid water with energy equal to 10 MeV, within 1 μm from the point of origin.

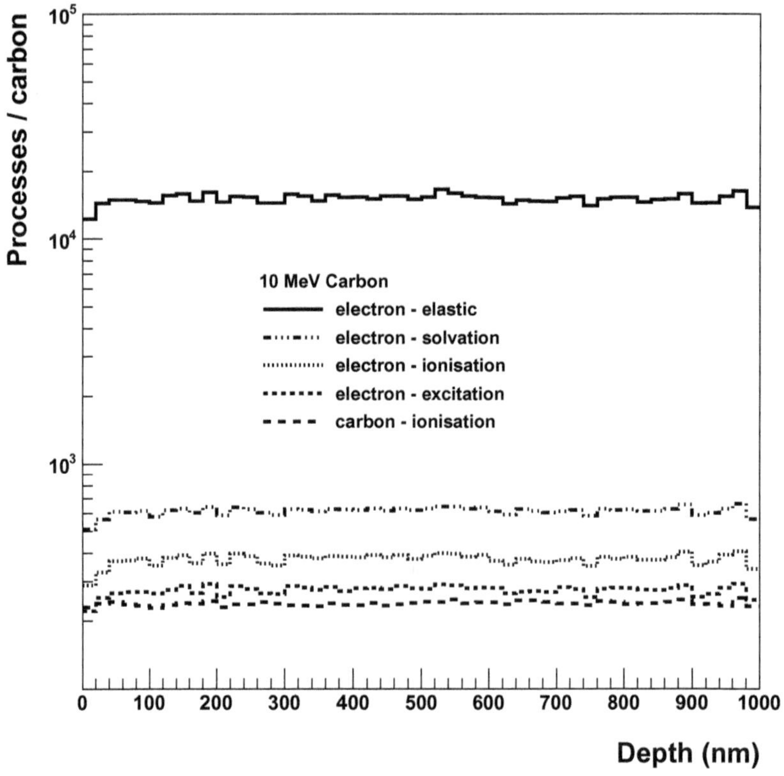

Fig. 2. Geant4-DNA processes for an incident ^{12}C ion of energy 10 MeV, in liquid water, within 1 μm from the point of origin.

2.4. *Validation of Geant4 physics component for ^{12}C ion therapy*

There have been a number of validation/benchmarking studies performed over the years using Geant4 with carbon ion energies relevant for ^{12}C therapy. For example, Lechner *et al.* [31], using water and polyethylene as the targets, validated Geant4 against experimental measurements of Bragg curves produced by ^{12}C ions with energy between 90 and 400 MeV/u. Böhlen *et al.* [13] validated the BIC and the QMD models in Geant4 (version 9.3) against experiments on the 400 MeV/u ^{12}C beam. Napoli *et al.* [69] compared the BIC and the QMD model from Geant4 (version 9.4) with experimental data for a thin carbon target using a 62 MeV/u ^{12}C beam. Dudouet *et al.* [70] validated the BIC, QMD and INCL++ models in Geant4 (version 9.6) through comparisons of the theoretical and experimental cross sections for 95 MeV/u ^{12}C ions incident upon the PMMA target. A more recent validation for a 400 MeV/u ^{12}C ion beam has been performed by Bolst *et al.* [71] using the version 10.2p2 of Geant4 and experimental data [72], made available by Haettner *et al.* [73].

In this sub-section, some of the recent results are reported. They are integrated into the Geant4 regression tests of relevance to medical physics [74]. Here, the version on 10.2p2 of Geant4 is adopted. Figure 3 compares the theoretical Bragg curves in water for a mono-energetic ^{12}C beam of energy 200 MeV/u (left) and 400 MeV/u (right) against experimental measurements from Haettner *et al.* [73].

The water phantom was simulated as a 60 cm thick slab with an area of $50{\times}50$ cm^2. The secondary particle production threshold was set to 0.1 mm. The energy deposited was scored within a 20×20 cm^2 area at the center of the beam with a voxel thickness of 0.1 mm along the direction of incidence of the beam. The lateral area corresponds to the physical dimensions of the ionization chamber used in the experiment. The thickness of the ionization chamber in the experiments was 3.7 cm [75]. The uncertainty of 1 mm quoted in the experiment refers to the uncertainty in the water equivalent thickness (WET) of materials positioned in front of the ionization chamber used.

The uncertainty of the Bragg peak position quoted in the simulation of 0.1 mm is due to the thickness of the used voxels. Because of the relatively large thickness of the experimental ionization chamber, this may cause a smearing of the shape of the Bragg peak. The electromagnetic physics was modeled using the "G4 Standard EM option 3 physics list" with the BIC, QMD, QMD-Frag and INCL++ models. Also employed here is the INCL++ model with ablation de-excitation (INCL-A). Overall, a good agreement can be seen between the experiment and the simulation results with all the mentioned fragmentation models. The agreement is within the errors of the experiment.

The work by Bolst and co-authors [71] benchmark three hadronic approaches available in Geant4: the BIC, QMD and INCL++ models. These models were benchmarked against experimental data [73] for a 400 MeV/u ^{12}C ion beam carried out at the GSI beamline. The experiment was performed using a combination of energy loss and time-of-flight measurements with a variable thickness of water target. The experimental data provide

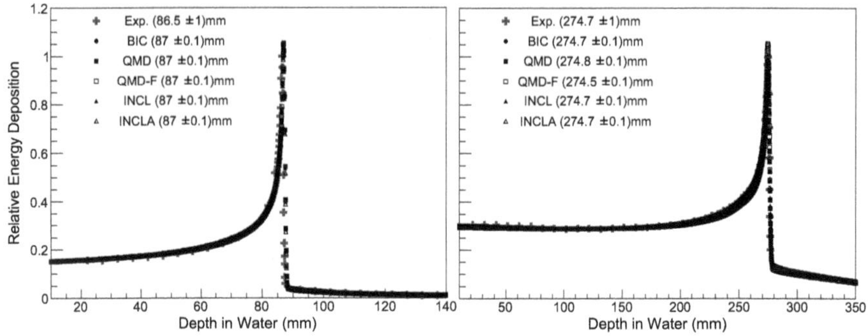

Fig. 3. Bragg curves in water produced by 200 MeV/u (left plot) and 400 MeV/u (right plot) ^{12}C ion beams. The Geant4 simulation results are compared with the reference experimental measurements [75]. The positions of the pinnacle Bragg peak are reported as calculated by the simulation using the alternative Geant4 fragmentation models, and as measured experimentally by Haettner *et al.* [75].

information on the yield of the fragments and their angular as well as energy distributions.

In the Geant4 simulation, electromagnetic interactions are modeled using the "G4 Standard EM option 3 physics list", as stated. To describe neutron interactions up to 20 MeV, the neutron "high precision" (HP) model was used. All the three fragmentation models are implemented using their default options. In addition to using their default options, the QMD and INCL++ models are also compared employing the alternative configurations denoted as QMD-F and INCL-A, respectively.

The QMD-F uses the "USE-Frag" option which alters the interaction criterion, while INCL-A uses the ablation model for de-excitation of nuclei, as mentioned earlier. The QMD-F option was observed not to significantly alter the angular or energy distributions of fragments compared to the default QMD model for these energies of ^{12}C ions and, thus, only the fragment yield results are shown for the QMD-F option.

For the energy distribution of fragments, the kinetic energy of the fragment is calculated by reference to the time-of-flight to match the assumption used in the experiment [73]. For angular and energy distributions, the total number of counts in the simulation is normalized to the corresponding total number in the experiment.

To make quantitative comparisons between the theoretical and experimental fragment distributions, the mean percentage difference $\langle \mathrm{PE}_Z \rangle$ is calculated for each fragment element using the following expression:

$$\langle \mathrm{PE}_Z \rangle = \frac{1}{m_Z} \sum_{j_Z=1}^{m_Z} \left(\frac{100\%}{n_{j_Z}} \sum_{i=1}^{n_{j_Z}} \left| \frac{\mathrm{Sim}_{i,} - \mathrm{Exp}_i}{\mathrm{Exp}_i} \right| \right). \tag{1}$$

Here, m_Z represents the total number of distributions compared for each element and n_{j_Z} is the number of data points in the j_Z^{th} distribution. For the fragment yields, there is a single distribution ($m_Z = 1$) for each fragment element with n_j being the number of thicknesses of water. Here, the version 10.2 patch02 of Geant4 was used. Presented here are the results from Ref. [71], including the new results obtained by means of the INCL++ model together with the ablation model.

Figure 4 shows the yield of different fragment elements produced per the incident primary ^{12}C ion. It can be seen that the yields differ quite significantly from model to model. Herein, each model reproduces the experiment within ~5-35%. A feature which can be seen with the experimental data is that it is characterized by larger errors upstream of the Bragg peak. It would be desirable to make new measurements where such errors are reduced. This would permit more stringent testings of various Monte Carlo models that predict different distributions of fragments outside the treatment field. The latter distributions are especially relevant for radiation-induced cancers in healthy tissue surrounding the treated tumor.

Table 3 summarizes the mean percentage difference for each element over all the thicknesses of water as calculated by Eq. (1). It can be seen that the QMD model, using the "Frag" option (QMD-F), is the best in reproducing the experimental data for the H as well as the He fragments. Moreover, the QMD-F model performs similarly to its competitors for the other elements with the exception of Li.

For this latter fragment, the BIC gives the best agreement with experimental data. Since H and He represent the two most dominant fragments produced, together with being the lightest fragments, they have the largest range and are more susceptible to scattering. These two factors mean that H and He contribute strongly to the dose delivered outside the treatment field. Thus, a more accurate reproduction of the H and He fragments represents a strong priority. From this discussion, it emerges that QMD-F may be the recommended model for accurate modeling of fragmentation yields using the version 10.2p02 of Geant4.

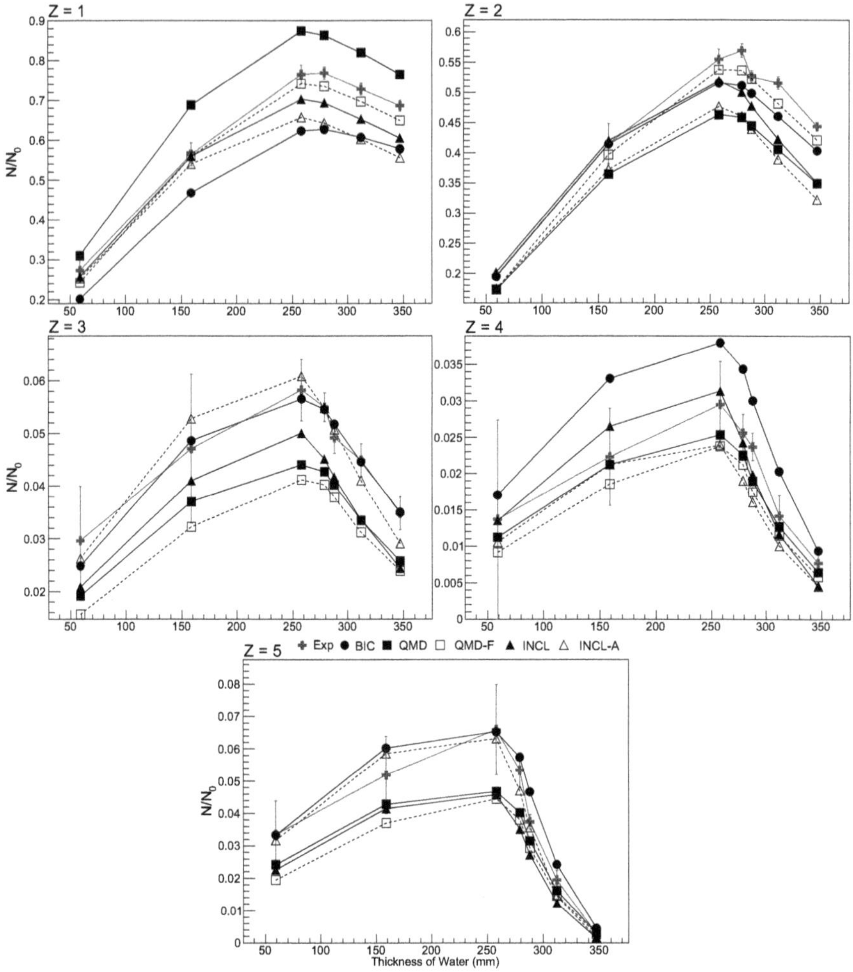

Fig. 4. Yields of fragments generated in a direction within an angle up to 10° from the direction of incidence of the primary ^{12}C ion beam. Geant4 simulation results are shown as calculated adopting the alternative fragmentation models. Experimental measurements are shown, as well.

Figure 5 compares the angular distribution calculated by means of the three different ion fragmentation models under study and the available experimental data. This is done for two different combinations of water thickness and fragment species. In total, there are 35 angular distributions for different target thicknesses and elements. The mean percentage difference, as calculated by Eq. (1), is summarized in Table 4.

Z	BIC	QMD	QMD-F	INCL	INCL-A
1	19 ± 2	14 ± 2	5 ± 2	8 ± 2	13 ± 2
2	6 ± 1	17 ± 1	5 ± 1	10 ± 1	18 ± 1
3	4 ± 7	25 ± 7	31 ± 7	21 ± 7	8 ± 7
4	33 ± 10	14 ± 10	22 ± 10	15 ± 10	25 ± 10
5	19 ± 8	20 ± 8	26 ± 8	33 ± 8	15 ± 8

Table 3. Mean percentage difference of the Geant4 ion fragmentation models compared to the experimental data [75].

It can be seen therein, that among the three Geant4 fragmentation models (BIC, QMD, INCL) and their variants, the INCL++ model reproduces best the measured angular distribution of the fragments. The exception is H ($Z = 1$) for which the QMD model gives the best reproduction of the experimental data. For Li ($Z = 3$) , Be ($Z = 4$) and B ($Z = 5$) ions, the QMD yields noticeably broader distributions compared to the experiment. The BIC model also gives broader distributions, but to a lesser extent.

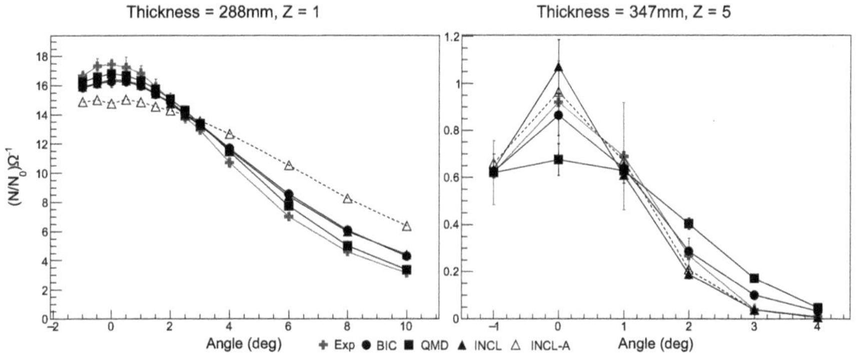

Fig. 5. Angular distributions for a selection of fragments types and thicknesses of water.

Figure 6 shows two examples of the kinetic energy distributions, calculated by means of Geant4, against the experimental reference data. In total, there are 151 kinetic energy distributions with different combinations of water thickness, angle and fragment species. The summary of the mean percentage errors is displayed in Table 5.

It is noticed that the INCL++ model reproduces the experiment considerably poorer than the BIC and the QMD models. The reason for this large

Z	BIC	QMD	INCL	INCL-A
1	14 ± 4	7 ± 4	15 ± 4	31 ± 4
2	24 ± 2	16 ± 2	7 ± 2	10 ± 2
3	29 ± 8	26 ± 8	16 ± 8	25 ± 8
4	47 ± 14	43 ± 14	18 ± 14	12 ± 14
5	131 ± 12	135 ± 12	27 ± 13	22 ± 12

Table 4. Mean percentage difference of the the Geant4 ion fragmentation models compared to the experiment for the angular distributions of each type of fragment. The water thickness and fragment species (Z = 1 and 5) are given on top of the related Fig. 5.

difference is that the INCL++ model systematically produces energy distributions that are peaked at lower energies compared to the other models, as shown in Fig. 6.

A similarly poorer performance of the BIC and the QMD is also encountered in the angular distributions. This is not unexpected since both models treat the fragmentation process using the Gaussian wave packets, while the INCL++ model describes the interaction as a Fermi free gas in a potential well.

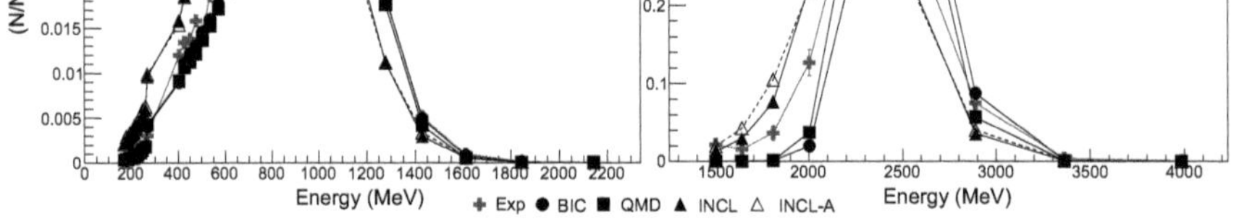

Fig. 6. Two examples of kinetic energy distributions as calculated by Geant4, against the reference experimental measurements. The water thickness, fragment species and angle of emission are reported on top of the figure.

In summary, it can be observed that Geant4 reproduces the Bragg curve of clinical ^{12}C ions with adequate accuracy. The fragmentation validation work showed that the QMD model with the "Frag" option reproduced the

Z	BIC	QMD	INCL	INCL-A
1	26 ± 6	21 ± 6	45 ± 6	71 ± 6
2	30 ± 7	33 ± 7	72 ± 7	69 ± 7
3	41 ± 8	43 ± 8	95 ± 8	116 ± 8
4	60 ± 9	55 ± 9	113 ± 9	102 ± 9
5	220 ± 11	196 ± 10	404 ± 10	435 ± 10

Table 5. Mean percentage error of the three fragmentation models compared to the experiment for the kinetic energy distributions of each type of fragment.

experimental fragment yields for both H and He with the best accuracy among the alternative fragmentation models offered by Geant4.

The INCL++ model reproduces best the angular distribution with respect to the experimental data with the exception of H for which the QMD model is the most superior. For kinetic energy distributions, the INCL++ model was seen to be the worst in reproducing the experimental data, systematically yielding the lower peaked distributions compared to the other models. For both angular and energy distributions, the BIC and the QMD models were seen to perform similarly.

2.5. *Characterization of the mixed radiation field of a clinical ^{12}C ion beam*

Here, we shall first deal with a setup of a biological beamline (Fig. 7) and the associated Geant4 simulations of the energy spectrum of the ^{12}C ion beam (Fig. 8). This will be followed by the Geant4 simulations of the total dose from a mono-energetic ^{12}C ion beam (Fig. 9) as well as from the individual components via the primary and secondary heavy particles (Fig. 10).

An ad-hoc Geant4 simulation has been developed to characterize the mixed radiation field produced by the ^{12}C ions at Chiba's Heavy Ion Medical Accelerator (HIMAC) beamline, as depicted in Fig. 7. For this simulation study, the version 10.2.patch02 of Geant4 is used alongside the BIC model for ion fragmentation.

Figure 7 shows the biological beamline at HIMAC which was modeled to produce a clinical spread out Bragg peak (SOBP). A thorough description of how the biological beamline was modeled can be found in Ref. [76], and only a brief description summary is provided here.

The ^{12}C beam exits the nozzle with a Gaussian energy distribution having the mean energy of 290 MeV/u and the standard deviation σ of 0.2%. The initial spot beam is shaped to a circle with a diameter up to 100 mm [77] using a pair of X and Y wobbler magnets operating using the single-ring wobbling method [78].

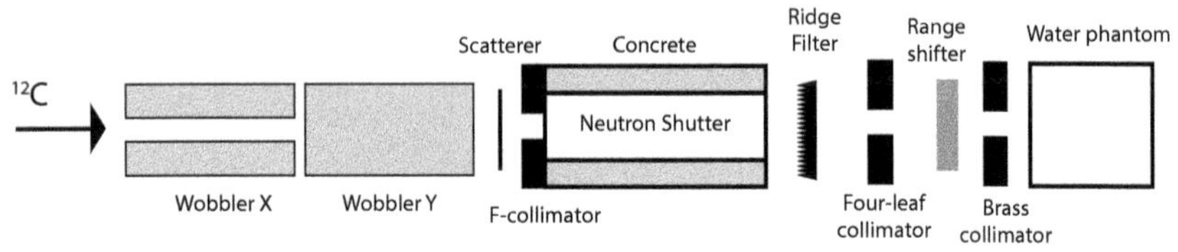

Fig. 7. Setup of the biological beamline at HIMAC, NIRS, Chiba, Japan, which was modeled in Geant4 to produce a clinical ^{12}C ion spread out Bragg peak. The diagram is not scaled. More details on the beamline can be found in Ref. [77].

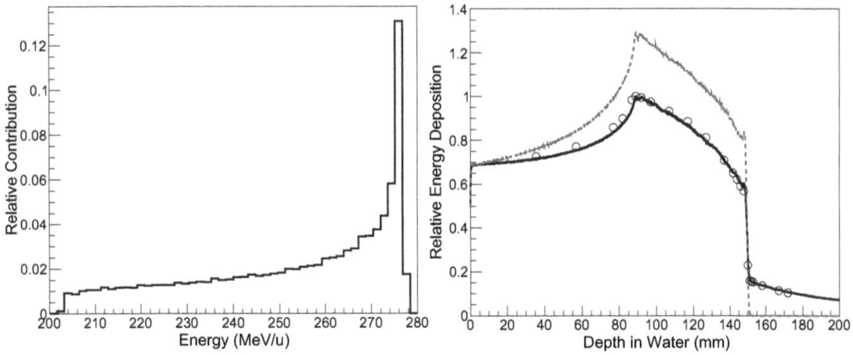

Fig. 8. Geant4 simulation results. Left: Energy spectrum of ^{12}C ions when reaching the surface of the phantom after passing through the beamline shown in Fig. 7. Right: SOBP produced by the HIMAC beamline in the water phantom; the dashed gray curve represents the SOBP calculated with electromagnetic physics only, while the solid black curve with the hadronic physics added. The open circle markers represent the data from the experimental measurements performed with an ionization chamber [14].

The uniformity of the dose profile is then improved by passing the ^{12}C ion beam through a tantalum scatterer with a thickness dependent upon the size requirements of the field. Because of the scatterer, the uniformity of the dose is within 2.5%.

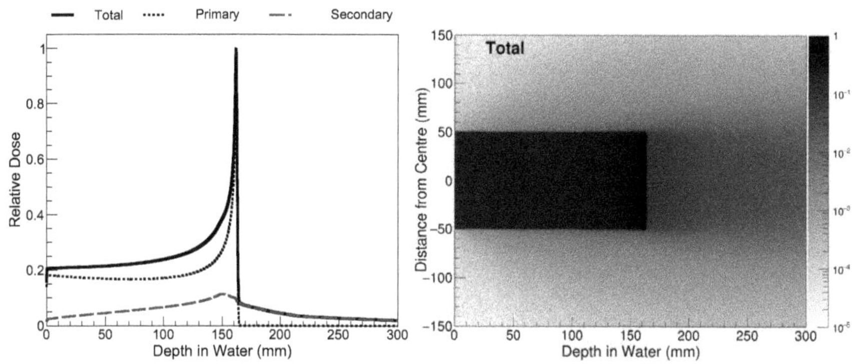

Fig. 9. Geant4 simulation results. Left: Bragg curve of a 290 MeV/u ^{12}C ion beam incident on a water phantom. Black curve: total dose. The dashed black and gray curves are the doses deposited by the primary beam and secondary particles in the water phantom, respectively. Right: the total dose distribution within the phantom. Both plots are normalized to the maximum total dose.

The beam then traverses a neutron shutter, which helps to remove the contribution of neutrons from the beam. Subsequently, it passes an aluminum ridge filter with 100 different thicknesses, producing a treatment field of several centimeters in depth.

The effect of the ridge filter can be observed on the left of Fig. 8, which shows the energy spectrum of the ^{12}C ion beam when it reaches the surface of the phantom, as calculated by Geant4. The beam is then shaped to the desired field size with a four leaf collimator and an XY collimator just before the isocenter.

The right side of Fig. 8 shows the depth-energy deposition curve in the water phantom when irradiated, using the setup shown in Fig. 7. In this figure, the gray curve represents the energy deposition when only the electromagnetic physics component is activated in the simulation. The black curve is the same physical quantity when the hadronic physics component is also taken into account.

It can be observed that the majority of the energy is deposited via ionization from the incident ^{12}C ion beam. The effect of the nuclear interactions is to decrease the energy deposition before the Bragg peak and increasing it after the distal edge. This is due to the production of fragments by the incident ^{12}C ions before the Bragg peak, which then propagate beyond the Bragg peak and scatter laterally, producing unwanted radiation doses outside the cancer target and, thus, altering the RBE.

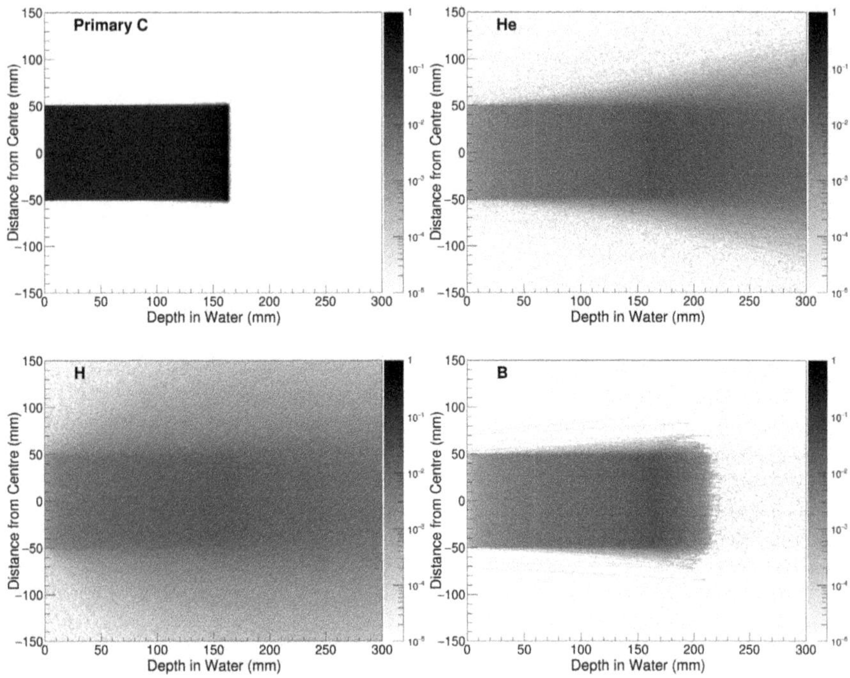

Fig. 10. Geant4 simulation results: dose distributions due to different particle species produced by the incident 290 MeV/u ^{12}C ion beam incident on the phantom target.

This effect can also be observed in Fig. 9, which shows (on the left) the dose produced by a mono-energetic ^{12}C ion beam (with an energy of 290 MeV/u) incident on a water phantom. In particular, the energy deposition originating from the incident ^{12}C ions and the secondary fragments are presented. The incident ^{12}C beam has a size of 10 × 10 cm^2.

Figure 10 shows the individual contribution to the dose deposited by the incident ^{12}C as well as the secondaries (i.e the H, He and B fragments). All the presented results are normalized to the maximum dose of the Bragg peak.

It is seen that the fragments scatter laterally and their range is extended beyond the Bragg peak. Such particles deposit energy out of the treatment field, possibly to organs at risk lying close to the targeted tumor and, hence, altering the RBE. Therefore, it is crucial that the fragments are modeled accurately in order to predict and minimize the dose to healthy tissue in accordance with the treatment plan.

3. Conclusions

This chapter reviews the capability of Geant4 for ^{12}C ion therapy. In summary, Geant4 offers a wide set of physics models to describe both electromagnetic and hadronic physics interactions, not only at millimeter scale of interest for dosimetry, but also at micro- and nano-scales. Geant4 is validated by both Geant4 collaborators and users. Its evolution for ^{12}C ion therapy is steadily monitored with the help of the regression testing suits.

Despite the extensive progress with the last ten years in the development of Geant4 physics capability for ^{12}C ion therapy, there are still some possible physics improvements which may significantly advance Geant4 in this domain. For example, new physics models of electromagnetic interactions of ^{12}C ions with liquid water or biological materials at low energies (below ~1 MeV/u) may provide a better accuracy of dose calculation and RBE determination at the distal edge of the Bragg peak, where organs at risk may be situated. For example, sophisticated methods such as the wave-packet continuum-discretization approach [79] to ion-atom collisions could be a powerful solution to describe the interactions of heavy ion projectiles with the biological medium.

In terms of hadronic physics capability, there are two domains of possible development. The first one is the development of better nucleus-nucleus hadronic physics models for energies below 400 MeV/u. It is important to note that this issue does not concern only Geant4, but any Monte Carlo code used in hadron therapy. Second, more accurate experimental measurements of fragment production should be carried out, eventually using different detector technology. The obtained new experimental data would be very useful in benchmarking Monte Carlo codes.

Finally, the literature is focused on the dosimetric validation of Monte Carlo codes for ^{12}C ion therapy in the presence of homogeneous phantoms alone. However, it would be important to validate Monte Carlo codes also in the presence of heterogeneity which, naturally, is part of the human anatomy. Additionally, given the current interest for in vivo imaging by positron emission tomography [80, 81], it would also be essential to validate the production of β^+ emitters.

References

[1] Health Policy Advisory Committee on Technology. Proton and Heavy Ion Therapy: An overview. `https://www.health.qld.gov.au/__data/assets/pdf_file/0033/643749/proton-hvy-ion-therapy.pdf` .

[2] Particle Therapy Co-Operative Group. https://www.ptcog.ch/ .
[3] H. Tsujii, Overview of Carbon-ion Radiotherapy, *J. Phys. Conf. Ser.* **777**, 012032 (2017).
[4] G. Battistoni, S. Muraro, P. R. Sala, F. Cerutti, A. Ferrari, S. Roesler, A. Fasso, and J. Ranft, The FLUKA code: description and benchmarking, *AIP Conf. Proc.* **896**, 31–49 (2007).
[5] A. Ferrari, P. Sala, A. Fasso, and J. Ranft. FLUKA: A multi-particle transport code (program version 2005) (2005).
[6] S. Agostinelli, J. Allison, K. Amako, J. Apostolakis, H. Araujo, P. Arce, M. Asai, D. Axen, S. Banerjee, G. Barrand, F. Behner, L. Bellagamba, J. Boudreau, L. Broglia, A. Brunengo, H. Burkhardt, S. Chauvie, J. Chuma, R. Chytracek, G. Cooperman, G. Cosmo, P. Degtyarenko, A. Dell'Acqua, G. Depaola, D. Dietrich, R. Enami, A. Feliciello, C. Ferguson, H. Fesefeldt, G. Folger, F. Foppiano, A. Forti, S. Garelli, S. Giani, R. Giannitrapani, D. Gibin, J. G. Cadenas, I. González, G. G. Abril, G. Greeniaus, W. Greiner, V. Grichine, A. Grossheim, S. Guatelli, P. Gumplinger, R. Hamatsu, K. Hashimoto, H. Hasui, A. Heikkinen, A. Howard, V. Ivanchenko, A. Johnson, F. Jones, J. Kallenbach, N. Kanaya, M. Kawabata, Y. Kawabata, M. Kawaguti, S. Kelner, P. Kent, A. Kimura, T. Kodama, R. Kokoulin, M. Kossov, H. Kurashige, E. Lamanna, T. Lampén, V. Lara, V. Lefebure, F. Lei, M. Liendl, W. Lockman, F. Longo, S. Magni, M. Maire, E. Medernach, K. Minamimoto, P. M. de Freitas, Y. Morita, K. Murakami, M. Nagamatu, R. Nartallo, P. Nieminen, T. Nishimura, K. Ohtsubo, M. Okamura, S. O'Neale, Y. Oohata, K. Paech, J. Perl, A. Pfeiffer, M. Pia, F. Ranjard, A. Rybin, S. Sadilov, E. D. Salvo, G. Santin, T. Sasaki, N. Savvas, Y. Sawada, S. Scherer, S. Sei, V. Sirotenko, D. Smith, N. Starkov, H. Stoecker, J. Sulkimo, M. Takahata, S. Tanaka, E. Tcherniaev, E. S. Tehrani, M. Tropeano, P. Truscott, H. Uno, L. Urban, P. Urban, M. Verderi, A. Walkden, W. Wander, H. Weber, J. Wellisch, T. Wenaus, D. Williams, D. Wright, T. Yamada, H. Yoshida, and D. Zschiesche, Geant4 –a simulation toolkit, *Nucl. Instrum. Meth. A.* **506**, 250 – 303 (2003).
[7] J. Allison, K. Amako, J. Apostolakis, H. Araujo, P. A. Dubois, M. Asai, G. Barrand, R. Capra, S. Chauvie, R. Chytracek, G. A. P. Cirrone, G. Cooperman, G. Cosmo, G. Cuttone, G. G. Daquino, M. Donszelmann, M. Dressel, G. Folger, F. Foppiano, J. Generowicz, V. Grichine, S. Guatelli, P. Gumplinger, A. Heikkinen, I. Hrivnacova, A. Howard, S. Incerti, V. Ivanchenko, T. Johnson, F. Jones, T. Koi, R. Kokoulin, M. Kossov, H. Kurashige, V. Lara, S. Larsson, F. Lei, O. Link, F. Longo, M. Maire, A. Mantero, B. Mascialino, I. McLaren, P. M. Lorenzo, K. Minamimoto, K. Murakami, P. Nieminen, L. Pandola, S. Parlati, L. Peralta, J. Perl, A. Pfeiffer, M. G. Pia, A. Ribon, P. Rodrigues, G. Russo, S. Sadilov, G. Santin, T. Sasaki, D. Smith, N. Starkov, S. Tanaka, E. Tcherniaev, B. Tome, A. Trindade, P. Truscott, L. Urban, M. Verderi, A. Walkden, J. P. Wellisch, D. C. Williams, D. Wright, and H. Yoshida, Geant4 developments and applications, *IEEE T. Nucl. Sci.* **53**, 270–278 (2006).

[8] J. Allison, K. Amako, J. Apostolakis, P. Arce, M. Asai, T. Aso, E. Bagli, A. Bagulya, S. Banerjee, G. Barrand, B. Beck, A. Bogdanov, D. Brandt, J. Brown, H. Burkhardt, P. Canal, D. Cano-Ott, S. Chauvie, K. Cho, G. Cirrone, G. Cooperman, M. Cortés-Giraldo, G. Cosmo, G. Cuttone, G. Depaola, L. Desorgher, X. Dong, A. Dotti, V. Elvira, G. Folger, Z. Francis, A. Galoyan, L. Garnier, M. Gayer, K. Genser, V. Grichine, S. Guatelli, P. Guéye, P. Gumplinger, A. Howard, I. Hivnáová, S. Hwang, S. Incerti, A. Ivanchenko, V. Ivanchenko, F. Jones, S. Jun, P. Kaitaniemi, N. Karakatsanis, M. Karamitros, M. Kelsey, A. Kimura, T. Koi, H. Kurashige, A. Lechner, S. Lee, F. Longo, M. Maire, D. Mancusi, A. Mantero, E. Mendoza, B. Morgan, K. Murakami, T. Nikitina, L. Pandola, P. Paprocki, J. Perl, I. Petrovi, M. Pia, W. Pokorski, J. Quesada, M. Raine, M. Reis, A. Ribon, A. R. Fira, F. Romano, G. Russo, G. Santin, T. Sasaki, D. Sawkey, J. Shin, I. Strakovsky, A. Taborda, S. Tanaka, B. Tomé, T. Toshito, H. Tran, P. Truscott, L. Urban, V. Uzhinsky, J. Verbeke, M. Verderi, B. Wendt, H. Wenzel, D. Wright, D. Wright, T. Yamashita, J. Yarba, and H. Yoshida, Recent developments in Geant4, *Nucl. Instrum. Meth. A.* **835**, 186 – 225 (2016).

[9] L. Waters, MCNPX user's manual, version 2.3.0, *LA-UR-02-260 Report* (2002).

[10] A. Dementyev and N. Sobolevsky, SHIELD – universal Monte Carlo hadron transport code: scope and applications, *Rad. Meas.* **30**, 553 – 557 (1999).

[11] I. Gudowska, N. Sobolevsky, P. Andreo, Dž. Belkić, and A. Brahme, Ion beam transport in tissue-like media using the monte carlo code SHIELD-HIT, *Phys. Med. Biol.* **49**, 1933–1958 (2004).

[12] K. Niita, T. Sato, H. Iwase, H. Nose, H. Nakashima, and L. Sihver, PHITS – a particle and heavy ion transport code system, *Rad. Meas.* **41**, 1080 – 1090 (2006).

[13] T. T. Böhlen, F. Cerutti, M. Dosanjh, A. Ferrari, I. Gudowska, A. Mairani, and J. M. Quesada, Benchmarking nuclear models of FLUKA and GEANT4 for carbon ion therapy, *Phys. Med. Biol.* **55**, 5833–5847 (2010).

[14] D. Bolst, L. T. Tran, L. Chartier, D. A. Prokopovich, A. Pogossov, S. Guatelli, M. I. Reinhard, M. Petasecca, M. L. Lerch, N. Matsufuji, V. L. Perevertaylo, C. Fleta, G. Pellegrini, M. Jackson, and A. B. Rosenfeld, RBE study using solid state microdosimetry in heavy ion therapy, *Radiat. Meas.* **106**, 512 – 518 (2017).

[15] A. Merchant, M. Newall, S. Guatelli, M. Petasecca, M. Lerch, V. Perevertaylo, G. Milluzzo, G. Petringa, F. Romano, G. Cirrone, G. Cuttone, M. Jackson, and A. Rosenfeld, Feasibility study of a novel multi-strip silicon detector for use in proton therapy range verification quality assurance, *Radiat. Meas.* **106**, 378 – 384 (2016).

[16] M. Asai, K. Amako, J. Apostolakis, H. Kurasige, T. Sasaki, and J.-P. Wellisch, *Design of Tracking and Generic Processes in Geant4*, In *Proc. of the Monte Carlo 2000 Conference, Lisbon, 23–26 October 2000*, pp. 1117–1122. Springer Berlin Heidelberg (2001).

[17] M. Pia, T. Basaglia, Z. Bell, and P. Dressendorfer, Geant4 in Scientific Literature, *IEEE Nucl. Sci. Symp. Conf. Rec.* (2009).

[18] www.opengatecollaboration.org/ .
[19] P. Arce, J. I. Lagares, L. Harkness, D. Pérez-Astudillo, M. Canadas, P. Rato, M. de Prado, Y. Abreu, G. de Lorenzo, M. Kolstein, and A. Díaz, *Nucl. Instrum. Meth. A.* **735**, 304 – 313 (2014).
[20] J. Perl, J. Shin, J. Schuemann, B. Faddegon, and H. Paganetti, TOPAS: An innovative proton Monte Carlo platform for research and clinical applications, *Med. Phys.* **39**, 6818–6837 (2012).
[21] http://geant4.web.cern.ch/geant4/UserDocumentation/UsersGuides/PhysicsReferenceManual/fo/PhysicsReferenceManual.pdf .
[22] M. Berger, Monte Carlo calculation of the penetration and diffusion of fast charged particles, *Method. Comput. Phys.* **1**, 135–215 (1963).
[23] J. Seco and F. Verhaegen, eds., *Monte Carlo Techniques in Radiation Therapy*. CRC Press (2013).
[24] F. Salvat, J. Fernandez-Varea, and J. Sempau, PENELOPE-2008: A Code System for Monte Carlo Simulation of Electron and Photon Transport, *OECD-NEA Report 6416, Issy-les-Moulineaux, France.* **6416** (2009).
[25] M. Bernal, M. Bordage, J. Brown, M. Davídková, E. Delage, Z. E. Bitar, S. Enger, Z. Francis, S. Guatelli, V. Ivanchenko, M. Karamitros, I. Kyriakou, L. Maigne, S. Meylan, K. Murakami, S. Okada, H. Payno, Y. Perrot, I. Petrovic, Q. Pham, A. Ristic-Fira, T. Sasaki, V. Stepán, H. Tran, C. Villagrasa, and S. Incerti, Track structure modeling in liquid water: A review of the Geant4-DNA very low energy extension of the Geant4 Monte Carlo simulation toolkit, *Phys. Medica.* **31**, 861–874 (2015).
[26] A. Valentin, M. Raine, J. E. Sauvestre, M. Gaillardin, and P. Paillet, Geant4 physics processes for microdosimetry simulation: Very low energy electromagnetic models for electrons in silicon, *Nucl. Instrum. Meth. B.* **288**, 66 – 73 (2012).
[27] A. Mantero, H. B. Abdelouahed, C. Champion, Z. E. Bitar, Z. Francis, P. Guéye, S. Incerti, and V. I. M. Maire, PIXE simulation in Geant4, *X-Ray Spectrom.* **40**, 135–140 (2011).
[28] S. Guatelli, A. Mantero, B. Mascialino, P. Nieminen, and M. G. Pia, Geant4 Atomic Relaxation, *IEEE T. Nucl. Sci.* **54**, 585–593 (2007).
[29] Dž. Belkić, Review of theories on ionization in fast ion-atom collisions with prospects for applications to hadron therapy, *J. Math. Chem.* **47**, 1366–1419 (2010).
[30] Dž. Belkić, (ed.), *Theory of Heavy Ion Collision Physics in Hadron Therapy.* In: *Advances in Quantum Chemistry*, Vol. **65** (Elsevier, Amsterdam, 2013).
[31] A. Lechner, V. Ivanchenko, and J. Knobloch, Validation of recent Geant4 physics models for application in carbon ion therapy, *Nucl. Instrum. Meth. B.* **268**, 2343–2354 (2010).
[32] ICRU Report 73-Stopping of Ions Heavier than helium, *J ICRU* (2005).
[33] ICRU Report 49-Stopping Power and ranges for protons and alpha particles, *J ICRU* (1993).
[34] V. Ivantchenko, M. Maire, L. Urbàn, V. Grichine, R. Kokoulin, H. Burkhardt, and P. Gumplinger. Overview and new developments in Geant4 electromagnetic physics. pp. 207–210 (2005).

[35] `https://twiki.cern.ch/twiki/bin/view/Geant4/LoweIonParameterized`
[36] R. Serber, Nuclear reactions at high energies, *Phys. Rev.* **72**, 1114–1115 (1947).
[37] D. Schardt, T. Elsässer, and D. Schulz-Ertner, Heavy-ion tumor therapy: Physical and radiobiological benefits, *Rev. Mod. Phys.* **82**, 383–425 (2010).
[38] D. Wright and M. Kelsey, The Geant4 Bertini Cascade, *Nucl. Instrum. Meth. A.* **804**, 175–188 (2015).
[39] G. Folger, V. Ivanchenko, and H. Wellisch, The Binary Cascade, *Eur. Phys. J. A.* **21**, 407–417 (2004).
[40] A. Boudard, J. Cugnon, J.-C. David, S. Leray, and D. Mancusi, New potentialities of the liège intranuclear cascade model for reactions induced by nucleons and light charged particles, *Phys. Rev. C.* **87**, 014606 (2013).
[41] J. J. Gaimard and K. H. Schmidt, A reexamination of the abrasion-ablation model for the description of the nuclear fragmentation reaction, *Nucl. Phys. A.* **531**, 709 – 745 (1991).
[42] J. Benlliure, A. Grewe, M. de Jong, K.-H. Schmidt, and S. Zhdanov, Calculated nuclide production yields in relativistic collisions of fissile nuclei, *Nucl. Phys. A.* **628**, 458–478 (1998).
[43] V. Grichine, A simple model for integral hadron-nucleus and nucleus-nucleus cross-sections, *Nucl. Instrum. Meth. B.* **267**, 2460 – 2462 (2009).
[44] V. Grichine, Geant4 hadron elastic diffuse model, *Comput. Phys. Commun.* **181**, 921 – 927 (2010).
[45] D. Johnson, Y. Chen, and S. Ahmad, Dose and linear energy transfer distributions of primary and secondary particles in carbon ion radiation therapy: A monte carlo simulation study in water, *J. Med. Phys.* **40**, 214–219 (2015).
[46] J. Griffin, Statistical Model of Intermediate Structure, *Phys. Rev. Lett.* **17**, 478–481 (1966).
[47] V. Weisskopf and D. Ewing, On the Yield of Nuclear Reactions with Heavy Elements, *Phys. Rev.* **57**, 472–485 (1940).
[48] K. Niita, S. ichiro Meigo, Y. Ikeda, and F. Maekawa, The GEM code – A Simulation Program for the Evaporation and Fission Process of an Excited Nucleus, *JAERI-Data/Code 2001-015, Japan Atomic Energy Research Institute (JAERI)* (2001).
[49] J. M. Quesada, M. A. Cortes-Giraldo, V. Ivanchenko, A. Ivanchenko, G. Folger, A. Howard, and D. Wright, Recent Developments in Pre-Equilibrium and De-Excitation Models in Geant4, *Prog. Nucl. Sci. Tech.* **2**, 936–941 (2011).
[50] N. Bohr and J. Wheeler, The Mechanism of Nuclear Fission, *Phys. Rev.* **56**, 426–450 (1939).
[51] J. Bondorf, A. Botvina, A. Iljinov, I. Mishustin, and K. Sneppen, Statistical multifragmentation of nuclei, *Phys. Rep.* **257**, 133–221 (1995).
[52] National Nuclear Data Center, Evaluated Nuclear Structure Data Files. URL `http://www.nndc.bnl.gov/ensdf` .
[53] A. Kraan, Range verification methods in particle therapy: underlying physics and Monte Carlo modeling, *Front. Oncol.* **5**, 150–177 (2015).

[54] M. Dingfelder, M. Inokuti, and H. Paretzke, Inelastic-collision cross sections of liquid water for interactions of energetic protons, *Radiat. Phys. Chem.* **59**, 255–275 (2000).
[55] D. Emfietzoglou, A. Pathak, G. Papamichael, K. Kostarelos, S. Dhamodaran, N. Sathish and M. Moscovitch, A study on the electronic stopping of protons in soft biological matter, *Nucl. Instrum. Meth. B.* **242**, 55–60 (2006).
[56] S. Uehara, H. Nikjoo, and D. Goodhead, Cross-sections for water vapour for the Monte Carlo electron track structure code from 10 ev to the mev region, *Phys. Med. Biol.* **38**, 1841–1858 (1993).
[57] N. Mott and H. Massey, *The theory of atomic collisions.* Oxford: Clarendon Press (1965).
[58] M. Bordage, J. Bordes, S. Edel, M. Terrissol, X. Franceries, M. Bardiès, N. Lampe, and S. Incerti, Implementation of new physics models for low energy electrons in liquid water in geant4-dna, *Phys. Medica.* **32**, 1833 (2016).
[59] D. Emfietzoglou, Inelastic cross-sections for electron transport in liquid water: a comparison of dielectric models, *Radiat. Phys. Chem.* **66**, 373–385 (2003).
[60] I. Kyriakou, S. Incerti, and Z. Francis, Technical Note: improvements in GEANT4 energy-loss model and the effect on low-energy electron transport in liquid water, *Med. Phys.* **42**, 3870–3876 (2015).
[61] Y. Kim and M. Rudd, Binary-encounter-dipole model for electron-impact ionization, *Phys. Rev. A.* **50**, 3954–3967 (1994).
[62] S. Incerti. Talk "Low-energy and DNA progress", 22nd Geant4 Collaboration Meeting, 25th-29th September, Wollongong, Australia .
[63] H. Tran, Z. E. Bitar, C. Champion, M. Karamitros, M. Bernal, Z. Francis, V. Ivantchenko, S. Lee, J. Shin, and S. Incerti, Modeling proton and alpha elastic scattering in liquid water in Geant4-DNA, *Nucl. Instrum. Meth. B.* **343**, 132–137 (2015).
[64] M. Dingfelder, L. Toburen, and H. Paretzke, An effective charge scaling model for ionization of partially dressed helium ions with liquid water, *Proc. Monte Carlo Topical Meeting.* **103**, 1–12 (2005).
[65] Z. Francis, E. Seif, S. Incerti, C. Champion, M. Karamitros, M. A. Bernal, V. N. Ivanchenko, A. Mantero, H. N. Tran, and Z. E. Bitar, Carbon ion fragmentation effects on the nanometric level behind the Bragg peak depth, *Phys. Med. Biol.* **59**, 7691–7702 (2014).
[66] W. Booth and I. Grant, Energy loss of oxygen and chlorine ions in solids, *Nucl. Phys.* **63**, 481–485 (1965).
[67] I. Plante and F. Cucinotta, Ionization and excitation cross sections for the interaction of HZE particles in liquid water and application to monte carlo simulation of radiation tracks, *New J. Phys.* **10** (2008).
[68] I. Plante and F. Cucinotta, Monte Carlo simulation of energy-deposit clustering for ions of the same LET in liquid water, *Phys. Med. Biol.* **57**, 209–224 (2012).
[69] M. D. Napoli, C. Agodi, G. Battistoni, A. A. Blancato, G. A. P. Cirrone, G. Cuttone, F. Giacoppo, M. C. Morone, D. Nicolosi, L. Pandola, V. Patera,

G. Raciti, E. Rapisarda, F. Romano, D. Sardina, A. Sarti, A. Sciubba, V. Scuderi, C. Sfienti, and S. Tropea, Carbon fragmentation measurements and validation of the Geant4 nuclear reaction models for hadrontherapy, *Phys. Med. Biol.* **57**, 7651–7671 (2012).

[70] J. Dudouet, D. Cussol, D. Durand, and M. Labalme, Benchmarking Geant4 nuclear models for hadron therapy with 95 Mev/nucleon carbon ions, *Phys. Rev. C.* **89**, 054616 (2014).

[71] D. Bolst, G. A. Cirrone, G. Cuttone, G. Folger, S. Incerti, V. Ivanchenko, T. Koi, D. Mancusi, L. Pandola, F. Romano, A. B. Rosenfeld, and S. Guatelli, Validation of Geant4 fragmentation for Heavy Ion Therapy, *Nucl. Instrum. Meth. A.* **869**, 68 – 75 (2017).

[72] `http://www.jcprg.org/master/exfor/D/d0589.txt` .

[73] E. Haettner, H. Iwase, M. Krämer, G. Kraft, and D. Schardt, Experimental study of nuclear fragmentation of 200 and 400 MeV/u ^{12}C ions in water for applications in particle therapy, *Phys. Med. Biol.* **58**, 8265–8279 (2013).

[74] S. Guatelli. Talk "Geant4 for medical physics: recent developments", 22nd Geant4 Collaboration Meeting, 25th-29th September, Wollongong, Australia. URL `https://indico.cern.ch/event/647153/contributions/2631985/attachments/1526575/2387183/guatelli_Geant4_User_Workshop_2017.pdf` .

[75] E. Haettner. Experimental study on carbon ion fragmentation in water using GSI therapy beams. Master's thesis, KTH Engineering Sciences (2006).

[76] D. Bolst, L. T. Tran, S. Guatelli, N. Matsufuji, and A. B. Rosenfeld, Modelling the biological beamline at himac using geant4, *J. Phys. Conf. Ser.* (2018).

[77] M. Torikoshi, S. Minohara, N. Kanematsu, M. Kormori, M. Kanazawa, K. Noda, N. Miyahara, H. Itoh, M. Endo, and T. Kanai, Irradiation system for HIMAC, *J. Radiat. Res.* **48**, A15–A25 (2007).

[78] Y. Shunsuke, K. Nobuyuki, K. Masataka, K. Tatsuaki, T. Yuka, T. Osamu, I. Yoshiharu, T. Mutsumi, K. Hajime, and T. Hideki, Evaluation of beam wobbling methods for heavy-ion radiotherapy, *Med. Phys.* **35**, 927–938 (2008).

[79] I. B. Abdurakhmanov, J. J. Bailey, A. S. Kadyrov, and I. Bray, Wave-packet continuum-discretization approach to ion-atom collisions including rearrangement: Application to differential ionization in proton-hydrogen scattering, *Phys. Rev. A.* **97**, 032707 (2018).

[80] W. Enghardt, P. Crespo, F. Fiedler, R. Hinz, K. Parodi, J. Pawelke, and F. Pönisch, Charged hadron tumour therapy monitoring by means of pet, *Nucl. Instrum. Meth. A.* **525**, 284 – 288 (2004).

[81] K. Parodi, H. Paganetti, H. A. Shih, S. Michaud, J. S. Loeffler, T. F. DeLaney, N. J. Liebsch, J. E. Munzenrider, A. J. Fischman, A. Knopf, and T. Bortfeld, Patient study of in vivo verification of beam delivery and range, using positron emission tomography and computed tomography imaging after proton therapy, *Int. J. Radiat. Biol.* **68**, 920 – 934 (2007).

Chapter 11

Theory of heavy ion collision physics for hadron therapy

Dževad Belkić[1,2]*

[1]*Nobel Medical Institute, Karolinska Institute, Department of Oncology-Pathology, P.O. Box 260, SE-171 76, Stockholm, Sweden*

[2]*Medical Radiation Physics & Nuclear Medicine, Karolinska University Hospital, SE-171 76 Stockholm, Sweden*

The salient aspects of the perturbative theoretical descriptions of ion-atom and ion-molecule rearrangement collisions at intermediate and high impact energies are succinctly illuminated with a focus on the use of the continuum distorted wave method in medical physics. This is done on the level of total cross sections as the input data for modeling energy loss needed in particle transport physics.

The emphasis is on one-electron capture by bare nuclei from multi-electron atomic and molecular targets. A set of the pertinent illustrations is given for total single charge-exchange in collisions of protons with atoms (H, O, Ne) and molecules (H_2, H_2O) by summing the contributions from all the shells in the case of multi-electron targets.

The independent electron model is employed for both atomic and molecular targets. Regarding molecular targets, the independent atom model is used. The Bragg and Gelius sum rules are discussed alongside the 'complete neglect of differential orbital' method as well as the one-center-expansions for description of the ground-state wave functions of molecular targets.

For capture cross sections, a simple atomic gas ^{20}Ne (neon), as the united atom limit of the ten-electron molecules H_2O (water), CH_4 (methane) and NH_3 (ammonia), is shown to be their good joint surrogate, i.e. a reasonable and useful first approximation.

The relevance of the proposed accurate and readily available molecular charge-exchange database in the independent atom model is illuminated, as it applies to hadron therapy.

*Email: Dzevad.Belkic@ki.se

Contents

1. Introduction

1.1. *The origins of hadron therapy*

The practical aspects of the concept of hadron therapy began in 1946 when Wilson published a short paper in a medical journal, Radiology [1] (see also [2]). Therein, he convincingly explained that protons and heavier atomic nuclei ($He^{2+}, ..., C^{6+}, ...$) could advantageously be used for radiotherapeutic treatments of patients with cancer. For this proposal, his motivation was in a far more favorable dose profile of atomic nuclei than those of electrons and photons. The latter two beams, within a non-conformal setting, deposit their energy throughout the traversed tissue, already from the entrance into the body of the patient all the way to the targeted tumor. In fact, electrons and photons alike deposit more energy at the entrance to the patient's body than at the tumor. This is opposite to what is actually needed and, as a result, the healthy tissue becomes damaged by non-conformal electron and photon therapies. This unwanted irradiation can induce tumor in normal tissue.

A precisely inverse dose profile is required, which would correspond to deposition of most energy at the tumor site, while simultaneously sparing the healthy tissue prior to reaching the target. This is exactly the case with protons and other heavier atomic nuclei. A dose profile of a proton beam is usually low and flat for most of the traversed path, which then abruptly ends with a pointed, narrow peak at the tumor site with no radiation spillover thereafter. This type of line-shape, representing the beam energy loss per distance has originally been established in the measurements of Bragg and Kleeman from 1904 [3] published the Philosophical Magazine, London (see also [4]). It is for this reason that the deposited narrow energy distribution of nuclei is called the Bragg curve, which ends with a sharp peak called the Bragg peak (BP).

The finding of Bragg and Kleeman that heavy particles (like alpha particles, He^{2+}, used in Refs. [3, 4]) become more efficient ionizers toward the end of their course is unprecedented. This is the case because the commonly accepted view at the time of this discovery was that ionization efficiency of alpha particles decreases with its range. Range (R), as the maximal distance attained by a particle traversing a medium, is determined by the initial impact energy. As per the explanation of Bragg and Kleeman [3, 4], alpha particles produced increased ionization near the end of the track (i.e. at the BP) because of their decreased velocity.

We can here support such a rationale for the emergence of the BP by the fact that the speed of the slowed down alpha particles will eventually match the orbital electronic velocity of some of the target atoms/molecules, at which point the cross section for charge exchange becomes maximal, yielding the Massey peak. As a result, the exhausted alpha particles are stopped after losing all their energy through one- and two-electron capture leading to formation of the single charged helium ions He^+ and neutral helium, He, respectively. Thus, on the side of electronic stopping powers, it is the resonance collisional effect that is one of the physical mechanisms responsible for the appearance of the Bragg peak (stopping power vs. distance), or equivalently, the Massey peak (cross section vs. impact energy).

Overall, it was the work of Bragg and Kleeman from Refs. [3, 4], which laid the ground for hadron therapy viz: the physical basis of the BP gives rise to a comparatively higher dose at the tumor site than anywhere else. This remarkable discovery laid dormant for 40 years before Wilson [1, 2] eventually decided to present it to physicians. Retrospectively, had Bragg and Kleeman [3, 4] published their article also in a dedicated medical journal, the idea about hadron therapy could have intrigued physicians for some four decades before the actual proposal of Wilson [1, 2]. Of course, it would have not been enough to merely get accustomed with the concept of the Bragg peak as early as 1904 [3], since the particle accelerators did not exist at that time. However, with the appearance of Wilson's article in 1946 [1], and the availability of accelerators, the time was just ripe to start implementing the Bragg peak in radiotherapy.

1.2. *Clinical significance of the Bragg curves (dose-effect distributions)*

The clinical significance of the Bragg peak is in that the area under this energy-loss line-shape is proportional to the dose delivered to the tumor. However, the tumor site is not a point which the top of the Bragg peak is supposed to cover. Rather, it is an extended area, which must be irradiated. A solution to this problem is in the fact that the position of the Bragg peak (i.e. the projectile's range R) is determined by the initial energy of the primary beam. Therefore, all that is needed is to attenuate the primary beam so as to lower the initial energy of the projectiles. This would move the peak height to a shorter distance. With such a beam energy modulation, the ensuing superposition of several successive Bragg peaks would yield the so-called spread out Bragg peak (SOBP). With the SOBP at hand, it becomes possible to cover the entire, extended region of the treated tumor.

Attractive as it certainly is, however, a downside to the SOBP should nevertheless be kept in mind. Namely, the SOBP significantly raises the doses deposited to the tissue prior to the tumor site. As an example, consider a tumor which is about 2.5 cm wide, residing between, say, the 11th and 13th cm from the entrance to the patient's body. To fully cover this tumor, at least 5 Bragg curves (with the diminishing positions of the BPs) would need to be generated by superimposing one on top of another. As a result, however, relative to the initial single BP, located at the range of 13 cm, the SOBP would deliver substantially higher doses (at least doubled) to the normal tissue prior to the tumor, i.e. at distances 0-11 cm from the entrance to the patient's body. Specifically, this SOBP would increase the dose to the healthy tissue by a factor of ~ 2 at 0 cm (entrance to the patient's body) and ~ 3 at 11 cm (beginning of the tumor location) [5, 6].

1.3. *Protons and heavier hadrons in radiotherapy*

Wilson's visionary connection of the Bragg curve alongside the Bragg peak with the main goal of radiotherapy (physician's controlling and possibly curing cancer while maximally sparing the normal tissue) was promptly and widely endorsed in the clinic. Following the initial research of physicists [7–13], already in the 1950s, patients with cancer began to be treated in the U.S. (Berkeley) and Sweden (Uppsala). However, due to discontinuation of funding, the Berkeley Project ended with no definite conclusion as to how to proceed further with hadron therapy. Before the Bevelac facility at the Lawrence Berkeley Laboratory was shutdown, about 3000 patients with cancer had been treated by both light (H^+, D^+, He^{2+}, C^{6+}) and heavy nuclei (Ne^{10+}, Ar^{18+}). One of the reasons for not fully materializing Wilson's goal in the clinic was placing more emphasis on heavier than on lighter nuclei and, thus, focusing on the oxygen enhancement ratio (OER). The unfortunate termination of the Berkeley Project put a damper on using hadrons heavier than protons as radiotherapeutic beams.

On the other hand, proton therapy not only survived, but also further expanded vigorously from year to year [14]. For ions other than protons, the situation also improved from the middle of the 1990s. This was due to the realization that lighter atomic nuclei (He^{2+}, C^{6+}, O^{8+}) should be more proactively explored with the emphasis on the relative biological effectiveness (RBE) in addition to the OER. Specifically, high energy carbon nuclei have been used in Japan [15] and Germany [16] to treat patients with deep-seated tumors. For e.g. 25 cm depth, protons and carbon ion would need to have the incident energy of about 200 and 4800 MeV, respectively [17–24]. Nowadays, several countries are either using lighter radiotherapeutic ions e.g. He^{2+}, C^{6+} or planning to build hospital-based accelerators for multiply charged nuclei. As of April 2019, the number of radiotherapeutic ion facilities in the world was 67 for H^+ (protons) and 13 for C^{6+} (carbon nuclei) [14]. Optimal choices of radiotherapeutic nuclei are guided by their biological effects on the tissue.

1.4. *Some possible side effects of hadron therapy*

Especially for deep-seated tumors, very energetic ions of a few hundreds of MeV per nucleon are clinically used to treat patients. At such high impact energies, nuclear reactions take place in the tissue yielding secondary heavy particles. In particular, nuclear transmutations produce energetic neutrons against which there could be no shielding in the patient's body [25]. Such neutrons can be produced by fragmentation of projectiles heavier than protons and from the tissue constituents. For example, at energies 200 or 400 MeV/amu needed to reach 25 cm of depth in a tissue-equivalent material readily produce 2 and 4 neutrons, respectively [26, 27]. This estimate is correct for the $^4He^{2+} + H_2O$ and $^{12}C^{6+} + H_2O$ collisions as we checked using the stopping power (S) of water molecule and the Bragg sum rule ($S_{H_2O} = S_O + 2S_H$).

At high energies necessary for treating deep-seated tumors, the main nuclear energy loss stems from generation of neutrons. The produced neutrons can cause late effects, including new tumors. This could occur years after the treatment and, thus, reliable risk estimates ought to be based on epidemiological studies with a long-term follow-up of patients. For particles heavier than protons, fragmentation tails appear beyond the prescribed range R. Such a radiation spillover can reach considerably farther distances beyond the tumor and, therefore, cause damage to normal surrounding tissue. Lighter secondary nuclei produced by transmutations of heavy ion primary beams have longer ranges.

Such secondaries can undergo more significant multiple scattering than the primaries. Moreover, due to their smaller masses, these secondaries can be deflected at larger scattering angles than the primaries. Dominant scattering of primary heavy nuclei is in the forward cone and this gives practically straight tracks and limited lateral spreading or diffusion. Overall, accurate estimates of doses deposited not only by the primary beams, but also by the secondary (tertiary and nuclei from higher-order generations) are necessary in reliable assessments of risks for therapy-induced cancers. Further, the secondary malignancy risk models have to be evidence-based by relying upon the most complete dose distributions (as per the just-stated argument) so as to have better chances for support by epidemiological studies when these become available.

If insufficiently accurate/complete dose distributions are used as the input data, no risk model would give a reliable prediction for secondary malignancies. For example, in 2006, two opposing risk predictions for secondary cancers have been published [28, 29] comparing proton to photon IMRT (intensity modulated radiotherapy). In Ref. [28], the estimate of the second cancer risk increase (for a factor ranging from 2 to 20) has been attributed to photon IMRT relative to proton therapy. By contrast, Ref. [29] reported approximately the same risk factor for protons and photon IMRT, as corroborated by a similar earlier finding from 2000 [30].

The two contradictory findings [28, 29] confused oncologists since at that time there were no epidemiological studies to support one or the other estimated risks. Eventually, in 2013 [31], the mentioned estimates from Ref. [29] have been supported by the first epidemiological study, which compared risks of second malignancy for two groups of patients, one treated with photon IMRT and the other with proton therapy under approximately the same conditions. With such an outcome, Schneider and Hälg [25] concluded: "the first epidemiological results by Chung *et al.* [31] suggest that the exaggerated secondary cancer risk estimates made by Hall [28] were misleading due to a questionable stray dose comparison."

1.5. *Role of basic sciences in hadron therapy*

Water, as a tissue-equivalent substance in about 70% of the human body, and deoxyribonucleic acid (DNA) are the two most important molecules to study within radiotherapy in general (and hadron therapy in particular). The same holds true for both accurate assessments of radiation damage of tumorous cells and repair of normal healthy cells after irradiation of the patient. This is where cellular radiobiology with radiobiological modeling, needed in dose planning systems [32–38], comes into focus while intertwining with radiophysics and radiochemistry in radiotherapy of deep-seated tumors by energetic hadrons. Nevertheless, this interdisciplinary basic science input into hadron therapy would not be complete without the needed accurate target definition (delineation) and reliable post-radiation follow-up of the treated patient so as to differentiate between e.g. recurrent tumor from radiation necrosis.

For this latter aspect of patient care, especially for early tumor detection (critical to the overall success of treatment), the multifaceted molecular imaging as a key diagnostic modality is indispensable. In particular, non-ionizing and non-invasive magnetic resonance spectroscopy (MRS) [39–42] is pivotal as it goes beyond anatomical/morphological imaging through peering into the structure of lesions by reconstructing the chemistry of the tissue, thus yielding the concentrations (abundance) of the diagnostically most important molecules. Such a panoply of aspects of basic sciences, ranging from theory and measurements on ion-molecule collisions to analysis/synthesis of the quantitative content of the irradiated tissue, constitute a key branch of the overall medical comprehensive strategy for fighting cancer as among the largest threats to public health worldwide.

1.6. *Physics modeling of energy loss of hadron beams in tissue*

Critically important for hadron therapy is physics modeling of energy loss of ion beams in their collisions with tissue molecules. For energetic beams, both the nuclear and electronic stopping powers are necessary. For electronic stopping powers, the theory of atomic and molecular collisions is used [43–48]. To this end, cross sections for various processes of electromagnetic interactions of nuclei with tissue (charge exchange, excitation, ionization as well as their combinations) are required.

In this regard, the continuum distorted wave (CDW) method [49, 50] for ion-atom and ion-molecule collisions is presently analyzed. Explicit computations of total cross sections for single charge-exchange with protons incident on one-to-ten-electron targets is carried out by using the general CDW code of Belkić *et al.* [51]. In this code, electron capture by bare nuclei from multi-electron target is modeled by a purely three-body problem within the frozen core approximation. In a multi-electron target, a core (as a set of all the passive electrons in the field of the parent nucleus) is viewed as being frozen by assuming that the orbitals of the non-captured electrons are the same before and after the transfer of the active electron to the impinging nucleus.

This is an idealization since, strictly speaking, no single electron transition is possible with a multi-electron target. The reason is that when an active electron transits from a multi-electron target to a bare projectile nucleus, the orbitals of the remaining electrons in the target are also perturbed. This perturbation is caused by the vacancy of the transferred active electron. Therefore, the binding energies of the non-captured electrons would be altered too. Hence, the final orbitals of the core would differ from their initial counterparts. Nevertheless, in practice, one supposes that the perturbation of the core is small relative to the direct interaction of the projectile nucleus with the target active electron.

This supposition amounts to ignoring the difference between the core orbitals after and before capture of the active electron. Hence the said notion known as the 'frozen core approximation'. In such a setting, the CDW method describes the initial state of a multi-electron atomic target (T) by using the Roothaan-Hartree-Fock (RHF) wave function built from the Slater-type orbitals (STO) with the electron orbital energies and the other parameters taken from the tables of Clementi and Roetti (CR) [52]. This is hereafter termed the CR-RHF wave function for a multi-electron atomic system. After the collision, the target remainder T^+ fades away from the RHF formalism for the CDW method. In other words, the wave function of T^+ in the exit channel does not appear at all in the integral for the transition amplitude in the CDW method.

Overall, the CDW code from Ref. [51] is restricted to atomic charge-exchange collisions involving multi-electron targets. Therefore, such a version of this theory will hereafter be referred to as the atomic continuum distorted wave (A-CDW) method. The simplest molecular CDW method is obtained directly from the code in Ref. [51] for the A-CDW method by using the independent atom model (IAM) [53–57], or equivalently, the Bragg additivity (the Bragg sum rule) [3, 4]: 'molecular cross section is the sum of the atomic cross sections each of which is multiplied by the number of atoms in the given molecule'. This molecular approximation will be called the Bragg – continuum distorted wave (B-CDW) method.

Atomic units will be used throughout the presentation unless stated otherwise.

2. Perturbative theories of fast ion-atom and ion-molecule collisions

At high impact energies, among the perturbative theoretical descriptions of ion-atom and ion-molecule rearrangement collisions, the CDW approximation is known to be adequate and reliable. This is true for both charge exchange [49] and ionization [58]. The focus of this chapter is on one-electron capture by bare nuclei from multi-electron atomic and molecular targets. The present illustrations are on single charge-exchange in both atomic ($H^+ + H$, $H^+ + Ne$, $H^+ + O$) and molecular ($H^+ + H_2$, $H^+ + H_2O$) collisions. Within the orbital RHF representation of multi-electron targets, advantage is taken of a reasonable similarity of the electronic configurations of the atomic gas (neon) and molecular (water vapor) targets. Both targets possess ten electrons and, therefore, we compare the measured cross sections on $H^+ + Ne \to H + Ne^+$ with their counterparts on $H^+ + H_2O \to H + H_2O^+$ to assess the extent of eventual agreement. Possible concordance of measured atomic and molecular cross sections would enable contrasting the A-CDW method on the $H^+ + Ne$ system to the corresponding experimental data on the $H^+ + H_2O$ system. Such a comparison is performed directly without any numerical adjustment, i.e. with no scaling nor normalization of theoretical to experimental data at any impact energies that are taken to run from 1 to 70000 keV.

2.1. *Independent atom model (the Bragg sum rule)*

The present theoretical formalism assumes the validity of the independent electron model (IEM) for both neon and water targets. In the IEM [50], capture of one electron occurs independently of what happens to the other target electrons. The implication is that the total capture cross section of one electron from the target is equal to the sum of the partial cross sections for each of the target orbitals. Additionally, for water molecules we use the IAM [53–57]. With the IAM, a molecule is viewed as a sum of its uncoupled constituent atoms. As a consequence, no molecular bonding, force nor interaction is included in the IAM. Thus, the IAM yields molecular cross sections as a simple, direct sum of the corresponding atomic cross sections. In such molecular cross sections, the sole weight coefficients multiplying the individual atomic cross sections are the gross atomic populations, i.e. the numbers of the given atoms in the molecule.

Thus, the IAM is recognized as the Bragg additivity rule [3, 4], which within the continuum distorted wave formalism yields the B-CDW method. As such, the Bragg sum rule gives the molecular cross sections for e.g. the $H^+ + H_2O \to H + H_2O^+$ collisions as the sum of the associated atomic cross sections for $H^+ + O \to H + O^+$ and $H^+ + H_2 \to H + H_2^+$. In its simplest form for diatomic molecular targets, the Bragg additivity approximates the molecular cross sections for e.g. $H^+ + H_2 \to H + H_2^+$ by the doubled atomic cross sections for $H^+ + H \to H + H^+$.

2.2. The Gelius sum rule

There are several ways to go beyond the direct summation of atomic cross sections (the Bragg additivity) to arrive at a molecular description. To this end, one can resort to e.g. the Gelius sum rule [59–61], which uses a molecular wave function for the target. Here, each molecular orbital (MO) is a superposition of atomic orbitals (AO). Thus, the Gelius model is within the general method known as the linear combination of atomic orbitals (LCAO). Then, a molecular cross section, obtained by means of the Gelius additivity, is not a bare addition, but rather a weighted sum of the constituent atomic cross sections. More precisely, a molecular cross section for each MO is a twofold sum, one over all the constituent atoms, and the other over all the AOs of each atom in the molecule. Each term in this double sum is the product of the corresponding parentage coefficient and the cross section for the given AO. The parentage coefficient is a number representing the electronic population of a fixed AO in the given MO. Therefore, the parentage coefficient can be deduced from the Mulliken population analysis [62–65].

The Gelius sum rule has first been applied to photoionization of molecules where it is alternatively called the Gelius intensity model [59–61]. It met with success despite ignoring all the interference terms among different atomic sites and different underlying MO-LCAOs in a molecular transition amplitude. Nevertheless, successful prediction of experimental data by the Gelius additivity can be explained by its applications to photoionization of molecules mainly for large electron and photon energies.

At sufficiently high energies, the de Broglie wavelength of the photoelectron is small relative to the average MO size. This implies that the cross section is predominantly determined by regions of very rapid variations of the MO wave function (which occurs close to the target nuclei). Neglect of molecular effects makes this model inapplicable at low photoelectron energies where the so-called shape resonances are encountered.

2.3. The complete neglect of differential overlap

A molecular concept of the type of the Gelius sum rule is incorporated in the complete neglect of differential overlap (CNDO) method [66]. As such, the CNDO method for a molecular target is based on the MO-LCAO formalism which, in practice, usually employs the CR-RHF wave function [52], but with the molecular instead of the atomic orbital energies for the electrons. These electronic energies can be optimized by some of the electronic structure codes, e.g. GAUSSIAN [67], as done in Refs. [68, 69]. In the CNDO method, a molecular cross section for a given molecular target is the sum of the associated atomic orbital cross sections multiplied by the pertinent parentage coefficients. The CNDO method has first been applied to ionization in energetic ion-molecule collisions in Refs. [70, 71] by using the plane wave Born (PWB) approximation. This was followed in Refs. [72–79] for both ionization and electron capture processes with molecular targets by employing the continuum distorted wave – eikonal initial state (CDW-EIS) approximation.

2.4. *The one-center expansion model for molecular wave functions*

An alternative approach to surpass the Bragg sum rule is to employ the one-center expansion (OCE) model [80–84], which is the self consistent field (SCF) version of the MO-LCAO method. The center in the one-electron molecular wave function from the OCE method is placed on the heaviest atom in the given molecule (which is oxygen in water). Here, each MO of the target molecule, expressed via a set of the LCAOs comprised of the STOs, is optimized by minimizing the expectation value of the exact, non-relativistic many-electron molecular Hamiltonian.

This gives the binding energies for each MO, as well as the expansion coefficients and the parameters of the equilibrium molecular spatial orientation (internuclear distance, angle). The OCE wave function is valid only for one fixed value of the internuclear distance and angle. Therefore, a molecular cross section computed with the OCE wave functions must be summed over all the orientations of the molecular target, whenever the experiments have not recorded any information about such orientations. This sum can be carried out analytically in some of the perturbative methods based on the OCE wave functions for molecular targets.

2.5. *Cross section sensitivity to different molecular wave functions*

The mentioned formalisms for description of the ground-states of molecular targets have previously been utilized for ionization and electron capture by protons and heavier nuclei in ion-molecule collisions at intermediate and high impact energies. These applications were concerned with several perturbative quantum-mechanical theories exported from ion-atom collisions to molecular physics, e.g. the PWB, boundary-corrected first Born

(CB1), CDW and CDW-EIS methods [69–79]. Also frequently employed were several other classical, perturbative and non-perturbative approximations [85–90].

Some of the previous theoretical studies [74, 78] have demonstrated that the total cross sections for charge exchange and ionization in ion-molecule collisions (integrated over the target molecular orientations and the projectile scattering angles) are not very sensitive to different target wave functions. Moreover, with or without the parameter optimization by e.g. the GAUSSIAN code [67], rather similar total cross sections have been obtained by the Bragg sum rules as well as the CNDO and OCE models for molecular targets. Based upon this past experience, it is deemed sufficient to opt for the IAM, i.e. the Bragg additivity in the present computation to illustrate the performance of the B-CDW method for molecular charge exchange $H^+ + H_2O \to H + H_2O^+$ which is of a special interest to radiotherapy.

Of course, it would make sense to apply the B-CDW method only if the A-CDW method is successful (within its validity domain) for each of the constituent atoms in the given molecular target. Thus, for the particular $H^+ + H_2O \to H + H_2O^+$ collision, it is necessary to recapitulate the well-known performance of the A-CDW method for the two separate component pathways, $H^+ + O \to H + O^+$ and $H^+ + H \to H + H^+$.

As has been found empirically [50], the A-CDW method is valid above the position of the Massey peak, i.e. at impact energies $E(\mathrm{keV/amu}) \geq 80\max\{|\mathrm{E_i}|, |\mathrm{E_f}|\}$ where E_i and E_f are the initial and final binding energies of the active electron, respectively. The present illustrations cover a much wider interval of E descending far below the Massey peak down to 1 keV for incident protons. This is generally unnecessary, but it is done here to clearly exhibit the expected departure of a perturbative second-order theory from measurements near and below the Massey peak, i.e. below the said estimate of the validity domain of the continuum distorted wave method [50].

2.6. *Explicit procedure for straightforward adaptations of the existing theoretical methods on ion-atom collisions to ion-molecule collisions*

For concreteness, we shall here single out the CDW method, although similar considerations also apply to the CDW-EIS, CB1 and other approximations. From the foregoing exposition, it is clear that a practical procedure exists for adapting the A-CDW method for charge exchange in ion-atom collisions to work also for ion-molecule collisions (the same holds true for ionization).

All that is needed are the matrix elements for the transition amplitudes in the case of multi-electron atomic targets impacted by ions. Moreover, for applications in hadron therapy, especially when addressing the equilibrium distributions of charged particles, besides bare nuclei incident on multi-electron atomic targets, the energy loss analyses require electron-bearing atomic projectiles, as well. The A-CDW code from Ref. [51] covers all these cases. It is based on the analytical expressions for the general set of quantum numbers in the initial and final bound state wave functions. In this code, a multi-electron atom is described by the analytically available one-electron CR-RHF wave function [52] in which each AO is a linear combination of the STOs as the basis set expansion functions.

Thus, the B-CDW method for charge exchange in ion-molecule collisions, as the simplest example of a molecular CDW theory, is obtained by adding the cross sections of the A-CDW method computed separately and independently for each of the atoms in a molecular target using directly the code from Ref. [51]. Overall, the B-CDW method does not use molecular wave functions at all. Rather, it employs exclusively atomic wave functions throughout, such that molecular cross sections are deduced by merely adding atomic cross sections for all the atoms present in the given molecular target.

On the other hand, the Gelius, CNDO and OCE models for ion-molecule collisions make the explicit use of molecular wave functions for molecular targets. These are one-electron molecular wave functions given by superpositions of precisely the same form of the STOs as encountered in the atomic CR-RHF wave functions used in ion-atom collisions. For example, the only essential difference between the molecular OCE wave function of Moccia [80–84] and the atomic CR-RHF [52] is in using different sets of the STOs with different binding energies and expansion coefficients. Of course, the difference in using the complex- and real-valued spherical harmonics in Refs. [52] and [83], respectively, is inessential. The reason is that the A-CDW code [51] with complex spherical harmonics can trivially be modified to accommodate wave functions with real spherical harmonics. Thus, the A-CDW method with its code from Ref. [51] can readily handle the molecular OCE wave functions [80–84]. This would yield the OCE – continuum distorted wave (OCE-CDW) method. Furthermore, in the OCE-CDW method, the integration of cross sections over all the molecular target orientations can be carried out analytically [91, 92].

In the orbital representation of a molecular treatment of an ion-molecule collision, the MOs of a molecular target are given by the LCAOs. Such an expansion assumes that the weight coefficients of all the AOs in the LCAOs are known, i.e. that the electronic populations are available. Even if this assumption is fulfilled, there will still remain a major task to obtain the molecular continuum wave function (which can be generated by using e.g. the well-known R-matrix method in a partial wave analysis). Once the bound and continuum MOs eventually become available, the matrix elements for the transition amplitude need to be calculated, including all the cross terms for the constituent AOs.

Rather than pursuing this demanding task, there is a much simpler way around these obstacles. It is where the Gelius sum rule [59–61] comes to the rescue by adding up the atomic cross sections (for different AOs in a fixed MO) multiplied by the parentage coefficients deduced from the Mulliken population analysis. In other words, all that is needed in the Gelius model is to know from which AOs an MO in a molecular target is built, and to have the electronic binding energies in all the MOs. This would be followed by computation of purely atomic cross sections for each AO from an MO. Afterward, such atomic cross sections would be multiplied by the parentage coefficients and added up to yield the molecular cross section for a fixed MO.

Finally, summing the latter molecular cross sections for all the MOs would yield the total cross sections for electron capture by an ion from any MO of a molecular target. In this procedure, the price paid by the Gelius model is the loss of all the interference or cross terms among different orbital in a molecule. Nevertheless, despite these obvious shortcomings, the success of the Gelius model is indeed astonishing, as shown in Refs. [59–61] and in many subsequent studies. The molecular orientation does not appear explicitly in the Gelius model for the CDW theory and, therefore, there are no integrations over these orientations to be performed. Here too the GAUSSIAN code [67] can be used to optimize all the electronic binding energies for the given molecular target. All told, the Gelius additivity can be implemented for charge-exchange in ion-molecule collisions by using again the A-CDW code [51] to generate the molecular cross sections with the indicated straightforward adjustments/modifications. The same holds true for the CNDO variant of molecular target wave functions [66, 70–72].

3. Illustrations

In the present illustrations, the main focus is placed onto single electron capture by protons from ten-electron atomic (neon, Ne) and molecular (water vapor, H_2O) targets. With such a goal, subsection 3.1 addresses only the corresponding experimental data for Ne and H_2O. Subsection 3.2 deals with neon alone as the target while comparing the A-CDW method with measurements. Subsection 3.3 confronts the A-CDW method on Ne with experiments on H_2O. Subsection 3.4 on the $H^+ + H$, $H^+ + H_2$ and $H^+ + O$ separate collisions juxtaposes the A-CDW and B-CDW methods to the corresponding measurements. Subsection 3.5 is on single charge-exchange in the $H^+ + H_2O$ collisions. For the studied collisional systems, there are experimental data from a large number of measurements [93–126]. Most of these measured cross sections (tabulated in Ref. [127]) will be used here for comparisons with theoretical predictions. In addition to the general charge-changing database with tables as well as graphs [127], there are also some other databases of recommended cross sections [128–131]. In all the present computations with the CDW method, the code of Belkić *et al.* [51] has been used.

3.1. *Measurements on electron capture by protons from neon and water*

Figure 1 displays the available and mutually concordant experimentally measured total cross sections for electron capture by protons from Ne and H_2O. Therein, cross sections on the ordinate and the impact energies on the abscissa are seen to vary within eight ($10^{-22} - 10^{-14}$ cm^2) and four ($10^0 - 10^4$) orders of magnitudes. The open and filled symbols are used for the neon and water target, respectively.

At the intermediate and high impact energies ($E \geq 150$ keV), the atomic (Ne) and molecular (H_2O) cross sections are in good agreement. Moreover, they follow the same steep declining pattern with augmentation of the impact energy. On the other hand, below 150 keV down to 2 keV, the Ne and H_2O data are observed to differ significantly, and the cross sections for Ne lie below those for water.

These two types of different behaviors of the measured cross sections on Ne and H_2O at lower and higher impact energies can be explained by the argument which runs as follows. At intermediate and high energies, fast projectiles pass almost instantly by Ne or H_2O without having the time to "peer" into the target inner structure.

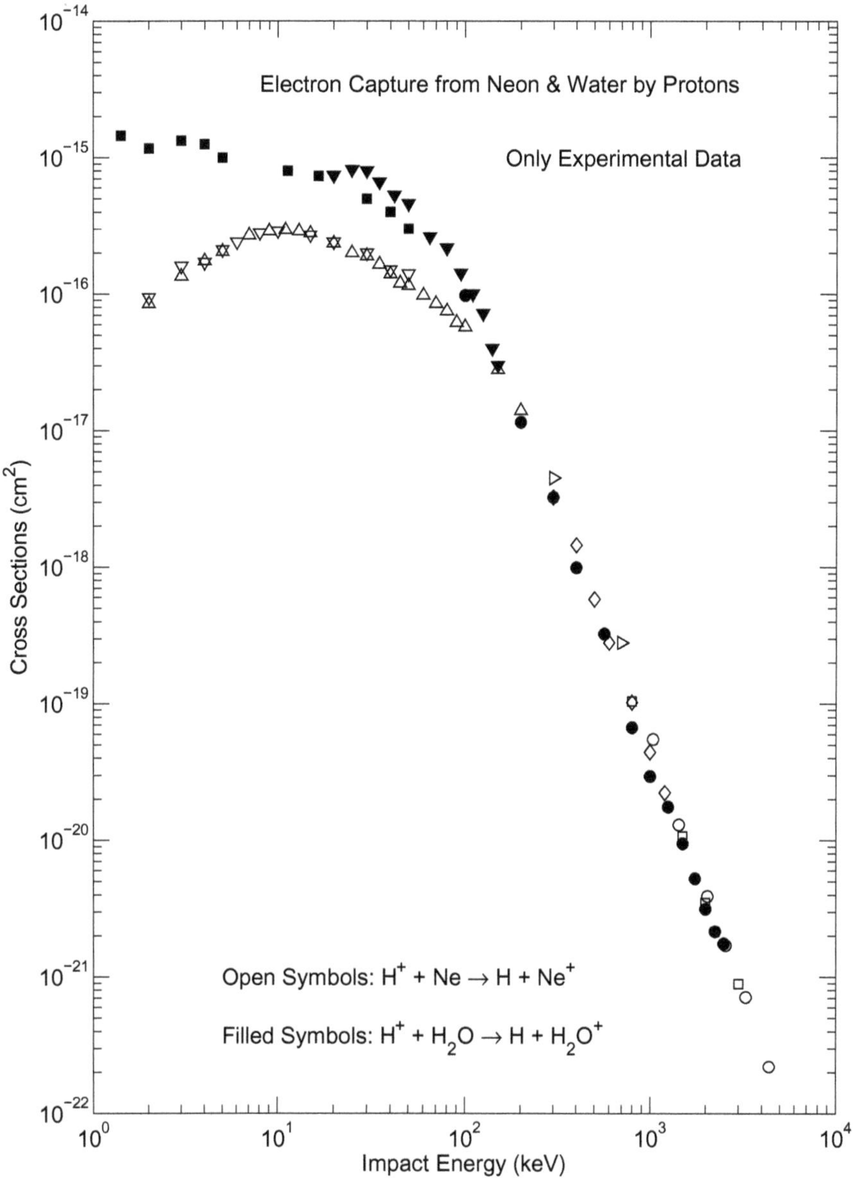

Fig. 1. Cross sections for electron capture by protons, H^+, from neon, Ne (open symbols) and water vapor, H_2O (filled symbols). Only experimental data are shown. Neon target: △ [94], ▽ [95], ○ [97], □ [98], ◊ [99], ▷ [101]. Water target: • [102], ■ [103], ▼ [106].

Therefore, to such energetic projectiles, all that matters is the total number of the target electrons. This statement is contingent upon the provision that the given atomic and molecular targets possess similar electronic configurations in the stationary discrete states.

This means that the atomic and molecular bound-state wave functions are assumed to be built from similar basis set functions that are usually the STOs. For example, the orbitals of the one-center molecular ground state wave function of H_2O, with the center placed on oxygen as the heavier atom [83], has the electronic configuration which is similar to that in neon.

With this provision, for fast projectiles, no significant consequence is expected to be caused by the differing internal structure of Ne and H_2O. This is precisely what is seen in Fig. 1 at $E \geq 150$keV for the neon and water targets that both have 10 electrons in total and quite similar electronic configuration.

However, an entirely different situation is encountered at lower impact energies at which slower projectiles spend enough time near the target. In such a case, these slow projectiles can discern and interact with the individual constituents of the target. Under this circumstance, it matters immensely whether the projectiles encounter atomic or molecular targets.

Molecules have more degrees of freedom than atoms. For example, vibrational and rotational molecular states are absent from atomic spectra. A projectile can capture an electron from some vibrational states and this collisional event is unique to a molecular target [132]. This would cause that at lower projectile energies molecular cross sections are very different from their atomic counterparts.

Such is the occurrence observed in Fig. 1 at $E < 150$ keV. Overall, Fig. 1 suggests that at $E \geq 150$ keV it could be possible to justify the use of the theoretical cross sections for electron capture in the $H^+ + Ne$ collisions as a surrogate of the corresponding predictions on the $H^+ + H_2O$ charge exchange.

3.2. *Theory vs. measurements on electron capture by protons from neon*

Figure 2 details the total cross sections in the A-CDW method for electron capture by protons from neon. According to the CR-RHF wave function [52], the ground state of a neon atom has the $1s, 2s$ and $2p$ fully occupied orbitals. Thus, the cross section from any orbital of the ground state of

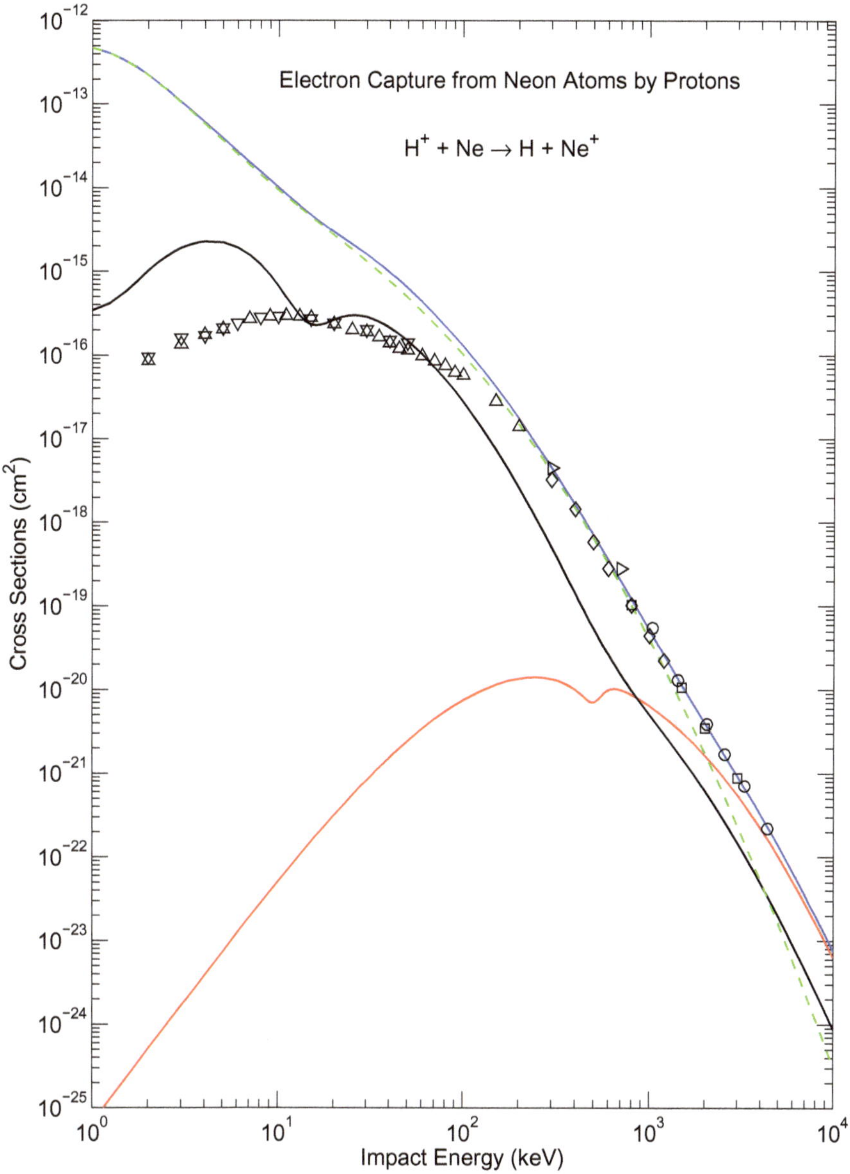

Fig. 2. Cross sections for electron capture by protons, H^+, from neon, Ne. Theoretical data, A-CDW method: full red line ($1s$ orbital of Ne), full black line ($2s$ orbital of Ne), dashed green line ($2p$ orbital of Ne), blue full line ($1s + 2s + 2p$ orbitals of Ne). Experimental data: △ [94], ▽ [95], ∘ [97], □ [98], ◊ [99], ▷ [101]. Here and in Figs. 3-11, the cross sections in the CDW method are computed only for ground-to-ground state capture and afterward multiplied by 1.202 to approximate the contributions from all the final excited states of H.

neon is given by the sum of the cross sections for the K($1s$) and L($2s + 2p$) shells of Ne($1s^2 2s^2 2p^6$). As stated, the underlying assumption here is the IEM, which assumes that capture of one electron from a given shell is independent of what happens to the other (passive) electrons in the target.

Very different lines are seen in Fig. 2 for electron capture by protons from the $1s, 2s$ and $2p$ orbitals of neon. Thus, below 850 keV, capture from the $1s$ orbital of Ne gives the smallest cross sections relative to the corresponding $2s$ and $2p$ contributions. Only above 2000 keV, capture from the $1s$ orbital begins to give a larger contribution than those from the $2s$ and $2p$ orbitals of neon. Around 450 keV, there is a dip in the curve in the A-CDW method for the cross section from the $1s$ orbital of neon. This is due to a destructive interference of the two Coulomb wave functions centered on the projectile and target nuclei in the A-CDW method. A dip is also encountered in the corresponding differential cross sections (angular distributions as a function of the scattering angles).

A similar dip, but at a much lower energy (near 15 keV), is also present in the cross sections for capture from the $2s$ orbital of Ne. Below 500 keV, the cross sections for capture from the $2s$ orbital of Ne are larger than those from the $1s$ orbital by a huge factor which reaches the value of 10^5 at 10 keV. On the other hand, below 2000 keV, the cross sections for capture from the $2p$ orbital of Ne yield the dominant contribution compared with the associated values for the $1s$ and $2s$ orbitals. For example, at 100 keV, the $2p$ to $2s$ cross section ratio is about five, whereas at 10 keV the corresponding $2p$ to $1s$ quotient is about 10^4.

The top line represents the total cross sections for capture from all the orbitals of neon as the sum of the cross sections from the $1s, 2s$ and $2p$ target orbitals. In particular, at $E \leq 1000$ keV, we see that the latter total cross sections are close to those from the $2p$ orbital alone. Most importantly, it is observed in Fig. 2 that at $E \geq 100$ keV, the top line ($1s + 2s + 2p$) in the A-CDW method is in excellent agreement with the corresponding experimental data for capture by protons from neon.

Figure 3 differs from Fig. 2 only in showing the partial cross sections for the summed contributions from the $2s$ to $2p$ orbitals of Ne corresponding to capture from the target L-shell. It is seen that below 2000 keV the L-shell capture dominates the K-shell capture. In fact, below 600 keV, the lines for the L-shell and the K+L shell of neon are indistinguishable from each other. Moreover, at impact energies $150 \leq E \leq 1500$ keV, the experimental data are nearly fully and excellently predicted by the L-shell capture alone. Above 1500 keV, the L-shell capture underestimates the measured cross sections.

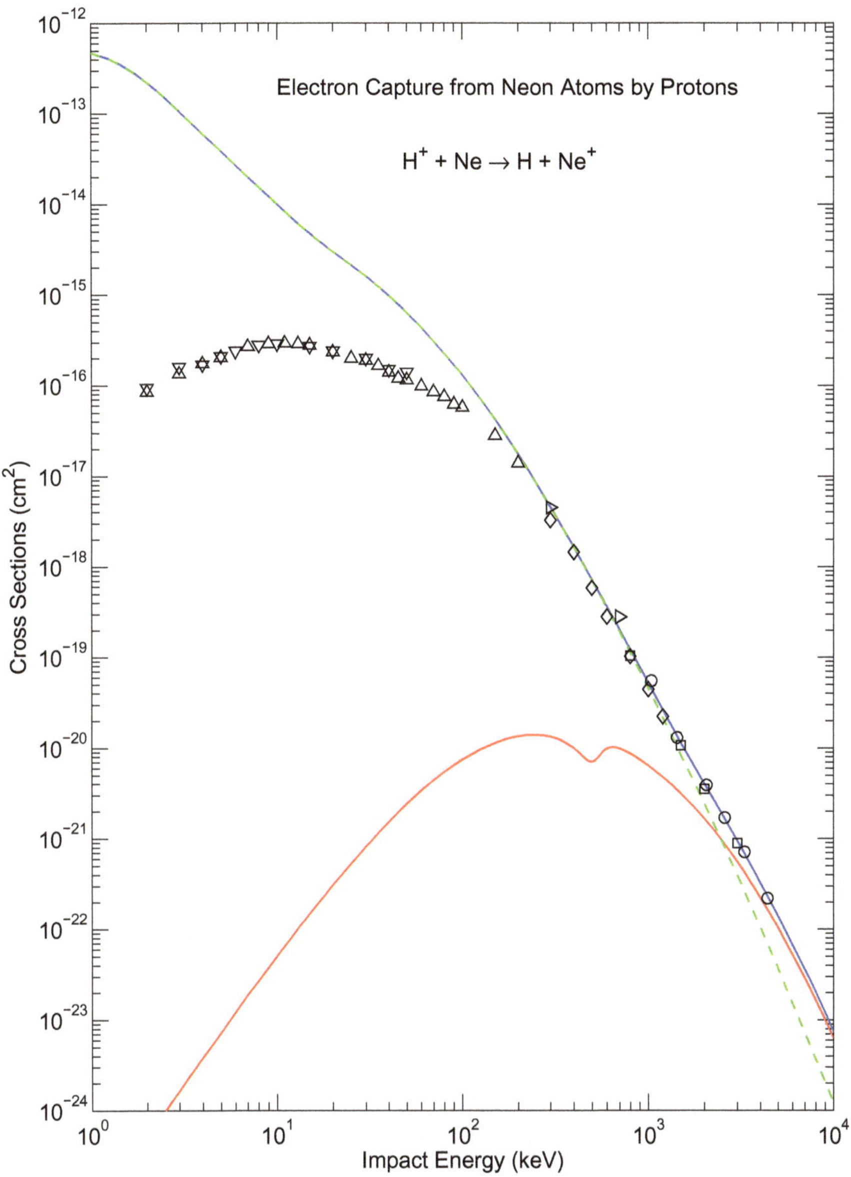

Fig. 3. Cross sections for electron capture by protons, H^+, from neon, Ne. Theoretical data, A-CDW method: full red line (K-shell of Ne), dashed green line (L-shell of Ne), blue full line (K+L shells of Ne). Experimental data: △ [94], ▽ [95], ○ [97], □ [98], ◊ [99], ▷ [101].

However, at $E \geq 1500$ keV, we see how the K-shell capture comes to the rescue such that the total cross sections (as the sum of the partial cross sections associated with the K and L shells of Ne) are again in excellent agreement with the experimental data. Note that the theoretical K+L shell capture exhibits a curvature change above 1500 keV. This slope alteration in the A-CDW method is due to the occurrence that above 1500 keV the K-shell capture becomes more noticeable. The fact that above 1500 keV the theoretical cross sections for the K+L shell capture are in accord with the measurements indicates that also the experimental data have the same slope change.

3.3. *Ten-electron atoms & molecules: theory on Ne vs. measurements on* H_2O

Figure 4 compares the A-CDW method for electron capture by protons from neon and the corresponding experimental data on water vapor. Therein, above 50 keV, it can be seen that the full line corresponding to the total (K+L shell) capture described by the A-CDW method for neon predicts reasonably well the associated cross sections measured on H_2O. This is consistent with Fig. 1 which shows concordance of the measured cross sections for electron capture in the $H^+ + Ne$ and $H^+ + H_2O$ collisions. The A-CDW method for Ne appears to be valid down to about 50 keV for charge exchange between protons and H_2O. In other words, the low-energy validity limit (~ 150 keV) for electron capture in the $H^+ + Ne$ collisions, seen in Fig. 3, is now decreased by a factor of three relative to the corresponding $H^+ + H_2O$ encounters.

Such a significant lowering of the smallest energy at which the A-CDW method remains valid is fortuitous. The reason is that when descending from 150 keV down to lower energies, the cross sections from the A-CDW method keep on rising due an over-account of continuum intermediate states whose normalization constant increases with diminished incident velocity. On the other hand, at $E \leq 150$ keV, the measured cross sections for water are larger than those for neon because of the availability of more open channels in the former relative to the latter target, respectively. As stated, such extra channels are e.g. capture from a vibrational or rotational states of H_2O that do not exist in Ne.

Hence, the said two different reasons for the rise of the cross sections in Fig. 4 at energies smaller than 150 keV make fortuitous the apparent agreement between an atomic theory and a molecular measurements. However,

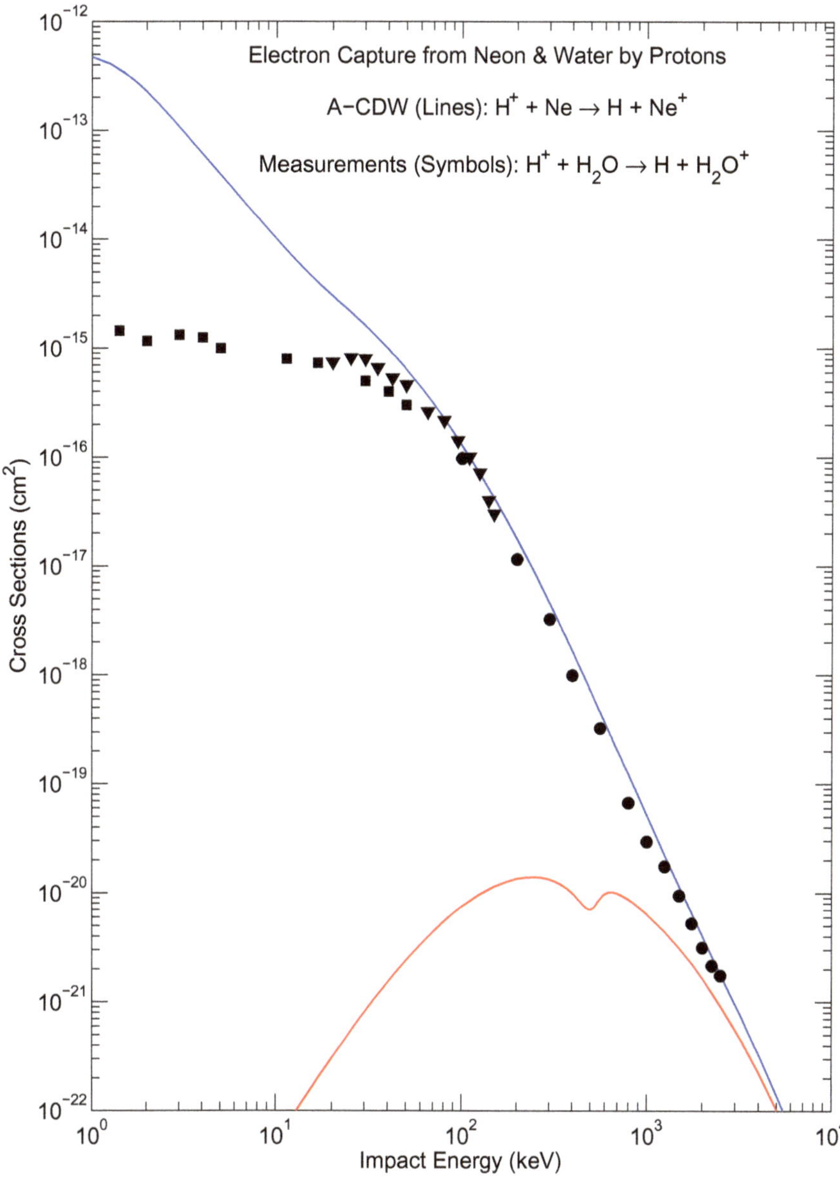

Fig. 4. Cross sections for electron capture by protons, H^+, from neon, Ne (full lines) and water vapor, H_2O (filled symbols). Theoretical data for neon target, A-CDW method: full red line (K-shell of Ne), blue full line (K+L shells of Ne). Experimental data for water target: • [102], ■ [103], ▼ [106].

above 150 keV, it can surely be stated, by reference to Fig. 4, that the A-CDW method for the atomic charge exchange $H^+ + Ne \rightarrow H + Ne^+$ represents a good surrogate for the molecular charge exchange $H^+ + H_2O \rightarrow H + H_2O^+$. Thus, starting already from the onset of the intermediate impact energy region, neon appears to be a good united atom limit of vapor-phase water (H_2O) as if oxygen (^8O) and H_2 (taken as 2 atomic hydrogens ^1H) were fused together in one atom with the common nucleus to become neon (^{20}Ne).

3.4. *One-, two- & eight-electron targets* $(\mathbf{H}, \mathbf{H_2}, \mathbf{O})$*: theory vs. measurements*

As a preparation for the application of the B-CDW method to water vapor (to be presented in subsection 3.5), we here consider atomic as well as molecular hydrogen (H, H_2) and atomic oxygen (O) as one-, two- and eight-electron targets, respectively. In Fig. 5, the cross sections for the process $H^+ + H(1s) \rightarrow H + H^+$ are shown. Here, both the measurements and the theory are for the atomic hydrogen H as the target. For this benchmark charge-exchange collision, with the exactly known bound state wave functions, the A-CDW method is seen in Fig. 5 to agree well with the experimental data at impact energies $E \geq 25$ keV.

Figure 6 compares the B-CDW method with the experimental data for $H^+ + H_2 \rightarrow H + H_2^+$. The cross sections in the B-CDW method are the doubled cross sections of those in the A-CDW method for the associated atomic collisions $H^+ + H \rightarrow H + H^+$ from Fig. 5. At energies $E \geq 30$ keV, the B-CDW method is observed in Fig. 6 to predict well the molecular experimental cross sections. All the experimental data in Fig. 6 are for molecular hydrogen H_2 (i.e. none of the displayed symbols correspond to the doubled cross sections on atomic hydrogen H as the target). Most of the measured cross sections from different studies are mutually concordant. The exceptions are the experimental data from Ref. [95] that are too low at intermediate energies and fall off too abruptly at higher energies, at variance with the other measured cross sections as well as with those from the B-CDW method.

The first instance when the IAM as the Bragg sum rule is often used is for diatomic molecules $(H_2, N_2, O_2, \ldots)$. This can be seen in many experimental studies as well as in several databases [127–131], where the measured cross sections for diatomic molecules are quoted, side-by-side, as cross section per molecule and cross section per atom with the latter being deduced

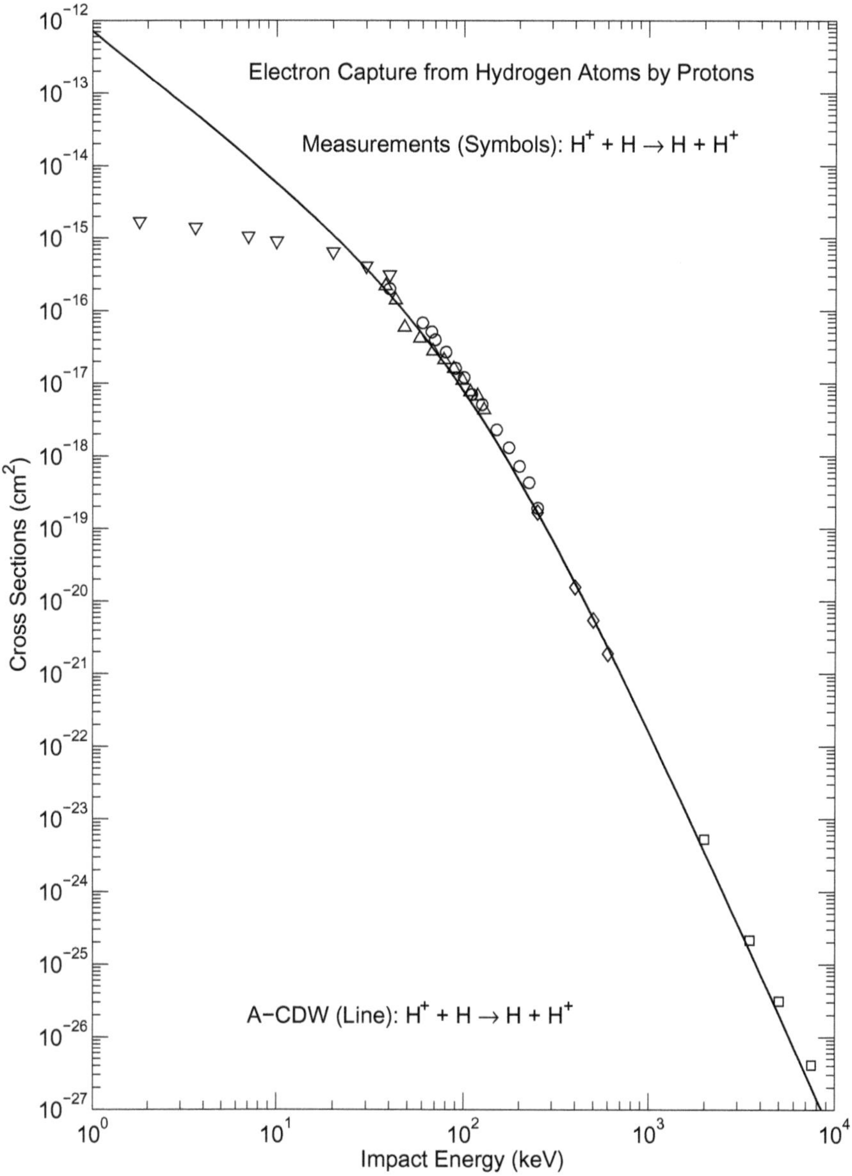

Fig. 5. Cross sections for electron capture by protons, H^+, from atomic hydrogen, H. Theoretical data: A-CDW method (full black line). Experimental data (open symbols): $\triangledown$ [108] $\triangle$ [109], $\circ$ [110], $\Diamond$ [115], $\square$ [116].

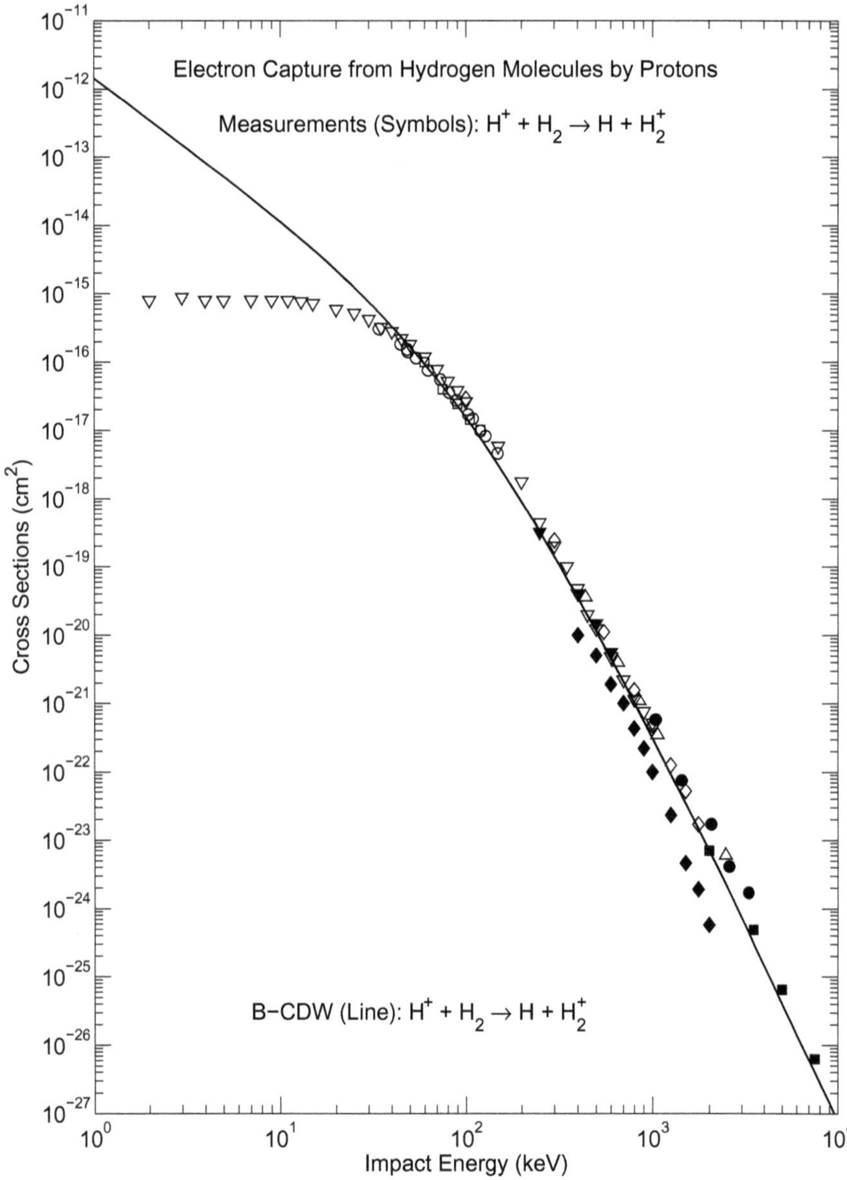

Fig. 6. Cross sections for electron capture by protons, H^+, from molecular hydrogen, H_2. Theoretical data: B-CDW method (full black line). Experimental data (open and filled symbols): ○ [118], ▽ [94], • [97], △ [121], ♦ [123], ◊ [102], ▼ [115], ■ [116].

by halving the former. By doubling the cross sections from the A-CDW method we have verified that the so deduced B-CDW method is also in good agreement with the experimental data for the $H^+ + N_2 \to H + N_2^+$ collisions. More details are given in Ref. [92] on this latter charge-exchange process, alongside several other applications of the B-CDW method to electron capture by protons and alpha particles from molecular targets of biomedical importance.

Figures 7 and 8 are of the type of Figs. 2 and 3 (neon) except that this time the target is atomic oxygen. In Fig. 7, the A-CDW method shows the dips in the cross sections for capture by protons from the target $1s$ and $2s$ orbitals. Capture from the $2p$ orbital of oxygen is a dominant process at 1-1000 keV. Above 1500 keV, capture from the $1s$ orbital yields the largest contribution to the total cross sections (summed over the K- and L-shells). It is this switch of dominance around 1500 keV (between the $2p$ and $1s$ orbitals) that causes the total cross section (K+L shells) to change the slope (the upper full line). Below 1000 keV and above 2000 keV, the lines for the $2p$ and $1s$ orbitals of oxygen, respectively, are close to the total cross sections for the K+L shell capture. As a matter of fact, the K-shell and K+L shell capture almost coincide above 4000 keV. It is also instructive, within the A-CDW method, to compare capture from the $2s$ and $2p$ orbitals of oxygen. Figure 7 shows that below 1000 keV, capture from the $2s$ orbital is negligible relative to that from the $2p$ orbital of oxygen. At higher energies, starting already with 2000 keV, the cross sections in the A-CDW method for capture from the $2p$ orbital fall off much faster than those for the $2s$ orbital.

As to the displayed measurements stemming from different studies, it is seen in Fig. 7 that at lower impact energies ($E \leq 100$) keV, experimental data due to Allison [94] and Gilbody *et al.* [109] are in a marked disagreement. Otherwise, all the other measured cross sections are in very good mutual accord within a huge energy range between 150 and 37700 keV. Acerbi *et al.* [125, 126] reported two sets of experimental data for single charge exchange processes $H^+ + N_2 \to H + N_2^+$, $H^+ + O_2 \to H + O_2^+$ and $H^+ + Ar \to H + Ar^+$. The first set was at 32500 and 37700 keV [125], whereas the second set [126] was at 24700, 27300, 32500 and 37700 keV. The preliminary measurement [125] is in agreement with the subsequent experiment [126]. Thus, it suffices to plot in Fig. 6 only the data of Acerbi *et al.* [126] at the mentioned four energies. These latter experimental data are perfectly reproduced by the A-CDW method. All the experimental data in Fig. 7 have been measured on molecular oxygen (O_2) as the target and

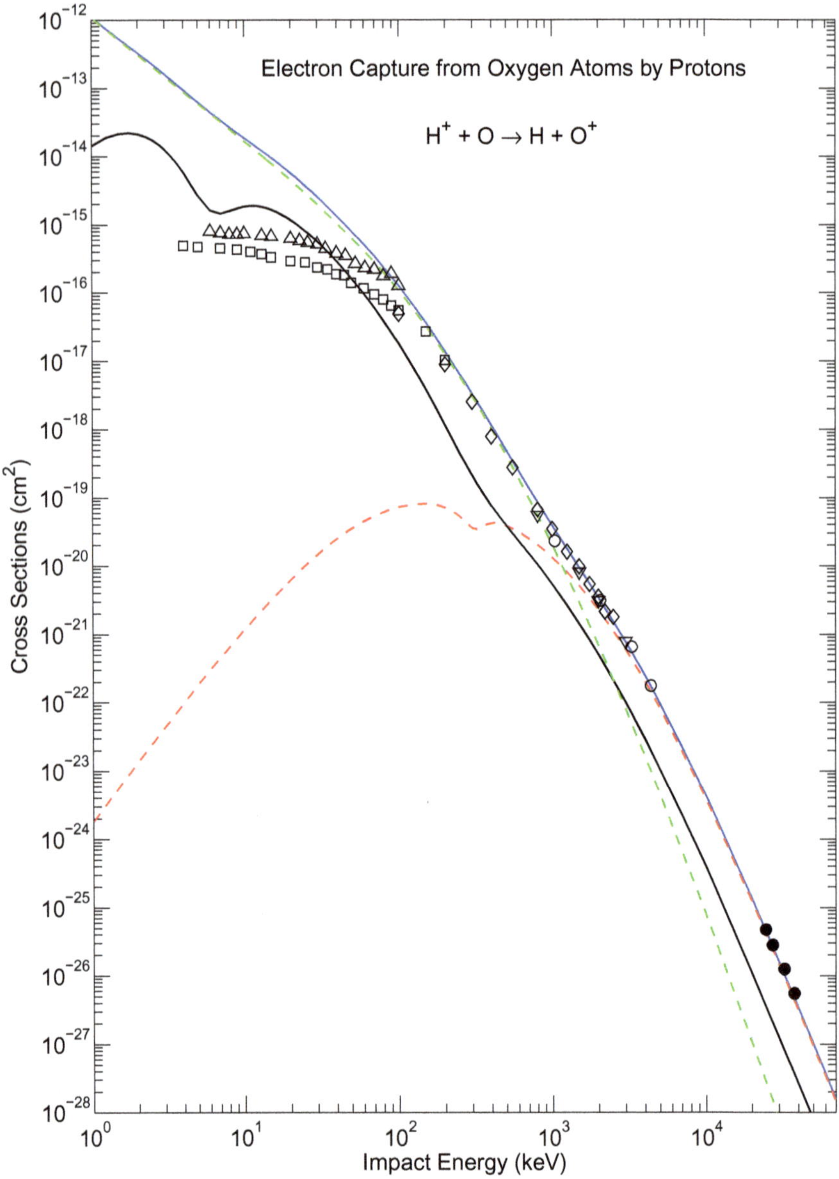

Fig. 7. Cross sections for electron capture by protons, H^+, from oxygen, O. Theoretical data, A-CDW method: dashed red line (1s orbital of O), full black line (2s orbital of O), dashed green line (2p orbital of O), blue full line (1s+2s+2p orbitals of O). Experimental data: □ [94], ◊ [102], ■ [126] ∘ [97], ▽ [98], △ [109].

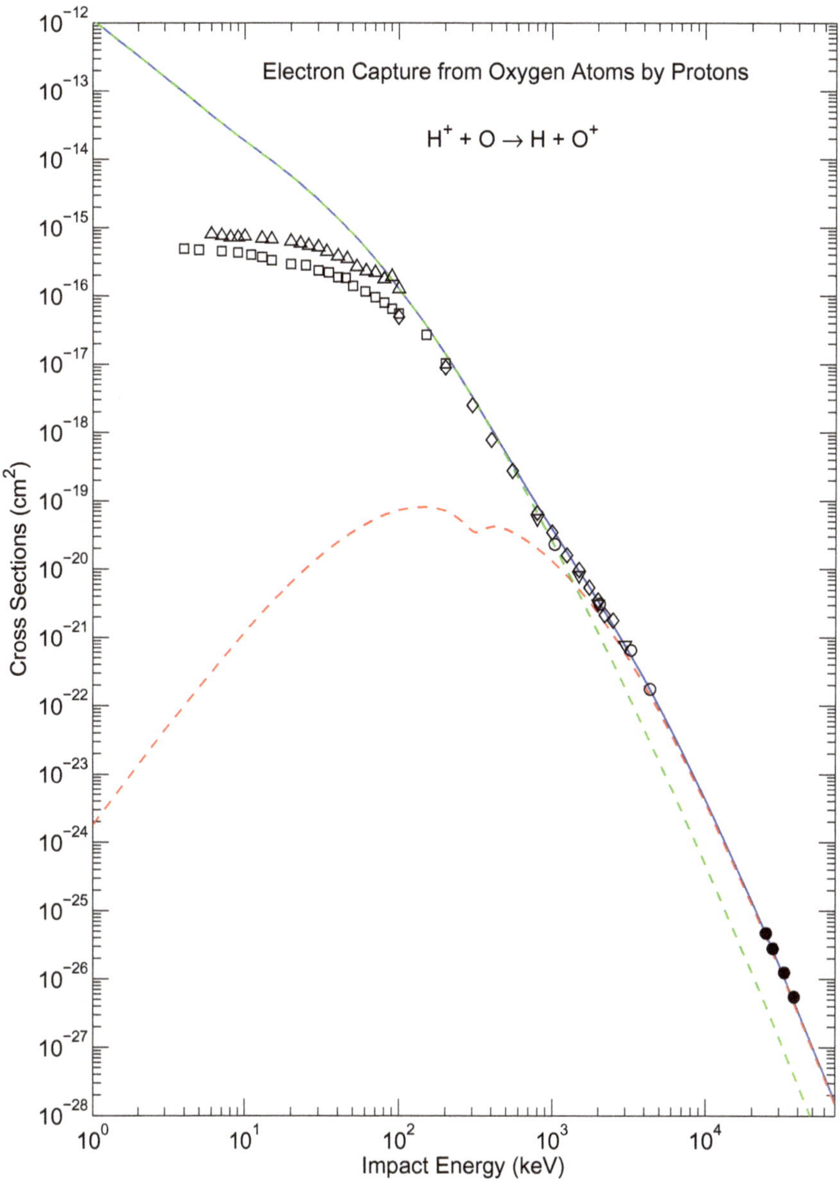

Fig. 8. Cross sections for electron capture by protons, H^+, from oxygen, O. Theoretical data, A-CDW method: dashed red line (K-shell of O), dashed green line (L-shell O), blue full line (K+L shells of O). Experimental data: □ [94], ◊ [102], ■ [126] ∘ [97], ▽ [98], △ [109].

are here plotted as the "cross sections per atom" (halved measured "cross sections per molecule"). Thus, as opposed to Fig. 6, the unaltered cross sections computed in the A-CDW method are depicted in Fig. 7, whereas this time it is the sets of the experimental data that are divided by 2.

Passing to Fig. 8, we show the separate K- and L-shell contributions to the total (K+L) capture process in the A-CDW method. It is observed that the K+L shell capture is exclusively determined by the L-shell capture below 500 keV. Conversely, above 4000 keV, the K-shell capture becomes practically the sole contributor to the total K+L shell capture cross sections in the $H^+ + O \to H + O^+$ collision.

3.5. *Molecular theories vs. measurements: both for protons on water*

In Fig. 9, we use the information from Figs. 6 (H_2) and 8 (O) through the Bragg additivity rule to obtain the total molecular cross sections in the B-CDW method (top full line) for the process $H^+ + H_2O \to H + H_2O^+$. At $E \geq 80$ keV, the B-CDW method is in excellent agreement with the corresponding molecular cross sections of Toburen *et al.* [102], Dagnac *et al.* [103] and Gobet *et al.* [104].

In this figure, also shown are the components of the additive total cross sections from the B-CDW method for the $H^+ + H_2O \to H + H_2O^+$ collision. These are the two independent charge-exchange processes: $H^+ + H_2 \to H + H_2^+$ and $H^+ + O \to H + O^+$. The latter component (oxygen, dashed line), as the heavier constituent, is seen to dominate at all the displayed energies (1-10000 keV). The former component (molecular hydrogen, bottom full line) is appreciable only below 100 keV. Above 100 keV, the composite process $H^+ + H_2O \to H + H_2O^+$ can be considered as being comprised solely of its purely atomic pathway where oxygen appears to be the only target via $H^+ + O \to H + O^+$, as if the H_2 part of the H_2O molecule were completely absent. This, in turn, would constitute a convincing justification of the hypothesis that a class of one-electron molecular wave functions of water with a single center placed on the heavier atom (oxygen, in H_2O^+) [83] could suffice for the OCE-CDW method which goes beyond the Bragg sum rule in the B-CDW method.

Figure 10 compares the molecular cross sections from the B-CDW method for molecular electron capture $H^+ + H_2O \to H + H_2O^+$ with the atomic cross sections from the A-CDW method for atomic charge-exchange $H^+ + Ne \to H + Ne^+$. As can be observed from this figure, the cross

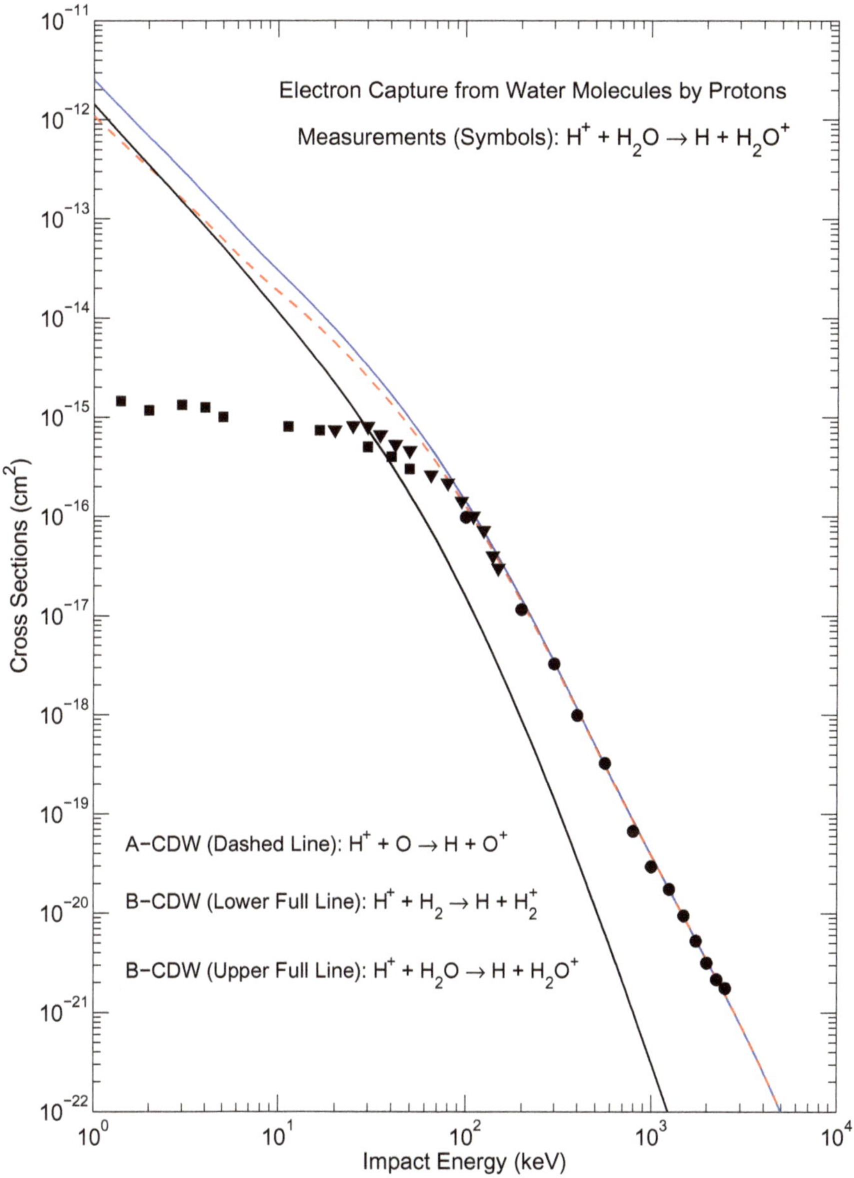

Fig. 9. Cross sections for electron capture by protons, H^+, from water vapor, H_2O. Theoretical data: A-CDW method (full black line, H_2 target), A-CDW method (dashed red line, K+L shells of O), B-CDW method (full blue line, K+L shells of H_2O). Experimental data for water target only: • [102], ■ [103], ▼ [106].

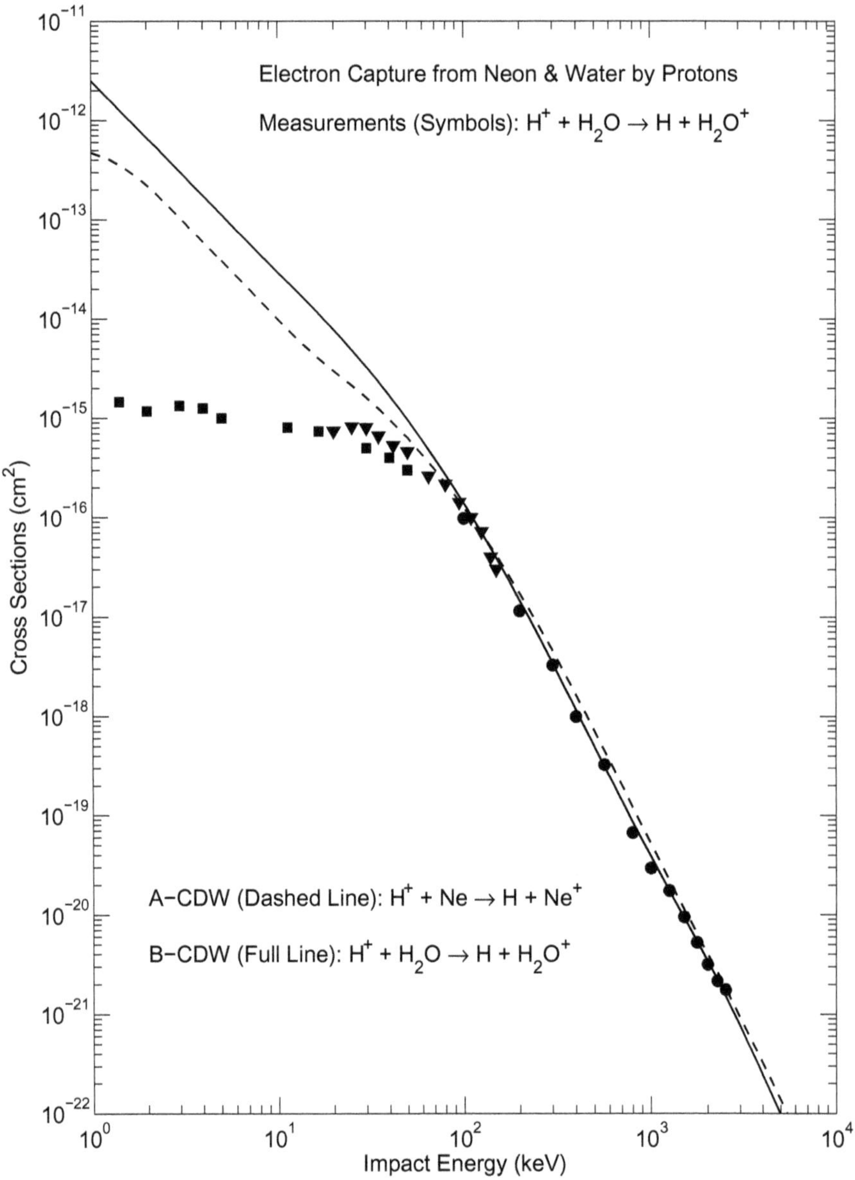

Fig. 10. Cross sections for electron capture by protons, H^+, from neon, Ne, and water vapor, H_2O. Theoretical data: A-CDW method (neon target), dashed line (K+L shells of Ne), B-CDW method (water target), full line (K+L shells of H_2O). Experimental data for water target only: • [102], ■ [103], ▼ [106].

sections from the B-CDW method for the $H^+ + H_2O \rightarrow H + H_2O^+$ collision are better aligned with most of the measured data on the H_2O target than the A-CDW method applied to electron capture by protons from neon, $H^+ + Ne \rightarrow H + Ne^+$.

Finally, in Fig. 11 for the $H^+ + H_2O \rightarrow H + H_2O^+$ collision, we depict the present cross sections from the B-CDW method and those from the PWB method [73] alongside the experimental data [102, 103, 106] for the same electron transfer process. In Ref. [73], the OCE molecular wave function of Moccia [83] has been used within the PWB method. Therein, it has been stated that the computations for the total cross sections were time-consuming due to multiple integrations (7 in total), including numerical quadratures over the Euler angles to average different orientations of water molecule target. This is in sharp contrast with the B-CDW method, whose programmed analytical T-matrix element provides extremely fast computations of total cross sections by only a single one-dimensional numerical integration over the transversal component of the momentum transfer [51].

In Fig. 11, the full and dashed lines are from the B-CDW and PWB methods, respectively. Here, the PWB method is seen to significantly overestimate the experimental data below 600 keV. Moreover, the PWB method exhibits an overly abrupt decrease of cross sections above 1000 keV, at variance with the measurements. This is opposed to a high degree of concordance above 70 keV between the B-CDW method and the measurements.

Overall, recapitulating Figs. 1-11, we can observe, by reference to the available experimental data, that both the A-CDW and B-CDW methods work very well at intermediate and high impact energies for electron capture by protons from the investigated atomic and molecular targets. Successful performance of the A-CDW method for many atomic targets different from those from the present work as well as for a number of projectiles heavier than protons has been established over the years in the abundant literature. The B-CDW method is also successful for electron capture by protons from molecular targets as shown in this chapter. We have explicitly checked that this also holds true for alpha particles impinging on a large number of molecules ranging from diatomic molecules, water, hydrocarbons, halides and DNA nucleobases [91, 92]. Moreover, the B-CDW (Bragg) and the OCE-CDW (Moccia) versions of the molecular CDW method have been shown in Refs. [91, 92] to yield quite similar total cross sections for single electron capture by nuclei from the mentioned variety of molecular targets. This makes the B-CDW method very useful for molecular collisions much as it is the case with the A-CDW method for atomic collisions.

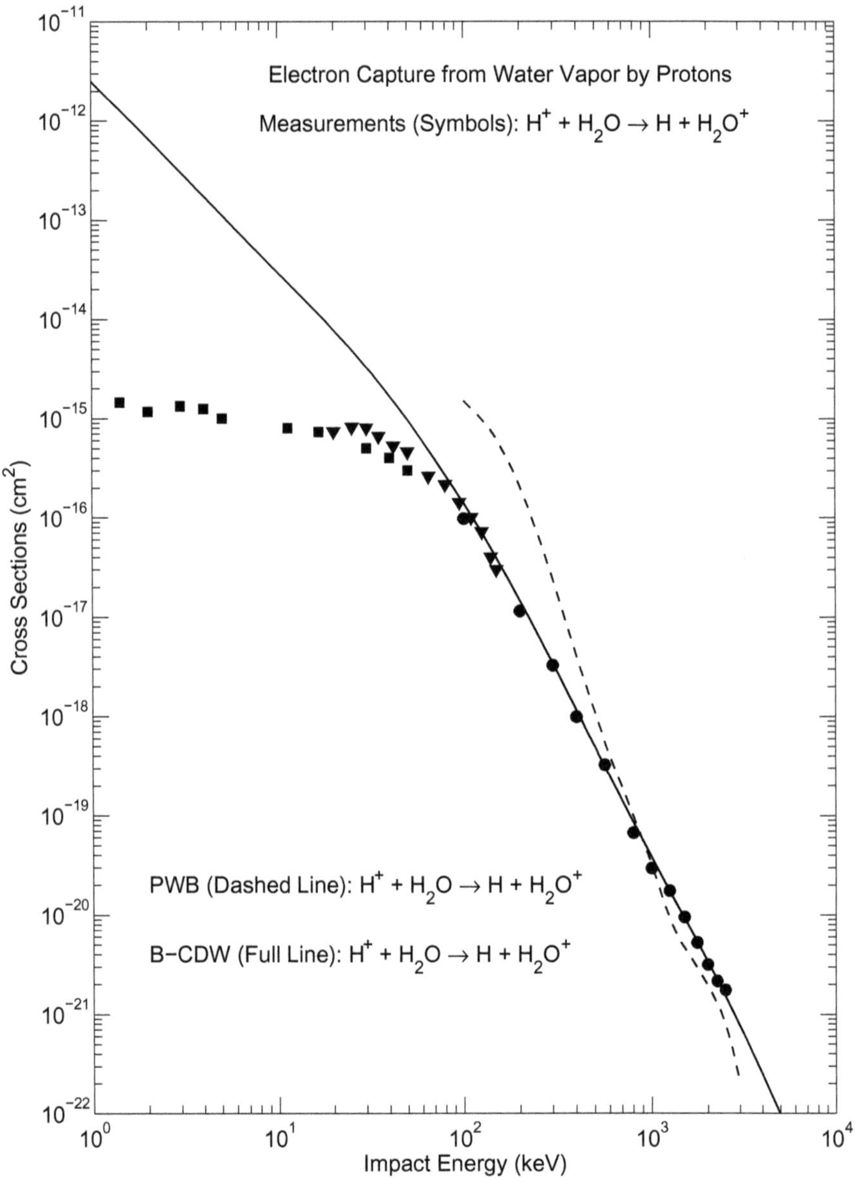

Fig. 11. Cross sections for electron capture by protons, H^+, from water vapor, H_2O. Theoretical data: the PWB method, dashed line (all the molecular orbitals of H_2O) [73], B-CDW method, full line (all the atomic orbitals, i.e. the K+L shells of H_2O). Experimental data for water target only: • [102], ■ [103], ▼ [106].

4. Discussion and conclusions

The present chapter reviews the distorted wave formalism of theory for fast ion-atom collisions relevant to hadron therapy. This framework is very efficient in providing total cross sections that are essential for describing energy loss of projectiles during their passage through tissue. Over decades, thorough testings of the computed cross sections from the leading quantum-mechanical distorted wave methods confirmed that they are in good agreement with the corresponding high-energy experimental data on atomic targets for electron capture and ionization.

Moreover, certain adequate extensions of these relatively straightforward yet powerful theoretical methods from atomic to molecular targets have been accomplished, as well. Such generalizations are also efficient in yielding total cross sections. This was made possible by way of the conventional STOs as the atomic basis functions within molecular orbitals for molecular targets.

We focus on quantum-mechanical *ab initio* methods from the first principles of physics, meaning that there are no free parameters whatsoever. This is distinctly advantageous relative to any of the existing phenomenological procedures (with no theoretical underpinning) based upon some *ad hoc* mathematical expressions. Such empirical recipes have no predictive power at all since they contain a number of fitting parameters that are adjusted specifically to the given set(s) of experimental data. In versatile applications (ranging from thermonuclear fusion to heavy ion radiation therapy), it is high time to make these fitting techniques completely obsolete by replacing them particularly with the boundary-condition-preserving quantum-mechanical *ab initio* methods of proven validity [20, 45, 48–50, 58].

Correct boundary conditions refer to the proper asymptotic behaviors of the scattering wave functions of the entire system (projectile + target), *and* their adequate connections to the corresponding perturbation potentials in the entrance and exit channels. Failure to satisfy these conditions can lead to utterly unphysical results. This has repeatedly occurred in the literature on electron transfer and ionization, as has been pointed out in Refs. [43, 45, 48, 50]. It is the physics of scattering events that imposes the mandatory fulfillment of these conditions so as to be able to distinguish the situations before and after the collision. The ultimate goal of research in particle scattering is to "peer" into the interaction region (small inter-aggregate separations) to learn what actually happened there by relying upon inferences from various observations made at asymptotically large distances. If proper care is not taken to fully separate these two domains with small and large distances (in both theory and experiment), there would be no justification in claiming that the observed effects are due to the events in the interaction region. Among several existing distorted wave theories with correct boundary conditions, we opted to analyze the continuum distorted wave method, CDW. It is a theoretically well-founded approximation, which is free from the so-called disconnected diagrams (associated with divergence of perturbation expansions). Crucially, the CDW method has the analytically calculable transition amplitudes that make the entire computations extremely expedient. This is of utmost practical importance, especially when extensively using the CDW method in ion transport physics for cross-disciplinary applications, including hadron therapy.

Among the irradiated targets, water molecules stand out as a tissue-equivalent substance of prime importance to hadron therapy. The reason is in the fact that water molecules are dominantly abundant constituents in most parts of the human body. In dose-effect considerations within hadron therapy, the irradiated critical target is widely considered to be DNA. Its double strand breaks can be lethal or can undergo miss-repair possibly leading to mutations. Irradiation can also yield secondary cancers in healthy, normal tissue. Risks for such radiation-induced cancers cannot be ruled out with certainty despite all the necessary precautions taken in the treatment of the patient.

The multi-faceted wealth of information from theory of atomic collisions should merge to hadron therapy through ion transport physics. In assessing the energy loss of ion beams, for the need of dose planning systems, the Bragg additivity rule is often used to deduce molecular cross sections through linear combinations of atomic cross sections. This is a procedure of considerable practical importance due to its simplicity and fast computations. Ion transport physics relies upon the stochastic [17, 18] and deterministic (based on e.g. the Fermi-Eyges function) [133–136] methods for modeling energy loss of primary beam particles (as well as secondaries, tertiaries, etc). Both methods use the collisional cross sections as the critical input data. Therefore, there would be no adequate ion transport physics without the most accurate cross section databases. Energy loss can be caused by electromagnetic and nuclear interactions of ions with tissue molecules. Geometry and composition of the targeted tissue molecules are too complex for the deterministic methods (Maxwell equation, Fokker-Planck equation, etc) to be of greater practical use in hadron therapy. For this reason, the stochastic methods are employed through Monte Carlo (MC) simulations by a number of ion transport codes (FLUKA, SHIELD-HIT, GEANT4, PHITS, MCNPX, etc) [18].

For hadron therapy of deep-seated tumors, very fast ion beams are necessary. For example, to reach the maximal distance or a range R of about 25 cm, e.g. carbon nuclei should have their impact energy of about 400 MeV/amu. Dose profiles for ions are characterized by the Bragg peak located at the range R at which most energy is deposited to the targeted site. Away from the Bragg peak, i.e. from the entrance to the patient's body to the tumor site, the ion dose profile becomes nearly flat. This latter plateau is due to a much smaller deposited energy compared to that in the immediate vicinity of the Bragg peak for ions. As such, ion beams, in fact, target mainly the tumor and largely spare the healthy tissue. In essence, ion dose profiles are diametrically opposite to those of electrons and photons. Ion beam paths or trajectories are almost straight lines because of heavy masses of projectiles. Much lighter electrons are scattered everywhere and, thus, have their trajectories literally all over the place due to their more frequent collisions with atoms and molecules of the medium. As a result, in non-conformal therapy, electron beams deposit their energy at every traversed distance, implying that healthy tissue is not spared from the unwanted radiation doses. In clinical practice, however, electron and photon therapies are used in a conformal setting which substantially improves dose profiles making them comparable to those of heavy ion beams.

Total cross sections (integrated over all the scattered angles) for atomic collisions are considered as a function of the impact energy of the projectiles. These cross sections attain their maximum values when the impact velocity is close to the velocity of the electron from the target orbit from which the active electron is captured/excited/ionized by the projectile. This velocity-matching resonance is known as the Massey peak. At impact energies larger than the location of the Massey peak height, total cross sections decrease monotonically. They rise steadily when approaching the Massey peak from lower impact energies. A mirror image of this pattern/profile/distribution is obtained by inverting the independent variable on the abscissa. Energy and distance are in the inverse proportionality relationship. To attain deeper distances, more energy is required. In the atomic units, a reciprocal energy is in the units of the inverse distance. This is used in atomic/molecular spectroscopy (especially within chemistry) where energy is often expressed in the units of cm^{-1}. Thus, plotting the total cross sections as a function of the reciprocal of the impact energy would give a dose profile reminiscent of the Bragg curve which represents the energy loss versus the traversed distance counted from the entrance to the patient's body all the way to the targeted tumor. Such an inversion of the abscissae (energy to distance) transparently pictures the one-to-one correspondence between the Massey and the Bragg peaks. This correspondence is deemed to be of help in relating the nomenclatures of atomic collision physics to those of ion transport physics.

At sufficiently high energies, ionizing collisions and neutron production yield the dominant contributions to electromagnetic and nuclear energy loss, respectively. Nuclear stopping powers needed for hadron therapy can readily be computed from the existing analytical expressions that are often phenomenologically established. Nevertheless, such empirical formulae are usually not employed in the conventional MC codes. Rather, the MC codes preferentially use the recommended tabular nuclear cross section data bases generated from a large set of mutually concordant measurements. It can then be stated with reasonable confidence that the nuclear cross section input data to MC simulations possess the needed adequacy as well as reliability.

On the other hand, the atomic/molecular ionization cross sections input data to MC codes are almost exclusively based upon the Bethe-Bloch formula for stopping power derived in the 1930s [137, 138]. There is no Bethe-Bloch formula for charge-exchange. Charge-exchange energy loss can be comparable to ionization energy loss especially around the Massey-Bragg peak [139]. In other words, reliable and more complete estimates for stopping powers must include the yield from electron capture processes. The Bethe-Bloch ionization formula is, in fact, a simplified high-energy limit of the corresponding non-relativistic first Born approximation for bare nuclei colliding with hydrogen-like atomic systems in their ground-states. An extension of this formula to relativistic impact velocities has been made only from the kinematical viewpoint. Some other corrections (the shell effect, molecular targets, density, etc) were also introduced phenomenologically.

However, in atomic collision physics, the first Born approximation itself has long been considered as unreliable for ionization around the Massey peak on the impact energy scale. The reason is in the fact that, from the onset, the first Born approximation neglects the two-center effect. This effect stems from the occurrence that the ionized electron moves in the field of two Coulomb centers due to the charges of the target and projectile nuclei [58]. The ensuing double continuum state generates a resonance for the coincident velocities of the projectile and emitted electron. The phenomenon is known as the electron capture to continuum, ECC, which appears as a strongly cusp-shaped line-shape in the energy distribution of the emitted electrons. The resulting resonantly enhanced cross section is a clear signature of an additional source of electron production in the traversed medium.

These supplementary electrons from the ECC mechanism, which is completely missing in the first Born and Bethe-Bloch methods, should also be included in the computation of electronic stopping powers. Because the scattered projectiles are fast, these additional electrons from the ECC phenomenon are energetic and, as such, many of them are the so-called δ−electrons, meaning that they are themselves capable of ionizing the tissue molecules. Due to their light mass and considerable energy, δ−electrons, both from the so-called binary encounters and the ECC source, can easily undergo multiple scattering. It is such a cumulative effect which produces the dominant 70% double strand breaks of DNA. The remaining 30% ionization yield of DNA is caused by the primary ion beam.

Heavy nuclei, as projectiles, are deflected mainly in the forward direction (through a fraction of milliradian) and, as such, do not directly exhibit multiple scattering effects. Nevertheless, they do induce multiple scattering indirectly through the electrons produced by the interactions of the projectiles with the targets. Moreover, besides light secondary particles (electrons), the primary beam can produce heavy secondary particles (nuclei). In addition to these latter secondaries, there could also be some heavy nuclei of higher-order generations (tertiaries, etc). Proton primaries produce other nuclei by fragmentation of the target nuclei alone. Heavier projectiles (e.g. nuclei of helium, lithium, carbon, oxygen, etc) can themselves be fragmented by nuclear transmutations in collisions with the tissue. Secondary nuclei are lighter than their primary counterparts. Therefore, scattering angles of secondary nuclei are larger than those of the primaries. For this reason, multiple scattering effects of the secondary nuclei, as well as of those from higher-order generations, need to be investigated in order to assess their influence on the nuclear stopping power. One of the ways to accomplish this is to use the Gaussian or Vavilov function [135, 136] to convolve the nuclear stopping power computed without the multiple scattering from any heavy nuclei.

The effects of the range and energy straggling are also important to take into account in computation of energy loss of heavy particles during their passage through tissue. These effects come from randomness of scattering phenomena. For example, for the given beam energy, the range R is not the same for all the projectile particles. This is the case because no realistic beam in any measurement is mono-energetic. A beam profile of heavy primaries is a Boltzmann distribution. Therefore, particles from different portions of the Boltzmann energy distribution of the primaries will have different ranges. Hence the range fluctuation or range straggling as if the particles "straggle" to reach their ranges. This effect too can be included in the description by convolving the stopping power without straggling by the Gaussian or Vavilov function [135, 136].

There are at least two major drawbacks with all the MC codes for hadron therapy (and beyond). They are exceedingly slow. Moreover, they are not yet updated for the state-of-the-art cross sections from atomic and molecular databases. Thus, ordinarily, for the given realistic geometry, e.g. SHIELD-HIT takes about ten hours to compute a single point on the energy loss curve (the Bragg curve) which, for completion, needs to be generated at a large number of distances traversed by an ion beam. The other MC codes perform similarly as to the central processing unit time, CPU. Such an exhaustive CPU time will likely be significantly shortened in the foreseeable future due to the ever improving upgrades in the computing power. One of the bottlenecks of all the MC codes is the procedure of sampling the events.

If the input data to MC codes are to be sampled from a given formula, all simulations would be time demanding. This can be improved considerably by first pre-computing cross sections from the available quantum-mechanical distorted wave methods, then storing the output findings in the memory, and finally using such tabular data as the inter-phase modules for the subsequent sampling in MC codes. This could mitigate the second mentioned drawback of MC codes by pre-computing cross sections from electromagnetic interactions and employing directly the resulting tables for sampling in the existing MC codes (as done with the nuclear cross section data bases that are currently based upon the recommended experimental data).

There is a both interesting and important relationship between total cross sections for atomic and the corresponding molecular targets. Here, the word "corresponding" refers to the atomic and molecular targets that have the same number of electrons (e.g. ten in the case of neon and water). This is presently illustrated by plotting together all the existing experimental data related to total cross sections for single electron transfer in the $H^+ + Ne$ and $H^+ + H_2O$ collisions. It is observed that at impact energies below and around the Massey peak, atomic and molecular cross sections differ considerably. However, they are quite close to each other after the Massey peak, i.e. at energies above 100 keV. Moreover, above 100 keV, these mutually concordant experimental data for the $H^+ + Ne$ and $H^+ + H_2O$ collisions are excellently reproduced by the associated theoretical cross sections from the purely atomic version of the continuum distorted wave method, A-CDW, for the $H^+ + Ne$ collisions.

This implies that at proton impact energies larger than or equal to 100 keV for one-electron capture from water vapor by protons, one can provisionally, as a first approximation, use the corresponding atomic cross sections from the A-CDW method for proton-neon collisions. Here, neon is viewed as the united atom limit of vapor-phase water molecule. Neon is also the united atom limit of some other ten-electron molecular species of great importance to radiological and auroral physics, e.g. methane (CH_4) and ammonia (NH_3) [140, 141]. For total charge-exchange cross sections, we found that ^{20}Ne is a good surrogate for CH_4 and NH_3 too [92].

Useful as this might appear, a still better description than treating water vapor as a neon-like molecule is provided by the simplest molecular variant of the continuum distorted wave theory based on the Bragg sum rule yielding the B-CDW method. We checked that the B-CDW method is also successful with projectiles heavier than protons and with heavier molecular targets e.g. hydrocarbons, halides as well as with the DNA and/or RNA nucleobases/components (uracil, adenine, guanine, thymine and cytosine) [92], where the acronym RNA stands for ribonucleic acid. Moreover, the B-CDW (Bragg's additivity) and OCE-CDW (Moccia's one-center molecular state) methods for proton-induced single-electron capture from DNA nucleobases have been found to predict very similar total cross sections [92].

This is of practical significance for hadron therapy at least for three reasons. First, a general program in the A-CDW approximation for capture cross sections from multi-electron atomic target is available [51]. It is fast due to programming the analytical expression for the transition amplitudes for any initial and final bound states. This would facilitate an efficient generation of databases for sampling in the MC simulations. Second, the A-CDW method has been established as being systematically reliable at intermediate and high impact energies (above 80 keV/amu). Therefore, these theoretical cross sections can be considered as the recommended atomic databases playing the role of its molecular counterpart by reliance upon the Bragg additivity rule in the B-CDW method. Third, too high cut-off impact energy (about 1000 keV/amu) is used in the Bethe-Bloch formula for MC simulations with some artificial fitting at lower energies towards the Massey peak. This should become obsolete with the cut-off impact energy lowered to 100 keV/amu by using the pre-computed and tabularly stored databases from the B-CDW method.

The present illustrations use the atomic and molecular CDW method for single electron capture alone. However, our conclusions in this chapter also hold true for ionization treated by the CDW method. Moreover, the same applies to the continuum distorted wave – eikonal initial state method, CDW-EIS, for both one-electron capture and single ionization. For example, we verified that ionization total cross sections in the CDW-EIS method are very similar for several ten-electron atomic and molecular targets, including neon and water, $H^+ + Ne \to H^+ + Ne^+ + e$ and $H^+ + H_2O \to H^+ + H_2^+O + e$, respectively.

Overall, the A-CDW method, which passed the test of time in purely atomic collisions at intermediate and high energies, is now conveniently emerging as a fully appropriate formalism for molecular collisions treated by the Bragg sum rule [3, 4] which, in turn, gives the B-CDW method. The results concordant with those reported in the present application of the B-CDW method also follow the formalisms with the explicit molecular orbital representations. These include the upgrades of the CDW method through the Gelius sum rule [59–61], the complete neglect of differential overlap [66] and the one-center expansion [80–84] (centered at the heaviest atom in a molecule) [91, 92]. Such an occurrence justifies the use of the B-CDW method for ion-molecule collisions when dealing with total cross sections (integrated over all the scattering angles).

In this review, we have not provided the mathematical expressions related to the continuum distorted wave theory. Its atomic version is well-known and the associated transition amplitudes are omnipresent [49–51]. Mathematical derivations in the extension of this approximation to molecular collisions are deferred to our two separate publications [91, 92]. By not overloading the presentation with formulae in this chapter, the room was opened to place the major focus onto the conceptual aspects of the addressed subject. Our goal was to show how, with a minimal effort, the door of ion-molecule collision theory could be made readily accessible to a number of atomic physics research groups that have not yet incorporated molecular targets into their current programs. This is a very important theme indeed, especially for applications in hadron therapy, which is definitely in need for expanding its unprecedented interdisciplinary character further to accommodate a fuller participation of the mainstream theoretical and experimental atomic/molecular collision physics. To this end, we tried to make the present exposition "as simple as it can be, but not simpler".

Acknowledgments

This work was supported by two Research Funds: Radiumhemmet via the Karolinska University Hospital and the Stockholm County Council (FoUU).

References

[1] R.R. Wilson, Radiological use of fast protons, *Radiology* **47**, 487 - 491 (1946).
[2] R.R. Wilson, Range, straggling and multiple scattering of fast protons, *Phys. Rev.* **71**, 385 - 386 (1947).
[3] W.H. Bragg, and R.D. Kleeman, On the ionization curves of radium, *Phil. Mag.* **8**, 726 - 738 (1904).
[4] W.H. Bragg, and R.D. Kleeman, On the α-particles of radium and their loss of range in their passing through various atoms and molecules, *Phil. Mag.* **10**, 318 - 340 (1905).
[5] A.M. Koehler, and W.M. Preston, Protons in radiation therapy. Comparative dose distributions for protons, photons and electrons, *Radiology* **104**, 191 - 195 (1972).
[6] A. Brown, and H. Suit, The century of the discovery of the Bragg peak, *Radiother. Oncol.* **23**, 265 - 268 (2004).
[7] C.A. Tobias, H.O. Anger, J.H. Lawrence, Radiological use of high energy deuterons and alpha particles, *Am. J. Roentg. Rad. Therapy & Nucl. Med.* **67**, 1 - 27 (1952).
[8] J.H. Lawrence, C.A. Tobias, J.L. Born, R. McCombs, J.L. Roberts, H.O. Anger, B.V. Low-Beer, and C. Huggins, Pituitary irradiation with high-energy proton beams: a preliminary report, *Cancer Res.* **18**, 121 - 134 (1958).
[9] D.E. Linstadt, J.R. Castro, and T.L. Phillips, Neon ion radiotherapy: Results of the phase I/II clinical trial, *Int. J. Radiat. Oncol. Biol. Phys.* **20**, 761 - 769 (1991).
[10] J.R. Castro, D.H. Char, P.L. Petti, I.K. Daftari, J.M. Quivey, R.P. Singh, E.A. Blakely, and T.L. Phillips, 15 Years experience with helium ion radiotherapy for uveal melanoma, *Int. J. Radiat. Oncol. Biol. Phys.* **39**, 989–996 (1997).
[11] B. Larsson, L. Leksell, B. Rexed, P. Sourander, W. Mair, and B. Andersson, The high-energy proton beam as a neurosurgical tool, *Nature* **182**, 1222 - 1223 (1958).
[12] B. Larsson, and B.A. Kihlman, Chromosome aberrations following irradiation with high-energy protons and their secondary radiation: A study of dose distributions and biological efficiency using root-tips of *vicia faba* and *allium cepa*, *Int. J. Rad. Biol.* **2**, 8 - 19 (1960).
[13] B. Larsson, Pre-therapeutic physical experiments with high energy protons, *Brit. J. Radiol.* **34**, 143 - 151 (1961).

[14] PTCOG, Particle therapy co-operative group; List of particle facilities in operation, https://www.ptcog.ch/index.php/facilities-in-operation (April 2019)

[15] T. Kanai, M. Endo, S. Minohara, N. Miyahara, H. Koyama-Ito, H. Tomura, N. Matsufuji, Y. Futami, A. Fukumura, T. Hiraoka, Y. Furusawa, K. Ando, M. Suzuki, F. Soga, and K. Kanachi, Biophysical characteristics of HIMAC clinical irradiation system for heavy-ion radiation therapy, *Int. J. Radiat. Oncol. Biol. Phys.* **44**, 201 – 210 (1999).

[16] G. Kraft, Tumor therapy with heavy charged particles, *Prog. Part. Nucl. Phys.* **45**, S473 – S544 (2000).

[17] Dž. Belkić, SHIELD-HIT: An optimal Monte Carlo code for simulation of transport of protons and heavier ions in tissue, In: *Proc. Int. Conf. Comp. Math. Meth. Sci. Eng.*, E. Brändas, J. Vigo-Aguiar (eds.), pp. 10–16 (Uppsala, Sweden, June 4-8, 2004).

[18] I. Gudowska, N. Sobolevsky, P. Andreo, Dž. Belkić, and A. Brahme, Ion beam transport in tissue-like media using the Monte Carlo code SHIELD-HIT, *Phys. Med. Biol.* **49**, 1933 – 1958 (2004).

[19] M. Torikoshi, Heavy-ion cancer therapy, *Laser Phys.* **16**, 654 – 659 (2006).

[20] Dž. Belkić, Review of theories on ionization in fast ion-atom collisions with prospects for applications to hadron therapy, *J. Math. Chem.* **47**, 1366 – 1419 (2010).

[21] D. Schardt, T. Elsässer, and D. Schulz-Ertner, Heavy-ion tumor therapy: physical and radiobiological benefits, *Rev. Mod. Phys.* **82**, 383 – 425 (2010).

[22] D.K. Ebner, and T. Kamada, The emerging role of carbon-ion radiotherapy, *Front. Oncol.* **6**, Art. ID 140 (2016).

[23] O. Sokol, E. Scifoni, W. Tianganelli, W. Kraft-Weyrather, J. Wiedemann, A. Maier, D. Boscolo, T. Friedrich, S. Brons, M. Durante, and M. Krämer, Oxygen beam for therapy: advanced biological treatment planning and experimental verification, *Phys. Med. Biol.* **62**, 7798 – 7813 (2017).

[24] M. Rovituso, and C. La Tessa, Nuclear interactions of new ions in cancer therapy: impact on dosimetry, *Transl. Cancer. Res.* **6**, S914 – S933 (2017).

[25] U. Schneider, and R. Hälg, The impact of neutrons in clinical proton therapy, *Front. Oncol.* **5**, Art. ID 235 (2015).

[26] M.A. Chaudhri, Production of secondary neutrons from tissue during radiotherapy, *Sixth Int. Congr. Rad. Oncol.* 30 January - 2 February 2001, Melbourne, Australia, Abstracts in the Supplement to: *Radiother. Oncol.* **58**, S22 (2001).

[27] M.A. Chaudhri, Production and potential implications of secondary neutrons within patients undergoing therapy with hadrons, *Cyclotrons and their Applications, Sixteenth Int. Conf.*, East Lansing, Michigan, 13-17 May 2001, Series: AIP Conf. Proc. Subseries: Accelerators, Beams and Instrumentation, F. Marti (ed.), **600**, 49 – 51 (2001).

[28] E.J. Hall, Intensity-modulated radiation therapy, protons, and the risk of second cancers, *Int. J. Radiat. Oncol. Biol. Phys.* **65**, 1 – 7 (2006).

[29] U. Schneider, A. Lomax, P. Pemler, J. Besserer, D. Ross, N. Lombriser, and B. Kaser-Hotz, The impact of IMRT and proton radiotherapy on secondary cancer incidence, *Strahlenther. Onkol.* **182**, 647 – 652 (2006).
[30] U. Schneider, A. Lomax, and N. Lombriser, Comparative risk assessment of secondary cancer incidence after treatment of Hodgkin's disease with photon and proton radiation, *Radiat. Res.* **154**, 382 – 388 (2000).
[31] C.S. Chung, T.I. Yock, K. Nelson, Y. Xu, N.L. Keating, and N.J. Tarbell, Incidence of second malignancies among patients treated with proton versus photon radiation, *Int. J. Radiat. Oncol. Biol. Phys.* **87**, 46 – 52 (2013).
[32] M. Krämer, O. Jäkel, G. Kraft, D. Schardt, and U. Weber, Treatment planning for heavy-ion radiotherapy: physical beam model and dose optimization, *Phys. Med. Biol.* **45**, 3299 – 3317 (2000).
[33] M. Astrahan, Some implications of linear-quadratic-linear radiation dose-response with regard to hypofractionation, *Med. Phys.* **35**, 4161 – 4172 (2008).
[34] Dž. Belkić, Survival of radiation-damaged cells via mechanism of repair by pool molecules: the Lambert function as the exact analytical solution of coupled kinetic equations, *J. Math. Chem.* **52**, 1201 – 1252 (2014).
[35] Dž. Belkić, Repair of irradiated cells by Michaelis-Menten enzyme catalysis: the Lambert function for integrated rate equations in description of surviving fractions, *J. Math. Chem.* **52**, 1253 – 1291 (2014).
[36] Dž. Belkić, and K. Belkić, Mechanistic radiobiological models for radiation damage, *Adv. Quantum Chem.* **70**, 163 – 263 (2015).
[37] Dž. Belkić, The Euler T and Lambert W functions in mechanistic radiobiological models with chemical kinetics for repair of irradiated cells, *J. Math. Chem.* **56**, 2133 – 2193 (2018).
[38] L. Bodgi, A. Canet, L. Pujo-Menjouet, A. Lesne, J.-M. Victor, and N. Foray, Mathematical models of radiation action on living cells: from the target theory to the modern approaches. A historical and critical review, *J. Theor. Biol.* **394**, 93 – 101 (2016).
[39] Dž. Belkić, *Quantum-Mechanical Signal Processing and Spectral Analysis* (Institute of Physics Publishing, Bristol, 2005).
[40] Dž. Belkić, and K. Belkić, *Signal Processing in Magnetic Resonance Spectroscopy with Biomedical Applications* (Taylor & Francis Group via CRC Press, New York, 2010).
[41] Dž. Belkić, and K. Belkić (eds.), *Magnetic Resonance Imaging and Spectroscopy*, Vol. **3**, *Comprehensive Biomedical Physics* (Elsevier, Amsterdam, 2014).
[42] Dž. Belkić, and K. Belkić, Review of recent applications of the conventional and derivative fast Padé transform for magnetic resonance spectroscopy, *J. Math. Chem.* **57**, 385 – 464 (2019).
[43] Dž. Belkić, *Principles of Quantum Scattering Theory* (Institute of Physics Publishing, Bristol, 2004).
[44] Dž. Belkić, *Quantum Theory of High-Energy Ion-Atom Collisions* (Taylor & Francis Group via CRC Press, New York, 2008).

[45] Dž. Belkić, I. Mančev, and J. Hanssen, Four-body methods for high-energy ion-atom collisions, *Rev. Mod. Phys.* **80**, 249 – 314 (2008).

[46] Dž. Belkić (ed.), *Fast Ion-Atom and Ion-Molecule Collisions*, In: *Interdisc. Res. Particle Coll. Quantitat. Spectrosc.* Vol. **1** (World Scientific Publishing, Singapore, 2013).

[47] Dž. Belkić (ed.), *Theory of Heavy Ion Collision Physics in Hadron Therapy*, In: *Advances in Quantum Chemistry*, Vol. **65** (Elsevier, Amsterdam, 2013).

[48] Dž. Belkić, Double charge exchange in ion-atom collisions using distorted wave theories with two-electron continuum intermediate states in one or both scattering channels, *J. Math. Chem.* **57**, 1 – 58 (2019).

[49] I. Cheshire, Continuum distorted wave approximation. Resonant charge transfer by fast protons in atomic hydrogen, *Proc. Phys. Soc.* **84**, 89 – 98 (1964).

[50] Dž. Belkić, R. Gayet, and A. Salin, Electron capture in high-energy ion-atom collisions, *Phys. Rep.* **56**, 279 – 369 (1979).

[51] Dž. Belkić, R. Gayet, and A. Salin, Computation of total cross sections for electron capture in high energy collisions: III, *Comput. Phys. Commun.* **32**, 385 – 397 (1984).

[52] E. Clementi, and C. Roetti, Roothaan-Hartree-Fock atomic wavefunctions: basis functions and their coefficients for ground and certain excited states of neutral and ionized atoms: $Z \leq 54$, *At. Data Nucl. Data Tables* **14**, 177 – 478 (1974).

[53] N.F. Mott, and H.S.W. Massey, *The Theory of Atomic Collisions* (Oxford University Press, Oxford, 1965).

[54] D. Raj, A note on the use of the additivity rule for electron-molecule elastic scattering, *Phys. Lett. A* **160**, 571 – 574 (1991).

[55] Y. Jiang, J. Sun, and L. Wan, Additivity rule for the calculation of electron scattering from polyatomic molecules, *Phys. Rev. A* **62**, 062712 (2000).

[56] P. Możejko, B. Żywicka-Możejko, and Cz. Szmytkowski, Elastic cross section calculations for electron collisions with XY_4(X = Si, Ge; Y=H,F,Cl,Br,I) molecules, *Nucl. Instr. Methods. Phys. Res. B* **196**, 245 – 252 (2002).

[57] P. Możejko, and L. Sanche, Cross section calculations for electron scattering from DNA and RNA bases, *Radiat. Environ. Biophys.* **42**, 201 – 211 (2003).

[58] Dž. Belkić, A quantum theory of ionization in fast collisions between ions and atomic systems, *J. Phys. B* **11**, 3529 – 3552 (1978).

[59] G. Gelius, C.J. Allan, G. Johansson, H. Siegbahn, D.A. Allison, and K. Siegbahn, The ESCA spectra of benzene and the isoelectronic series, thiophene, pyrrole and furan, *Phys. Scr.* **3**, 237 – 242 (1971).

[60] U. Gelius, Molecular spectroscopy by means of ESCA: experimental and theoretical studies (PhD Thesis), *Acta Universitatis Upsaliensis* (University of Uppsala, Uppsala, Sweden, 1973).

[61] U. Gelius, Recent progress in ESCA studies of gases, *J. Electr. Spectr. Related Phenom.* **5**, 985 – 1057 (1974).

[62] R.S. Mulliken, Electronic population analysis on LCAO-MO molecular wave functions: I, *J. Chem. Phys.* **23**, 1833 – 1840 (1955).

[63] R.S. Mulliken, Electronic Population Analysis on LCAO-MO Molecular Wave Functions: II; Overlap populations, bond orders and covalent bond energies, *J. Chem. Phys.* **23**, 1841 – 1846 (1955).
[64] R.S. Mulliken, Electronic Population Analysis on LCAO-MO Molecular Wave Functions: III; Effect of hybridization on overlap and gross atomic AO populations, *J. Chem. Phys.* **23**, 2338 – 2342 (1955).
[65] R.S. Mulliken, Electronic Population Analysis on LCAO-MO Molecular Wave Functions: IV; Bonding and antibonding in LCAO and valence-bond theories, *J. Chem. Phys.* **23**, 2343 – 2346 (1955).
[66] K. Siegbahn, C. Nodling, G. Johansson, H. Hedman, P.F. Hedén, K. Hamrin, U. Gelius, T. Bergmark, L.O. Werme, R. Manne, Y. Baer, *ESCA Applied to Free Molecules* (North-Holland, Amsterdam, 1969).
[67] M.J. Frisch, G.W. Trucks, H.B. Schlegel, G.E. Scuseria, M.A. Robb, J.R. Cheeseman, G. Scalmani, V. Barone, B. Mennucci, G.A. Petersson, H. Nakatsuji, M. Caricato, X. Li, H.P. Hratchian, A.F. Izmaylov, J. Bloino, G. Zheng, J.L. Sonnenberg, M. Hada, M. Ehara, K. Toyota, R. Fukuda, J. Hasegawa, M. Ishida, T. Nakajima, Y. Honda, O. Kitao, H. Nakai, T. Vreven, J.A. Montgomery, Jr., J. E. Peralta, F. Ogliaro, M. Bearpark, J.J. Heyd, E. Brothers, K.N. Kudin, V.N. Staroverov, R. Kobayashi, J. Normand, K. Raghavachari, A. Rendell, J.C. Burant, S.S. Iyengar, J. Tomasi, M. Cossi, N. Rega, J. M. Millam, M. Klene, J.E. Knox, J.B. Cross, V. Bakken, C. Adamo, J. Jaramillo, R. Gomperts, R.E. Stratmann, O. Yazyev, A.J. Austin, R. Cammi, C. Pomelli, J.W. Ochterski, R.L. Martin, K. Morokuma, V.G. Zakrzewski, G.A. Voth, P. Salvador, J.J. Dannenberg, S. Dapprich, A.D. Daniels, O. Farkas, J.B. Foresman, J.V. Ortiz, J. Cioslowski, and D.J. Fox, Gaussian 09, Revision A.02, *Gaussian, Inc.* (Wallingford CT, 2009).
[68] Ph. Bernhardt, and H.G. Paretzke, Calculation of electron impact ionisation cross sections of DNA using the Deutsch-Märk and binary-encounter-Bethe formalism, *Int. J. Mass. Spectrom.* **223/224**, 599 – 611 (2003).
[69] M.E. Galassi, C. Champion, P.F. Weck, R.D. Rivarola, O.A. Fojón, and J. Hanssen, Quantum-mechanical predictions of DNA and RNA ionization by energetic proton beams, *Phys. Med. Biol.* **57**, 2081 – 2099 (2012).
[70] B. Senger, and R.V. Rechenmann, Angular and energy distributions of δ–rays ejected from low–Z molecular targets by incident protons and α particles, *Nucl. Instr. Meth. Phys. Res. B* **2**, 204 – 207 (1984).
[71] B. Senger, Calculated molecular double-differential cross-sections for ionization under proton impact, *Z. Phys. D* **9**, 79 – 89 (1988).
[72] G.H. Olivera, A.E. Martinez, R.D. Rivarola, and P.D. Fainstein, Theoretical calculation of electronic stopping power of water vapor by proton impact, *Radiat. Phys.* **144**, 241 – 247 (1995).
[73] S. Houamer, Yu.V. Popov, C. Dal Cappello, and C. Champion, Charge transfer in collisions of protons with water molecule at high projectile energies and small scattering angles, *Nucl. Instr. Meth. Phys. Res. B* **267**, 802 – 806 (2009).

[74] C. Champion, H. Lekadir, M.E. Galassi, O.A. Fojón, R.D. Rivarola, and J. Hanssen, Theoretical predictions for ionization cross sections of DNA nucleobases impacted by light ions, *Phys. Med. Biol.* **55**, 6053–6067 (2010).

[75] M.E. Galassi, P.N. Abufager, P.D. Fainstein, and R.D. Rivarola, Single-electron capture in collisions of protons with molecules of biological interest, *Phys. Rev. A* **81**, Art. ID 062713 (2010).

[76] C. Champion, P.F. Weck, H. Lekadir, M.E. Galassi, O.A. Fojón, B. Abufager, R.D. Rivarola, and J. Hanssen, Proton-induced single electron capture on DNA/RNA bases, *Phys. Med. Biol.* **57**, 3039 – 3049 (2012).

[77] J.M. Monti, C.A. Tachino, J. Hanssen, O.A. Fojón, M.E. Galassi, C. Champion, and R.D. Rivarola, Distorted wave calculations for electron loss process induced by bare ion impact on biological targets, *Appl. Radiat. Isot.* **83**, 105 – 108 (2014).

[78] L. Gulyás, S. Egri, H, Ghavaminia, and A. Igarashi, Single and multiple electron removal and fragmentation in collisions of protons with water molecules, *Phys. Rev. A* **93**, Art. ID 032704 (2016).

[79] M.A. Quinto, P.R. Montenegro, J.M. Monti, O.A. Fojón, and R.D. Rivarola Electron capture by swift ions from molecules of biological interest, *J. Phys. B* **51**, Art. ID 165201 (2018).

[80] R. Moccia, One center expansion self-consistent field molecular orbital electronic functions for XH_n molecules, *J. Chem. Phys.* (Letter to the Editor) **37**, 910 – 911 (1962).

[81] R. Moccia, One center basis set SCF MO's: I; HF, CF_4 and SiH4, *J. Chem. Phys.* **40**, 2164 – 2176 (1964).

[82] R. Moccia, One center basis set SCF MO's: II; NH_3, NH_4^+, PH_3, PH_4^+, *J. Chem. Phys.* **40**, 2176 – 2185 (1964).

[83] R. Moccia, One center basis set SCF MO's: III; H_2O, H_2S and HCl, *J. Chem. Phys.* **40**, 2186 – 2192 (1964).

[84] R. Moccia, Optimization of the basis functions in SCF MO calculations: optimized one-center SCF MO basis set for HCl, *Theoret. Chim. Acta* (Berlin) **8**, 8 – 17 (1967).

[85] C. Illescas, L.F. Errea, L. Méndez, B. Pons, I. Rabadán, and A. Riera, Classical treatment of ion – H_2O collisions with a three-center model potential, *Phys. Rev. A* **83**, Art. ID 052704 (2011).

[86] P.M. Gabás, L.F. Errea, C. Illescas, L. Méndez, and I. Rabadán, Ab initio treatment of charge transfer in ion-molecule collisions based on one-electron wave functions, *Phys. Rev. A* **85**, Art. ID 012702 (2012).

[87] L.F. Errea, C. Illescas, L. Méndez, and I. Rabadán, Ionization of water molecules by proton impact: Two perturbative studies of the electron-emission spectra, *Phys. Rev. A* **87**, Art. ID 032709 (2013).

[88] M. Murakami, T. Kirchner, M. Horbatsch, and H.J. Lüdde, Single an multiple electron removal processes in proton-water-molecule collisions, *Phys. Rev. A* **85**, Art. ID 052704 (2012).

[89] M. Murakami, T. Kirchner, M. Horbatsch, and H.J. Lüdde, Quantum-mechanical calculation of multiple electron removal and fragmentation cross sections in He^+ – H_2O collisions, *Phys. Rev. A* **86**, Art. ID 022719 (2012).

[90] H. Luna, W. Wolff, E.C. Montenegro, A.C. Tavares, H.J. Lüdde, G. Schenk, M. Horbatsch, and T. Kirchner, Ionization and electron capture cross sections for single- and multiple-electron removal from H_2O by Li^{3+} impact, *Phys. Rev.* A **93**, Art. ID 052705 (2016).
[91] Dž. Belkić, Single electron capture by fast protons from water vapor, *J. Math. Chem.* **57** (in press 2019).
[92] Dž. Belkić, Single electron capture by fast protons from diatomic and hydrocarbon molecules, *J. Math. Chem.* **57** (in press 2019).
[93] P.M. Stier, and C.F. Barnett, Charge exchange cross sections of hydrogen ions in gases, *Phys. Rev.* **103**, 896 – 907 (1956).
[94] S.K. Allison, Experimental results on charge-changing collisions of hydrogen and helium atoms and ions at kinetic energies above 0.2 keV, *Rev. Mod. Phys.* **30**, 1137 – 1168 (1958).
[95] J.F. Williams, and D.N.F. Dunbar, Charge exchange and dissociation cross sections for H_1^+, H_2^+ and H_3^+ ions of 2 to 50 keV energy incident upon hydrogen and the inert gases, *Phys. Rev.* **149**, 62 – 69 (1966).
[96] F.J. de Heer, J. Schutten, and H. Moustafa, Ionization and electron capture cross sections for protons incident on noble and diatomic gases between 10 and 140 keV, *Physica* **32**, 1766 – 1792 (1966).
[97] U. Schryber, Elektroneneinfang schneller protonen in gasen, *Helv. Phys. Acta* **40**, 1023 – 1051 (1967).
[98] S.L. Varghese, G. Bissinger, J.M. Joyce, and R. Laubert, Atomic total electron-capture from C-, O-, F- and S-bearing molecule gases for MeV/u H^+ and He^+ projectile, *Phys. Rev. A* **31**, 2202 – 2209 (1985).
[99] D.P. Almeida, N.V. de Castro Faria, F.L. Freire, E.C. Montenegro, and A.G. de Pinho, Collisional formation and destruction of fast negative hydrogen ions in He, Ne and Ar targets, *Phys. Rev. A* **36**, 16 – 36 (1987).
[100] V.V. Afrosimov, R.N. Il'in, and E.S. Solov'ev, Electron capture by protons in inert gases, *Sov. Phys. - Techn. Phys.* **5**, 661 – 666 (1960).
[101] V.S. Nikolaev, I.S. Dmitriev, Ya.A. Teplova, and Yu.A. Fainberg, Charge-exchange cross sections of protons, *Vestnik Moskovskoga Universiteta*, Seriya 3, *Fizika, Astronomiya* **35**, 84 – 85 (1994).
[102] L.H. Toburen, M.Y. Nakai, and R.A. Langley, Measurements of high-energy charge-transfer cross sections for incident protons and atomic hydrogen in various gases, *Phys. Rev.* **171**, 114 – 171 (1968).
[103] R. Dagnac, D. Blanc, and D. Molina, A study on the collision of hydrogen ions H_1^+, H_2^+ and H_3^+ with a water-vapour target, *J. Phys. B* **3**, 1239 – 1251 (1970).
[104] F. Gobet, B. Farizon, M. Farizon, M.J. Gaillard, M. Carré, M. Lezius, P. Scheier, and T.D. Märk, Total, partial and electron-capture cross sections for ionization of water vapor by 20-150 keV protons, *Phys. Rev. Lett.* **86**, 3751 – 3754 (2001).
[105] F. Gobet, S. Eden, B. Coupier, J. Tabet, B. Farizon, M. Farizon, M.J. Gailllard, S. Ouaskit, M. Carré, T.D. Märk, and P. Scheier, Ionization of water by (20-150)-keV protons: separation of direct-ionization and electron-capture processes, *Phys. Rev. A* **70**, Art. ID 062716 (2004).

[106] F. Gobet, S. Eden, B. Coupier, J. Tabet, B. Farizon, M. Farizon, M.J. Gailllard, S. Ouaskit, M. Carré, and T.D. Märk, Electron-loss and target ionization cross sections for water vapor by 20-150 keV neutral atomic hydrogen, *Chem. Phys. Lett.* **421**, 68 – 71 (2006).

[107] M.E. Rudd, T.V. Goffe, R.D. DuBois, and L.H. Toburen, Cross sections for ionization of water vapor by 7-4000 keV protons, *Phys. Rev. A* **31**, 492–494 (1985).

[108] W.L. Fite, R.F. Stebbings, D.G. Hummer, and R.T. Brackmann, Ionization and charge transfer in proton-hydrogen atom collisions, *Phys. Rev.* **119**, 663 – 668 (1960).

[109] H.B. Gilbody, and G. Ryding, An experimental study of charge transfer in proton-atomic hydrogen collisions at impact energies above 40 keV, *Proc. Roy. Soc. A* **291**, 438 – 443 (1966).

[110] A.B. Wittkower, G. Ryding, and H.B. Gilbody, An experimental study of charge transfer in proton-hydrogen collisions using furnace target method, *Proc. Phys. Soc.* **89**, 541 – 546 (1966).

[111] G.W. McClure, Electron transfer in proton-hydrogen atom collisions: 2-117 keV, *Phys. Rev.* **148**, 47 – 54 (1966).

[112] G.W. McClure, Ionization and electron transfer in collisions of two hydrogen atoms: 1.25-117 keV, *Phys.Rev.* **166**, 22 – 29 (1968).

[113] J.E. Bayfield, Electron capture into the metastable 2s state in collisions of protons with hydrogen atoms, *Phys. Rev. Lett.* **20**, 1223 – 1226 (1968)

[114] J.E. Bayfield, Measurement of the total cross section for charge transfer into the metastable state H($2s$) for proton collisions with atomic hydrogen, *Phys. Rev.* **185**, 105 – 112 (1969).

[115] P. Hvelplund, and A. Andersen, Electron capture by fast H^+, He^+ and He^{++} ions in collisions with atomic and molecular hydrogen, *Phys. Scr.* **26**, 375 – 380 (1982).

[116] W. Schwab, G.B. Baptista, E. Justiniano, R. Schuch, H. Vogt, and E.W. Weber, Measurement of the total cross sections for electron capture of 2.0-7.5 MeV H^+ in H, H_2 and He, *J. Phys. B* **20**, 2825 – 2834 (1987).

[117] J.P. Keene, Ionization and charge exchange by fast ions of hydrogen and helium, *Phil. Mag.* **40**, 369 – 385 (1949).

[118] F.L. Ribe, Electron-capture cross sections for protons passing through hydrogen gas, *Phys. Rev.* **83**, 1217 – 1225 (1951).

[119] C.F. Barnett, and H.K. Reynolds, Charge exchange cross sections of hydrogen particles in gases at high energies, *Phys. Rev.* **109**, 355–359 (1958).

[120] U. Schryber, Elektroneneinfang schneller protonen in gasen, *Helv. Phys. Acta* **32**, 562 – 564 (1966).

[121] L.M. Welsh, K.H. Berkner, S.N. Kaplan, and R.V. Pyle, Cross sections for electron capture by fast protons in H_2, He, N_2 and Ar, *Phys. Rev.* **158** 85 – 92 (1967).

[122] J.F. Williams, Single-electron capture and loss cross sections for 2-50 keV hydrogen atoms incident upon hydrogen and inert gases, *Phys. Rev.* **153**, 116 – 120 (1967).

[123] J.F. Williams, Measurements of charge-transfer cross sections for 0.25 to 2.5 MeV protons and hydrogen atoms incident upon hydrogen and helium gases, *Phys. Rev.* **157**, 97 – 100 (1967).

[124] J.M. Sanders, S.L. Varghese, C.H. Fleming, and G.A. Soosai, Electron capture by protons and electron loss from hydrogen atoms in collisions with hydrocarbons and hydrogen molecules in the 60-120 keV energy range, *J. Phys. B* **36**, 3835 – 3846 (2003).

[125] E. Acerbi, M. Castiglioni, G. Dutto, F. Resmini, C. Succi, and G. Tagliaferri, Electron capture by 33 and 38 MeV protons in gases, *N. Cim. B* **50**, 176 – 178 (1967).

[126] E. Acerbi, B. Candoni, M. Castiglioni, G. Dutto, G. Fait, F. Resmini, and C. Succi, Electron-capture cross-sections in various gases by protons with energies from 25 to 38 MeV, *N. Cim. A* **64**, 1068 – 1081 (1969).

[127] N.V. Novikov, and Ya.A. Teplova *Charge-Changing Cross Sections in Ion-Atom Cross Collisions*, http://cdfe.sinp.msu.ru/services/cccs/cccs.htm, Scobeltsyn Institute of Nuclear Physics, Lomonosov Moscow State University, Moscow, Russia (2009).

[128] C.F. Barnett, J.A. Ray, E. Ricci, M.I. Wilker, E.W McDaniel, E.W. Thomas, and H.B. Gilbody, *Atomic Data for Controlled Fusion Research* (Oak Ridge National Laboratory, Tennessee, USA), Vol. **1**, ORNL-5206 (1977).

[129] C.F. Barnett, H.T. Hunter, M.I. Kirkpatrick, I. Alvarez, C. Cisneros, and R. Phaneuf, *Atomic Data for Fusion*, Vol. **1** (Oak Ridge National Laboratory, ORNL-6086/V1, 1990)

[130] M.E. Rudd, R.D. DuBois, L.H. Toburen, C.A. Ratcliffe, and T.V. Goffe, Cross sections for ionization of gases by 5-4000 keV protons and electron capture by 5-150 keV protons, *Phys. Rev. A* **28**, 3244 – 3257 (1983).

[131] B.G. Lindsay, and R.F. Stebbings, Charge transfer cross sections for energetic neutral atom data analysis, *J. Geophys. Res.* **110**, A12213 (2005).

[132] R. Cabrera-Trujillo, L. Hernández, E.M. Hernández, and G. Hinojosa, Single electron capture cross sections for protons colliding with neon amd methane targets: Effects of the initial vibrational state of CH_4, *J. Phys. B* **49**, Art, ID 035201 (2016).

[133] The ionization loss of energy in gases and in condensed materials, *Phys. Rev.* **57**, 485 – 493 (1940).

[134] L. Eyges, Multiple scattering with energy loss, *Phys. Rev.* **74**, 1534 – 1535 (1948).

[135] M. Hollmark, J. Uhrdin, Dž. Belkić, I. Gudowska, and A. Brahme, Influence of multiple scattering and energy loss straggling on the absorbed dose distributions of therapeutic light ion beams: I; Analytical pencil beam model, *Phys. Med. Biol.* **49**, 3247 – 3265 (2004).

[136] M. Hollmark, I. Gudowska, Dž. Belkić, A. Brahme, and N. Sobolevsky, An analytical model for light ion pencil beam dose distributions: multiple scattering of primary and secondary ions, *Phys. Med. Biol.* **53**, 3477 – 3491 (2008).

[137] H. Bethe, Zur theorie des durchgangs schneller korpuskularstrahlen durch materie, *Ann. Phys.* (Leipzig) **5**, 325 – 400 (1930).
[138] F. Bloch, Zur bremsung rasch bewegter teilchen beim durchgang durch materie, *Ann. Phys.* (Leipzig) **16**, 285 – 320 (1933).
[139] Yu.A. Belkova, and Yu.A. Teplova, Charge exchange effects in stopping of light ions in solids, *Bull. Russ. Acad. Sci.* **70**, 977 – 980 (2006).
[140] R.P. Singhal, D.E. Rio, P.F. Schippnick, and A.E.S. Green, Electron energy degradation in neon, *Radiat. Res.* **95**, 32 – 44 (1983).
[141] C.C. Montanari, and J.E. Miraglia, Neonization method for stopping, mean excitation energy, straggling, and for total and differential ionization cross sections of CH_4, NH_3, H_2O and HF by impact of heavy projectile, *J. Phys. B* **47**, Art. ID 015201 (2015).

Index

Lightning Source UK Ltd.
Milton Keynes UK
UKHW020738171019
351773UK00001B/27/P

9 789811 211607